PHYTOREMEDIATION RHIZOREMEDIATION

FOCUS ON BIOTECHNOLOGY

Volume 9A

Series Editors

MARCEL HOFMAN
Centre for Veterinary and Agrochemical Research, Tervuren, Belgium

JOZEF ANNÉ
Rega Institute, Catholic University of Leuven, Belgium

Volume Editors

MARTINA MACKOVA
Institute of Chemical Technology,
Prague, Czech Republic

DAVID DOWLING
Institute of Technology,
Carlow, Ireland

THOMAS MACEK
Institute of Organic Chemistry and Biochemistry,
Academy of Sciences of the Czech Republic,
Prague, Czech Republic

COLOPHON

Focus on Biotechnology is an open-ended series of reference volumes produced by Springer in co-operation with the Branche Belge de la Société de Chimie Industrielle a.s.b.l.

The initiative has been taken in conjunction with the Ninth European Congress on Biotechnology. ECB9 has been supported by the Commission of the European Communities, the General Directorate for Technology, Research and Energy of the Wallonia Region, Belgium and J. Chabert, Minister for Economy of the Brussels Capital Region.

Phytoremediation Rhizoremediation

Edited by

MARTINA MACKOVA
Institute of Chemical Technology,
Prague, Czech Republic

DAVID DOWLING
Institute of Technology,
Carlow, Ireland

and

THOMAS MACEK
Institute of Organic Chemistry and Biochemistry,
Academy of Sciences of the Czech Republic,
Prague, Czech Republic

Springer

A C.I.P. Catalogue record for this book is available from the Library of Congress.

ISBN-13 978-90-481-7238-2
ISBN-10 1-4020-4999-4 (e-book)
ISBN-13 978-1-4020-4999-4 (e-book)

Published by Springer,
P.O. Box 17, 3300 AA Dordrecht, The Netherlands.

www.springer.com

Printed on acid-free paper

TABLE OF CONTENTS

INTRODUCTION

JOHN FLETCHER
University of Oklahoma, Department of Botany, Norman, USA

To appreciate the history of phytoremediation and its importance, it is helpful to start with a brief review of common remediation strategies used around the world to treat soil contaminated with toxic metals and/or organic chemicals. Three widely used strategies are: 1) immobilization or retention of toxicants within a confined area (i.e. the soil at the site of their release or in contaminated soil placed in a landfill, 2) removal of contaminants from the soil, 3) destruction of organic pollutants by chemical, physical, or biological means. These strategies either individually or in combination with each other have been routinely implemented by the remediation industry to successfully treat contaminated soil. Unfortunately, implementations have often required extensive earth moving, expensive equipment, and costly construction; all features that have aroused public resistance on occasion and have sometimes been tagged, rightly or wrongly, as more threatening to the environment than the contaminants themselves. In any event, public concern for the implementation features listed above have been instrumental in keeping pressure on the remediation industry to develop more cost effective and friendly methods, including bioremediation.

Bioremediation started over 50 years ago with research examining the fate of pesticides in agricultural soils. In view of the wide range of catabolic reactions mediated by bacterial enzymes, it is not surprising that from the beginning bioremediation research focused on bacteria. The capacity for bacteria to degrade xenobiotics was so impressive that other living organisms were virtually ignored for 30 years. As a result bioremediation became thought of as degradation of organic contaminants by bacteria even though the bio prefix suggested involvement of all life forms. Early investigators in plant remediation work were confronted with an attitude held by some persons that if remediation of a contaminant could not be achieved by bacteria with their diversified array of catabolic enzymes it sure couldn't be achieved with plants. This attitude is perhaps why investigators striving to call attention to the unique remediation features of plants felt obliged to establish a separate remediation field, phytoremediation, and include several subdivisions (i.e. phytoextraction, rhizoremediation, etc.).

Unique remediation features possessed by plants are easily illustrated by returning to the 3 common remediation strategies listed earlier to treat contaminated soil: 1) immobilization, 2) removal, and 3) destruction. Partial immobilization of water soluble contaminants is brought about by plant transpiration (soil water taken up, transported, and evaporated from leaf surfaces) since the process removes soil water that would

M. Mackova et al. (eds.), Phytoremediation Rhizoremediation, 1–3.

otherwise cause contaminant leaching and movement. Removal of toxic metals from contaminated soil occurs when inorganic ions are taken up by plant roots and translocated through the stem to aboveground plant parts. Regarding contaminant destruction, plants, because of their autotrophic nature, were rarely examined for catabolic properties until phytoremediation emerged and studies conducted then with nonphotosynthetic tissue culture cells and axenic roots clearly showed that plant enzymes degrade some organic pollutants. The use of plants to foster the degradation or organic soil contaminants has been further advanced by studies showing that soil microflora under the chemical influence of plant roots (rhizosphere zone) can be important in xenobiotic metabolism. The catabolic activity within the rhizosphere has been attributed to both bacteria and fungi whose presence and enzymatic expression are believed to be modulated by organic chemicals released from both living and dead roots. Since both root physiology and biosynthetic pathways vary considerably among plant species it is anticipated that rhizoremediation properties will also vary among plants. The most useful species for rhizoremediation may be previously unexplored species with no commercial importance prior to their use in rhizoremediation. Both the direct and indirect degradation of soil contaminants can potentially occur at the lowest depth of root penetration, a special feature of plant remediation. Thus, through the efforts of a relatively small group of scientists working around the world over the last 20 years, phytoremediation has become a well established, multifaceted technology capitalizing on three plant properties: transpiration, ion uptake, and metabolism with the later having both direct and indirect influences.

As phytoremediation technology has evolved it has become increasingly apparent that no single plant species excels in all three plant remediation properties, nor does any single species show maximum uptake of all toxic metals or foster degradation of all organic contaminants. Therefore, successful treatment of soils with mixed waste requires a combination of plant species with appropriate remediation properties, and also the inclusion of plant species hosting rhizosphere communities (bacteria and fungi) active against specific contaminants that are present. Thus, a major contribution that has emerged from the field of phytoremediation is the biosystems approach to soil remediation where the joint actions of several different organisms functioning in unison or in sequence are used to treat contaminated soil. The concept of a plant driven remediation system may be foreign in some remediation circles but it is certainly consistent with the fundamental principals of terrestrial ecology where the central role of plants as the primary producers and greatest users of water is well established.

Accepting the central role played by plants in biosystem remediation, raises many unanswered questions on how to assemble and manage the most effective biosystem. What are the best plant species to use since the physiology, biochemistry, and rhizosphere of very few of the thousands of native species have been studied? How does the microflora of a dead root and its degradative properties compare with that of a living root? Are genetically altered organisms necessary to degrade some soil contaminants? Do some plant roots release surfactants? Can plant species that move deep groundwater to surface root zones be capitalized on in rhizoremediation? Should plants be introduced as single annual crops or as perennial communities? Should plant succession be encouraged? As such questions are addressed and new remediation technology emerges it is very likely that phytoremediation employing ecologically and physiologically sound biosystems will be accepted as a necessary and first step in successful ecological restoration of contaminated habitats.

THE CHEMICAL ECOLOGY OF POLLUTANT BIODEGRADATION

Bioremediation and phytoremediation from mechanistic and ecological perspectives

ANDREW C. SINGER
Centre for Ecology & Hydrology–Oxford, Mansfield Rd, Oxford OX1 3SR, United Kingdom, E-mail: acsi@ceh.ac.uk

1. Introduction

As the yachtswoman Dame Ellen MacArthur returned to the south coast of Britain in early 2005 after a record 71-day solo circumnavigation of the globe on a trimaran, she noted pointedly, "It's funny when you smell the land and you have not smelled it for two months". MacArthur's comment reflects the multitude of odours originating from *terra firma* and highlights an important and underappreciated feature of our world – a dizzying abundance and diversity of chemicals surround us and in some subtle, as well as some very direct ways, dictate the actions and reactions of all life.

Among the numerous sources of chemicals in our environment, molecules of plant origin are arguably the most abundant and best characterised. This chapter aims to highlight the ecological functions of plant-derived chemicals and discuss their roles in both multi-trophic interactions and (pollutant-degrading) enzyme evolution. Evidence to support these positions has largely been generated in the past decade and will be reviewed in the later part of the chapter.

Rhizodeposition, the release of carbon compounds from living plant roots into the surrounding soil, is dominated by low molecular mass solutes such as sugars, amino acids and organic acids. There are numerous studies which aim to understand the regulation and ecological significance of rhizodeposition, for which the reader is directed to three excellent reviews [1-3]. Although rhizodeposition plays a central role in establishing and sustaining a soil system, this chapter will focus on a class of compounds, secondary plant metabolites (SPMe), that are nearly four-orders of magnitude more diverse than the typical rhizodeposits. Over 100 000 low-molecular-mass SPMe have been described with an estimated 400 000 yet to be discovered [4]. Many of these SPMe contain one of the following chemical structural backbones: isoprene, phenylpropene, alkaloid or fatty acid/polyketide (Figure 1) [5].

M. Mackova et al. (eds.), Phytoremediation Rhizoremediation, 5–21.

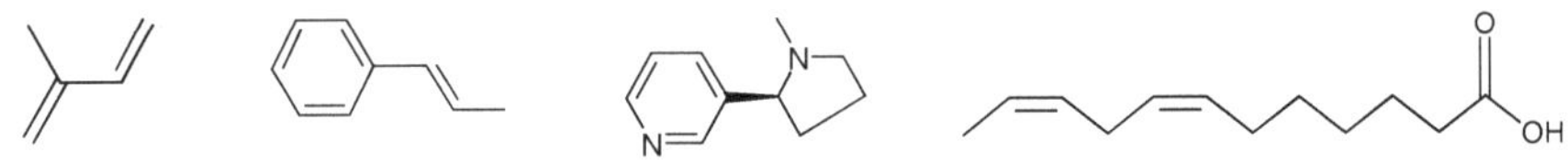

Isoprene Phenylpropene Alkaloid Fatty acid/polyketide

Figure 1. Typical skeletal backbones for the majority of secondary plant metabolites.

Although referred to as "secondary metabolites", implying a function of only secondary importance to the plant, SPMe fulfill a range of vital functions: (1) antimicrobial activity; (2) insect and microbial attraction; (3) insect and microbial deterrent; (4) plant-plant signal; (5) stress response; and (6) germination and growth inhibition [6].

Volatile low-molecular mass SPMe, consisting of a range of functional groups (hydrocarbons, alcohols, aldehydes, ketones, ethers and esters), are integral in how plants interact with their environment. Volatile emissions from flowers and fruits, for example, provide clues to animals, pollinators and seed disseminators, while those from vegetative tissues contribute to plant defence systems by repelling microorganisms and animals or attracting herbivore predators, thereby protecting the plant through tritrophic interactions [7].

1.1. THE NATURE OF THE PROBLEM

The Twenty-Fourth Report by the Royal Commission on Environmental Pollution stated that there are between 30 000 and 100 000 chemicals on the market [8]. Every year, approaching 2000 novel xenobiotic chemicals are added to this list, the vast majority of which have not been tested for even the most basic indications of environmental hazard. It is now recognised that this policy has been responsible for a number of environmental catastrophes such as: (1) reproduction failures in songbirds resulting from the organochlorine pesticide 4,4'-(2,2,2-trichloroethane^{-1},1-diyl)bis(chlorobenzene) (DDT) which was highlighted by Rachel Carson's landmark book, *Silent Spring* in 1962 [9]; (2) bioaccumulation of the organochlorine polychlorinated biphenyl (PCB) and reproduction failures at all levels of the food web from fish to eagles and humans [10]; and (3) depletion of the ozone layer induced by the release of chlorofluorocarbons (CFC) [11].

DDT, the chemical for which Carson is most noted for highlighting, was banned in the United States at the end of 1972, eight years after her untimely death from cancer. Although the DDT ban spread to many temperate countries, few tropical countries acceded to the ban, largely due to the pesticide's efficacy to control the spread of malaria and other insect-borne diseases. DDT has been shown to dissipate much more rapidly in tropical than temperate soils [12]. The mechanism for the latter is partly attributed to increased temperature-mediated volatility, but more importantly increased microbial biodegradation. The mechanisms underpinning chemical persistence in the environment are complex but are thought to be heavily influenced by the rarity of the chemical's structure and substituents.

My contention is that the chemical ecology of a site should also be considered as an important variable in determining a chemical's persistence. In this chapter, the term chemical ecology is used as defined by the International Society of Chemical Ecology, "the chemical mechanisms which help control intra- and interspecific interactions among living beings". Owing to key differences in the local chemical ecology, a recalcitrant molecule in one locale might be readily biodegraded in another. In part, the local flora provides the evolutionary mechanism for the development and modification of SPMe-degrading enzymes in the metagenome, which, it is argued, is fortuitously responsible for the presence of pollutant-degrading enzymes, *a priori* the chemicals synthesis by chemists.

1.2. TRITROPHIC TRINITY

Although it has been widely proposed that pollutant-degrading enzymes evolved from isozymes in response to industrial production and environmental release of xenobiotics, the *a priori* existence of readily mutable pollutant-degrading isozymes remains largely absent from the literature [13-16]. This chapter will contribute to the dialogue on the source and developmental mechanism of these pollutant-degrading enzymes.

The evolution of plants and their natural enemies was, arguably, responsible for generating much of the Earth's biological diversity [17]. A corollary was proposed [18], stating that the synergistic and antagonistic relationships between plants, micro-organisms, and insects (P-M-I) are responsible for the diversity of SPMe. A second corollary then suggested that the P-M-I tritrophic interactions serve as one of the main driving forces of pollutant-degrading enzyme evolution [19].

1.2.1 Theories on the evolution of pollutant-degrading isozymes

In "The Fractal Geometry of Nature", Mandelbrot highlights the fractal structure of many natural systems [20]. In this chapter, catabolic enzymatic systems are proposed to conform to a fractal architecture. Elucidation of the organisation and evolution of catabolic systems will aid in the investigation into the origins of pollutant-degrading enzymes.

The classic example of a fractal structure can be found in the form of a tree (Figure 2). The long tree trunk provides the foundation from which a repeating series of shorter branches are serially connected. At the metaphorical "leaves" of the tree lie molecules which necessitate an individualised enzymatic step before moving into more central metabolic pathways located at the base of the tree (e.g. citrate cycle, glycolysis). Their location in the periphery of the tree can be attributed to their relatively unusual chemical structure, substituent, or both. Whereas chemicals metabolised in well connected, more centralised locales of the tree (nearer the "trunk"), consist of relatively more common chemical structures and substituents [21]. Elaborating upon Firn and Jones (2000), it is proposed that substrate specificity of biodegradative enzymes is proportional to the distance of the enzymatic reaction from the "leaves", i.e. enzymes with a low substrate specificity are located in the "leaves" while enzymes with a high substrate specificity are typical of central metabolic enzymatic reactions (i.e. "trunk"). This metabolic architecture can be helpful in developing a mechanistic understanding of the evolution of pollutant-degrading enzymes. For example, in the event a microorganism encounters a novel molecule, it might perish if the molecule is toxic, as are many SPMe, or a mutant enzyme might emerge from the population, enabling the molecule's detoxification or

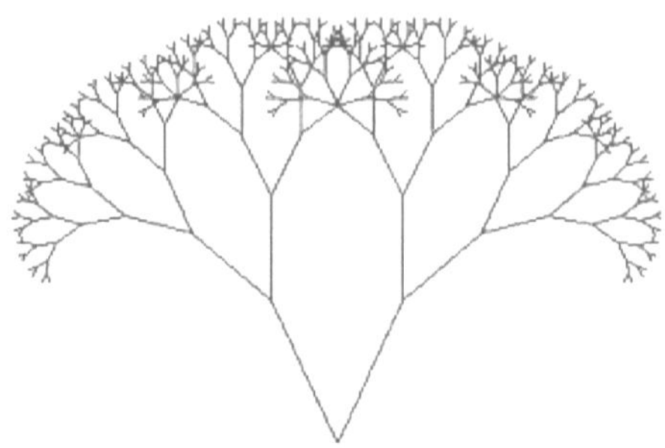

Figure 2. Fractal structure of catabolic enzymatic system within microorganisms. Novel xenobiotic and natural chemical structures are catabolised in the outer branches, funnelling metabolites to more central, substrate specific enzymatic steps.

metabolism. The mutant enzyme would develop in the "leaves" as it is the location of enzymes which are responsible for interacting with the environment (e.g. detoxification, communication, assimilation). Moreover, as the novel chemical is metabolised it may continue down the "tree" into central metabolic pathways, or it may persist without any further metabolism [14, 22]. Persistent metabolites would be further evidence of enzymatic activity in the "leaves", as catabolism of molecules with relatively novel structures or substituents: (1) might produce equally novel metabolites requiring additional modification to indigenous enzymes for further metabolism; (2) require low substrate specific enzymes for primary and secondary catabolism, which is a characteristic of enzymes residing in the "leaves".

Networked databases, such as MetaRouter and the University of Minnesota Biocatalysis/Biodegradation Database, provide a useful framework to visualise pollutant catabolic pathways (http://pdg.cnb.uam.es/MetaRouter/index.html) [23], UM-BBD; http://umbbd.ahc.umn.edu/) [24] and discern their fractal, interwoven structure. Integrating this information into a "suprametabolism" network (i.e. incorporating all pathways), will enable predictions to be made on the fate of both current and future environmental pollutants [22, 24].

2. Chemical ecology of pollutant degradation

In the early 1990's, researchers began to theorise about the "natural substrate" of pollutant-degrading enzymes. Among the first pollutants to be scrutinised were PCBs. Higson [25] and Furukawa [26] postulated that lignin may be the natural substrate for the PCB-degrading enzyme. *Rhodococcus erythropolis* TA421, a microorganism isolated from a wood-feeding termite ecosystem, was shown to degrade the recalcitrant pollutant PCB [27, 28]. The association of TA421 with a wood-feeding termite provided the researchers with an opportunity to link lignin-degrading ability with the capacity to catabolise PCBs. Maeda *et al.*, confirmed the presence of three PCB-degrading genes (*bph*C) in TA421, each with a different, yet narrow, substrate specificity [27], which might correlate with the three monomers of lignin (Figure 3). This, however, has yet to be tested.

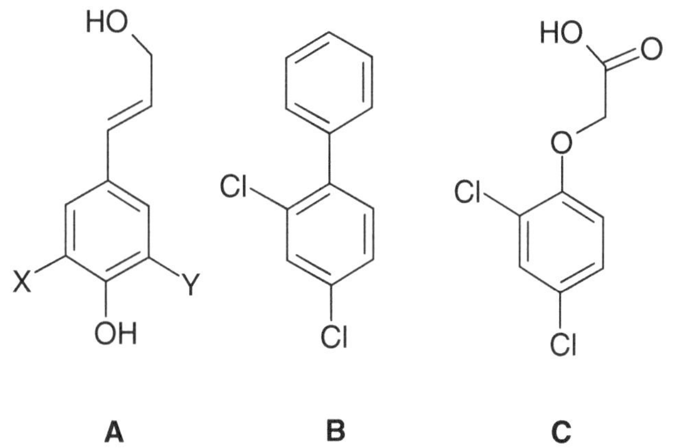

Figure 3. Structure similarity between: (A) Lignin monomer structure, where (1) X = Y = H (p-coumaryl alcohol), (2) X = OMe; Y = H (coniferyl alcohol), (3) X = Y = OMe (sinapyl alcohol) [29]; (B) 2,4-dichlorobiphenyl, PCB congener; and (C) 2-4-dichlorophenoxyacetate.

Among the first studies that specifically investigated the link between plant-derived chemicals and pollutant remediation was that by Donnelly *et al.*, The authors demonstrated that a range of flavonoids could support the growth of PCB-degrading microorganisms [30]. The best growth substrate and concentration were determined for each of three PCB-degrading microorganisms: *Ralstonia eutrophus* strain H850; *Burkholderia cepacia* LB400; and *Corynebacterium* sp. MB1. Each bacterium was subjected to a congener depletion assay [31], which was designed to show in a 24-hour period the extent of congener degradation after growth on a particular flavanoid. Naringin proved the best growth substrate for H850 and supported its greatest metabolic activity on PCBs. Myricetin induced the greatest PCB degradation by LB400, which catabolised 16 of the 19 congeners tested. Strain MB1 degraded thirteen PCB congeners in the presence of coumarin in excess of the biphenyl controls.

The researchers suggested that fine plant roots may ultimately serve as a "naturally occurring injection system," capable of dispensing phenolic plant-derived compounds into the rhizosphere. They advocated the use of these exudates as a means to support the growth of PCB-degrading microorganisms. A major step in the development of the chemical ecology of pollution was provided by Focht (1995) who proposed that plant terpenes, rather than biphenyl [32], might be the natural substrates for PCB catabolising enzymes. Subsequent studies by Hernandez *et al.*, showed that soils enriched with orange peel, ivy leaves, pine needles or eucalyptus leaves resulted in 10^5 times more biphenyl (unchlorinated PCB) utilizers (10^8 g^{-1}) than their unsupplemented control (10^3 g^{-1}), which suggested that terpenes found in these plants might be natural substrates for biphenyl-utilizing bacteria. Notably, complete disappearance of Aroclor 1242 was observed in soils amended with orange peel, ivy leaves, pine needles and eucalyptus leaves [33]. The authors examined the efficacy of terpene-degrading isolates to biotransform Aroclor 1242 in broth. Three bacteria isolated from the experimental soil with the capacity to utilize cymene as a sole carbon source exhibited enhanced (20-80%) transformation of Aroclor 1242, in comparison with glucose-grown cultures. Four of five limonene-utilising isolates exhibited elevated (43-83%) Aroclor 1242 transformation compared with the controls. In conclusion, the authors speculated that biphenyl may provide soil microorganisms with a relatively labile source of carbon, which is rapidly utilized by fast-growing soil microorganisms (copiotrophs). They argued that slow-growing microorganisms (oligotrophs), which rely on low concentrations and slowly delivered secondary plant metabolites, might be more effective in degrading PCBs [33]. Evidence for the interaction between pollutant degradation and the availability of labile carbon sources is highlighted in several papers referenced in this chapter.

In a similar vein, Dzanto and Woolston, demonstrated removal of 10, 21 and 24% PCB (Aroclor 1248) from soils supplemented with pine needles, biphenyl and orange peel, respectively, compared with control soil, although the differences were not deemed mathematically significant ($P > 0.05$) [34]. The authors suggested that further promotion might be achieved by biostimulating the rhizosphere with specific inducing substrates for the target pollutant. This approach has since been demonstrated successfully in two classic studies by Narasimhan *et al.* [35] and Kupier *et al.* [36, 37], as discussed elsewhere [19].

The first methodical screening and isolation of active inducing compounds in plants was achieved by Gilbert and Crowley. The authors examined a number of plant extracts (spearmint, pennyroyal, basil, barley, green bean, dill, avocado litter and garden compost) to determine if any stimulated the degradation of 4-4'-dichlorobiphenyl by a known PCB-degrading bacterium, *Arthrobacter* sp. strain B1B [38]. Spearmint extract resulted in approximately 33% of the metabolites produced by the known PCB-inducing compound, biphenyl. Subsequent analysis of this extract identified carvone as the principal component responsible for catabolism induction. Ten terpenoids of similar structure to carvone were assayed for their ability to induce PCB (50 mg l^{-1}) degradation: *p*-cymene, isoprene, (S)-(+)-carvone, (R)-(–)-carvone, (S)-(–)-limonene, (R)-(+)-limonene, carvacrol, cumene, trans-cinnamic acid and thymol. These plant-derived compounds are commonly found in dill and caraway seed, spearmint, pine needles, citrus, juniper, oregano, thyme and numerous other aromatic plants. With the exception of cumene, trans-cinnamic acid and thymol, all terpenes enhanced 4-4'-dichlorobiphenyl metabolism, while *p*-cymene and isoprene accelerated catabolism in comparison with biphenyl ($P < 0.05$). The workers highlighted that not only was PCB degradation induced by nonaromatic compounds, but also among the most effective was isoprene, which lacks a ring structure. It was proposed that the relatively high antimicrobial activities of terpenes might induce a P450-like detoxification and fortuitous degradation of the PCBs. Cytochrome P450 enzymes are a large family of enzymes that have been shown to oxidize terpenes, such as camphor (P450cam), as well as pollutants, such as polycyclic aromatic hydrocarbons (PAHs; e.g. naphthalene and pyrene [39]), chlorinated phenols [40], and biphenyls [41].

p-Cymene is among the more frequently investigated SPMe in pollutant-degradation studies, and is arguably among the more effective. It is a natural aromatic hydrocarbon that occurs in the oils of over 100 gymnospermic and angiospermic plants, including eucalyptus, cumin, thymine, cypress, coriander, sage, star anise and cinnamon [42, 43]. Its efficacy might stem from: (1) its structural similarity to many pollutants (e.g. toluene, xylene, ethylbenzene, biphenyl, chlorobenzene); and (2) a common evolutionary origin of the genes encoding the catabolic pathways [43].

Encouraged by carvone induction of PCB degradation by *Arthrobacter sp.* strain B1B, Park *et al.,* demonstrated expression of the *bph*C gene (2,3-dihydroxybipheyl 1,2-dioxygenase) in the PCB degrader *Ralstonia eutrophus* H850 following induction by (R)-(–)-carvone (50 mg l^{-1}) [44]. The researchers concluded that carvone might induce a different degradative pathway, potentially generating different congener specificity to that of biphenyl-induced cells. Jung *et al.,* examined the efficacy of carvone or limonene to induce the *bph*C gene of *R. eutrophus* H850 in soil. Although biphenyl was capable of inducing the *bph*C gene up to 4 days after addition to the soil, neither carvone nor limonene were able to maintain the induction [45]. The authors concluded that the presence of potential inducing compounds *in situ* does not necessarily ensure that induction will occur, and that a greater understanding of induction is needed before field implementation [45].

Using a similar approach to Jung *et al.*, Oh *et al.,* examined the ability of terpene to prolong the survival of a known PCB-degrading bacterium, *Pseudomonas pseudoalcaligenes* KF707, in soil [46]. The addition of 50 mg l^{-1} *p*-cymene or 50 mg l^{-1} α-terpinene increased KF707 survival by 10- to 100-fold compared with biphenyl-supplemented and control mesocosms. *Rhodococcus* sp. strain T104, a PCB-degrading bacterium, was shown to catabolise biphenyl as well as the SPMe limonene, cymene, pinene and abietic acid as sole sources of carbon. Limonene was capable of inducing the biphenyl degradation pathway. The bacterium contains three genes, T1, T3 and T5, which potentially code for aromatic-degrading compounds. T1 was induced by limonene and cymene, and to a much lower extent, biphenyl. Notably, glucose exerted a similar degree of induction to that of limonene. Cymene was the strongest inducer of T3, while limonene and cymene induced T5 more strongly than biphenyl and glucose [47-49]. Kim *et al.,* further demonstrated that T104 is responsible for three distinct catabolic pathways for phenol, biphenyl and limonene the last of which can induce both the upper and lower pathways for biphenyl degradation [49]. Therefore, the authors concluded that microorganisms might harbour several mechanisms for degrading structurally similar compounds [49].

Rhodococci play an important role in the carbon cycle due to their ability to degrade many semi-recalcitrant organic compounds. One of the three linear plasmids from a well-studied PCB-degrading bacterium, *Rhodococcus* sp. strain RHA1, was sequenced to elucidate the number, structure and regulation of the open reading frames [50]. The smallest of the linear plasmids is divided into three clusters, one of which contains limonene degradation genes, which are potentially responsible for its ability to grow on limonene, as well as carveol and carvone as sole sources of carbon. Interestingly, the plasmid contains three cytochrome P450-encoding genes. Earlier, the bacterium had been shown to possess multiple isozymes (three *bph*-type ring-hydroxylating dioxygenases and seven *bph*-type ring cleavage enzymes) for PCB degradation. The identification of aromatic- and terpene-degrading genes as well as cytochrome P450 on the same plasmid suggests that the bacterium employs both broad and narrow substrate range enzymes to: (a) make maximum utilization of available carbon sources, particularly those deemed recalcitrant by specialised bacteria; and (b) detoxify compounds that may associate with, otherwise, labile carbon sources. Thus, RHA1 is an excellent model microorganism to examine the link between SPMe and pollutant degradation.

Through the use of a chromosomally-encoded lacZ reporter, Master and Mohn gained insight into the differential induction of *bph*A, the large subunit of the biphenyl dioxygenase, in two PCB-degrading bacteria, *Pseudomonas* sp. strain Cam-1 and *Burkhoderia xenovorans* LB400 [51]. The latter exhibited constitutive expression of *bph*A in the presence of twelve different inducers, including many plant-derived compounds such as pinene, limonene, cymene, cumene, carvone and salicylate. Due to its constitutive PCB-degrading capacity, however, the authors suggested a cautious interpretation of the efficacy of the inducing compounds. In contrast, the biphenyl-induced strain Cam-1 demonstrated a *bph*A activity six times greater than the basal level in cells at 30°C in the presence of pyruvate, indicating the need for induction prior to bioaugmentation of PCB-contaminated soil. Of the twelve SPMe examined, only salicylate induced Cam-1 *bph*A activity to levels greater than basal levels recorded for pyruvate-exposed cells.

Tandlich *et al.*, used carvone and limonene to stimulate biodegradation of Delor 103 (a commercial mixture of PCBs) by *Pseudomonas stutzeri.* An expansion of PCB congener removal was achieved after supplementation with 10 mg l^{-1} carvone compared to glucose-grown control cells [52]. It is interesting to note that the spectrum of congeners degraded decreased with the addition of 20 mg l^{-1} carvone, which suggested that terpene induction may be compound- and concentration-specific. Limonene and glycerol-cultured cells increased the range of congeners degraded as well as the total PCB catabolised compared to the controls. Increased congener depletions were recorded with elevated limonene concentrations from 10 to 20 mg l^{-1}, while a decline in PCB degradation resulted with the co-addition of biphenyl and carvone or limonene. Specifically, 90% of a tri-ortho-substituted PCB congener was removed by biphenyl-induced cells while no removal was observed in the presence of carvone. Furthermore, biodegradation in the presence of glycerol or xylose, with carvone or limonene addition, increased the suite of congeners degraded.

Nishio *et al.,* demonstrated the broad substrate specificity for *p*-cymene monooxygenase (CMO) found in the soil microorganism *Pseudomonas putida* F1 (PpF1). The bacterium can grow on *p*-cymene as a sole carbon and energy source by employing a different degradative pathway compared with cultivation on the structurally similar pollutant, toluene. CMO was shown to actively biotransform 4-ethyltoluene, styrene, *m*- and *p*-xylene, 4-chlorostyrene, 4-(methylthio)toluene, 3-chlorotoluene, 4-chlorotoluene, 4-fluorotoluene and 4-nitrotoluene [53]. Interestingly, the highest biotransformation rate was found not with cymene but with 4-chlorostyrene, which shares the same chemical substructure with flavones such as anthocyanidin and isoflavone as well as the lignin monomer *p*-coumaryl alcohol (Figure 4; [53]).

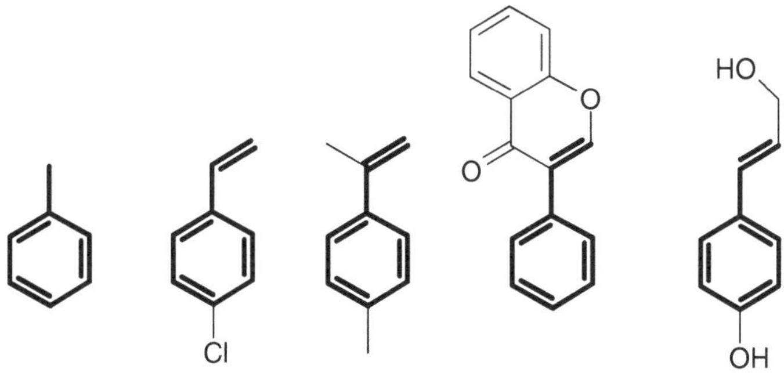

A B C D E

Figure 4. Structural similarities are bolded between: (A) toluene (pollutant); (B) 4-chlorostyrene (pollutant); (C) p-cymene (SPMe); (D) isoflavone (SPMe); and (E) p-coumaryl alcohol – lignin monomer.

Qui *et al.*, assessed the influence of the addition of two flavonoids, morin and flavone, on benz[*a*]pyrene (B[*a*]P) degradation in rhizosphere soil [54]. The soils were exposed to 0, 0.1, 1, 10, 100 µmoles of the flavonoids for 60 days. Both morin and flavone-supplemented soils recorded decreased mineralization of ^{14}C-B[*a*]P with flavonoid concentrations as low as 10 µmoles. Flavone-supplemented soils lowered B[*a*]P bioavailability as monitored by decreased recovery from serial extractions with hexane, water and ethyl acetate. The researchers suggested that morin might have either inhibited the enzyme system responsible for B[*a*]P degradation or was preferentially used as a carbon source by the native B[*a*]*P*-degrading population. Due to a decrease in its extractability, flavone might have stimulated B[*a*]P transformation only and not mineralization, and so resulted in the sequestration, sorption or humification of the metabolite. The authors supported the need for an understanding of the complicated and potentially confounding effects of root exudation, secondary plant metabolite selection, and the specific soil system on the rate and extent of pollutant degradation in soils treated by phytoremediation [54].

Recent studies of the consumption of atmospheric methane in forest soils has indicated that monoterpenes can inhibit methanotrophy (40-100%), with (–)-α-pinene the most effective [55]. The degree of inhibition was found to be species and monoterpene specific in mono-cultures (*Methylosinus trichosporium* OB3b), for unsaturated, cyclic hydrocarbon forms such as (–)-α-pinene, (S)-(–)-limonene, (R)-(+)-limonene and χ-terpinene) [56]. Amaral and Knowles applied an aqueous extract of two depths of forest soil to examine if natural substrates inhibit methanotrophy [55]. They observed a concentration-dependent and transient inhibition after the addition of 0-5 cm depth soil extracts, whereas extracts from deeper soil (5-12 cm) proved non-inhibitory. Consistent with the current literature, monoterpene depositions from plant leaves might accumulate within the upper soil horizon and result in methanotrophy inhibition. Owing to its global implications, this system provides an interesting and environmentally important model to study the chemical ecology of terpene- and pollutant-degrading genes.

2.1. STRUCTURAL- AND STEREO-ISOMERS

Many of the environmental pollutants controlled under international agreements, such as the United Nations Economic Commission for Europe Persistent Organic Pollutants Protocol (U.N.E.C.E. P.O.P.s Protocol) and the United Nations Environment Programme PoPs Convention, are mixtures of structural- and stereoisomers (e.g. aldrin, chlordane, dieldrin, DDT, heptachlor, hexabromobiphenyl, hexachlorocyclohexane, PCBs, dioxin). Detailed investigation of differential biological activity on, and biodegradation of, these complex isomeric mixtures of PoPs, universally demonstrates highly variable activities and persistence [57]. Isomers are molecules with the same chemical formula. Structural-isomers have different bonding patterns whereas stereo-isomers have identical bonding patterns but differ only in the geometric position of the bond. Hence, it is misleading to discuss the efficacy of a remediation approach when addressing structural- or stereo-isomeric mixtures without acknowledging the potential for differential isomeric activity. Similarly, when investigating the recalcitrance of inducing pollutant degradation with SPMe, one must be cautious of the differential effects of structural- and stereo-isomers. Two studies are presented here as evidence of the differential effects of structural- and stereo-isomers in both the pollutant and the secondary plant metabolite.

Strong evidence for the induction of alternative PCB catabolic pathways using SPMe within many of the well-known PCB-degrading bacteria was demonstrated by Singer *et al.* [58] through the use of stereoselective degradation. Five PCB-degrading bacteria, *Ralstonia eutrophus* H850, *Burkholderia xenovorans* LB400 ([59]), *Rhodococcus globerulus* P6, *Rhodococcus* sp. strain ACS and *Arthrobacter* sp. strain B1B were assessed for their ability to differentially degrade four atropoisomeric PCBs (one tetrachlorobiphenyl and three pentachlorobiphneyls). Catabolism was assessed for each bacterium after growth on tryptic soy broth and in the presence of biphenyl, (S)-(+)-carvone or *p*-cymene. Stereoselectivity varied with respect to strain, congener and co-substrate. The authors concluded that the inducing compounds might facilitate alternative PCB-degradation pathways within the bacterium, thereby accounting for the observed stereoselective degradation pattern. The stereoselective degradative pattern for each enzyme can exist owing to the enzyme and the chemicals' chirality. Hence, changes in metabolic pathway might be detectable through the use of chiral chemical substrates as they might be differentially degraded by each enzymatic system.

2.2. RHIZOSPHERE ECOLOGY

Yu *et al.*, reported the recovery, by three to four orders of magnitude, of more resin acid degraders (tricyclic terpenoids originating from softwood trees) in hydrocarbon-contaminated soils than in pristine Arctic tundra soil [60]. Notably, the soil samples were collected thousands of kilometres from the nearest source of resin acids (conifer forest) and contained no native resin acids. The bacteria isolated in the study, *Pseudomonas* and *Sphingomonas*, are hydrocarbons degraders, which suggested that their ability to mineralise resin acid and xenobiotics may not be purely coincidental. The results from Yu *et al.*, were particularly interesting in light of a publication by Button who discovered that over 10% of the bacteria in a litre of seawater near Seward, Alaska (similar Arctic region to that studied by Yu *et al.*,), catabolised terpenes [61]. The author postulated that very heavy precipitation on the conifer forest of the Pacific Northwest carries the canopy drip and guttation fluid into the surface water and, ultimately, the sea. However, due to the Alaska Coastal Current, the dissolved terpenes are carried into the estuaries upstream, thereby sustaining a terpene-based food web [61]. The distribution of large quantities of SPMe in the Arctic region may provide the elusive mechanism Yu *et al.* [60] sought for the presence of resin acid (and hydrocarbon) degraders.

2.2.1 Induction by plant phenolics and root recycling

It has been proposed that fine plant root recycling can provide the stimulus needed to sustain pollutant-degrading microorganisms in the rhizosphere [62, 63]. The researchers demonstrated that a majority of fine roots (<1 mm diameter) from mulberry (*Morus* sp.) die at the end of a 6-month growing season. Flavones, such as morusin, morusinol and kuwanon C, contribute to approximately 4% of the fine root biomass (dry weight) after a full growing season. The authors have demonstrated that a wide range of flavones sustain the growth of the PCB-degrading bacterium *Burkholderia xenovorans* LB400 and concluded that a continual supply in the rhizosphere, through fine root recycling, might facilitate the structure and function of the microbial populations facilitating degradation of otherwise recalcitrant pollutants [30, 62]. The biphenyl dioxygenase of *Pseudomonas pseudoalcaligenes* KF707, a well studied PCB-degrading microorganism, has also been shown to catalyze both flavone and 5,7-dihydroxyflavone [64], and as discussed earlier, was shown to exhibit protracted survival in soil supplemented with *p*-cymene or α-terpinene [46].

The fine root recycling hypothesis was evaluated by Parrish *et al.*, who, following application of an herbicide to kill the roots of fescue (*Festuca arundinacea* Schreb.) and yellow sweet clover (*Melilotus officinalis* Lam.), assessed the rate and extent of PAH degradation in the plant rhizospheres. Although they demonstrated differences in the extent of PAH removal by the two species, there was no enhancement of PAH removal due to "induced root death" [65].

Addressing a similar question, Shaw and Burns demonstrated that *Trifolium pratense* exudates and a supplement of roots grown in non-sterile soil, increased the maximum 2,4-dichlorophenoxyacetic acid (2,4-D) degradation rate and decreased the lag time to the maximal 2,4-D degradation rate [66]. Notably, both these promotions also resulted with supplementation of autoclaved roots. Conversely, gnotobiotic hydroponic and sand-grown roots did not increase the rate of 2,4-D degradation, which suggested that the stimulatory component was both a function of the plant and cultivation medium. The authors also found evidence that unfractionated legume rhizodeposits enhanced 2,4-D mineralization. The implication was that flavonoids, as major signalling components of the rhizobia-legume symbiosis [1], might select for microorganisms capable of detoxifying and utilising the flavanoid signals or their metabolites [66]. For example, cinnamic acid is one of the possible metabolites of flavanoid degradation and has been shown to induce *TfdA*, the gene responsible for the first step of 2,4-D catabolism [66]. 2,4-D is structurally analogous to *p*-coumaryl alcohol, a lignin monomer, which has been proposed to be a natural inducer of PCB degradation (Figures 3 and 4) [26].

2.2.2 Salicylate

Akin to *p*-cymene, salicylate is another SPMe that has been studied extensively, not only for its efficacy to stimulate pollutant degradation but also in relation to its role as a plant-plant signalling compound [18]. Salicylate has been shown to induce biphenyl, xylene and toluene degradation in *Pseudomonas paucimobilis* Q1 [67] and PAH degradation in *P. saccharophila* P15 and *P. putida* 17484 and PpG1 [68-70]. Filonov *et al.*, demonstrated preferential expression of the *ortho*-pathway for catechol cleavage (a metabolite of PAHs and salicylate), as well as the presence of silent genes for the *meta*-catechol cleavage pathway. Recognition of this alternative pathway extends the range of substrates utilized although at a potential cost of cell death if the microorganism is exposed to particular halogenated isomers [71].

A considerable body of results has amassed that demonstrates the funnelling of PAH metabolites through one of two intermediate pathways, salicylate and phthalate [72]. However, the discovery of naphthalene-, phenanthrene-, anthracene-, chrysene-, fluorine-, pyrene-degrading bacteria, which do not grow on salicylate or phthalic acid, suggests that a variety of pathways and inducers exist for the degradation of PAHs [72-74].

2.2.3 Pollutant-degrading pathway repression

Rentz *et al.*, examined the effect of hybrid willow (*Salix alba* × *matsudana*) root exudates on the phenanthrene-degrading activity of *P. putida* 17484. Although salicylate was expected to increase phenanthrene degradation, it was repressed by approximately 21% of its maximum [70]. The researchers concluded that the prevalence of alternative carbon sources in the rhizosphere exerted catabolite repression [3]. However, in this and a previous study with *Pseudomonas fluorescens* HK44, it was suggested that increased numbers of total heterotrophs and pollutant-degrading bacteria, as well as increased metabolic activity, can, potentially, compensate for catabolite repression [70, 75]. Global carbon source regulation was implicated by a decline in phenanthrene degradation in cells exposed to 2.0 mM acetate, lactate, pyruvate, glucose and glutamate. The amino acids aspartic acid and glutamate, quantified as up to 3.9% of the total organic carbon of willow root exudates, might have contributed to the repression. In previous studies, it has been demonstrated that the availability of amino acids in the concentration range of 0.001 to 0.1% can suppress the *Pseudomonas*-derived *DmpR*-Po σ^{54}-dependent regulatory system, and so delay expression of the (methyl)phenol catabolic enzyme. The authors emphasised that the appropriate transcriptional response to specific signals in their environment are contingent on the physiological status of the cell [76,77]. Notably, the σ^{54} promoter for the toluene/xylene catabolic TOL plasmid has also been shown to be growth-phase regulated in rich media [78]. Hence, the efficacy of SPMe induction will likely be dependent on the availability of carbon sources (e.g. amino acids) and the growth stage of the catabolic microorganism (e.g. stationary phase). Coordinated expression of pollutant-degrading genes upon entry into stationary phase was also demonstrated by Denef *et al.*, in the PCB-degrading bacterium *B. xenovorans* LB400 [79]. Rentz *et al.*, were careful to note that root-derived substrate repression is likely to vary among different microbial strains and plant species [70]. Yoshitomi and Shann confirmed this in a study that involved continuous application of corn (*Zea mays* L.) root exudates to pyrene contaminated soil for 90 days. The researchers observed enhanced pyrene mineralization in root-exudate supplemented as compared with controls [80].

3. The tortoise and the hare: Exponential silencing

Repression of a microbial catabolic gene in log-phase growth on nutrient rich medium is termed exponential silencing [81]. This phenomenon has been studied in only a few microorganisms (*Pseudomonas putida* pWWO [82], *Acinetobacter sp.* ADP1 [83] and *Burkholderia xenovorans* [79]). The insights gained from understanding this process have immediate implications towards rhizostimulation and the chemical ecology of pollutant remediation.

Exponential silencing might suggest that copiotrophic rhizosphere-competent bacteria will preferentially exploit labile carbon substrates (e.g. pyruvate, malate, citrate, succinate) before degrading less labile molecules, such as pollutants (e.g. toluene, xylene, biphenyl). However, on entering stationary phase, the copiotroph experiences a general stress response, which up-regulates σ^{54}-dependent promoters and activates enzymes (e.g. monooxygenases, dioxygenases) with broad substrate specificity thus enabling utilization of semi-recalcitrant, lower energy-yielding carbon sources, which in some cases, might (fortuitously) be a pollutant. Conversely, oligotrophs, which arguably rely more on lower energy-yielding carbon sources might thereby avoid exponential silencing and thereby carry out a significant proportion of what is termed "natural attenuation". In this way, the tortoise (oligotroph) could provide more extensive pollutant removal than the hare (copiotrophs). If validated, laboratory studies demonstrating the efficacy of copiotroph-mediated pollutant attenuation might be *in vitro* anomalies, unrepresentative of complex soil microbial systems.

This chapter has highlighted the value of consolidating interdisciplinary knowledge to generate new hypotheses for pollutant degradation. Due to the increasing literature base in all fields of science, it is now possible (and necessary) to initiate interdisciplinary collaboration between microbiology, ecology, biochemistry, botany and entomology, to resolve this complex problem.

References

[1] Dakora, FD and Phillips, DA (2002) Root exudates as mediators of mineral acquisition in low-nutrient environments. Plant Soil 245: 35-47
[2] Jones, DL; Hodge, A and Kuzyakov, Y (2004) Plant and mycorrhizal regulation of rhizodeposition. New Phytol 163: 459-480
[3] Hutsch, BW; Augustin, J and Merbach, W (2002) Plant rhizodeposition - an important source for carbon turnover in soils. J Plant Nutr Soil Sci - Z Pflanzenernaehr Bodenkd 165: 397-407
[4] Hadacek, F (2002) Secondary metabolites as plant traits: Current assessment and future perspectives. CRC Crit Rev Plant Sci 21: 273-322
[5] Dixon, R (2001) Natural products and plant disease resistance. Nature (London) 411: 843-847
[6] Farmer, EE (2001) Surface-to-air signals. Nature (London) 411: 854-856
[7] Dudareva, N and Negre, F (2005) Practical applications of research into the regulation of plant volatile emission. Curr Opin Plant Biol 8: 113-118
[8] Anonymous (2003) Chemicals in Products: Safeguarding the Environment and Human Health. Cm 5827, in Royal Commission on Environmetal Pollution, Twenty-Fourth Report
[9] Carson, R (1962) Silent Spring. Houghton Mifflin Company. New York
[10] Jensen, S (1966) Report of a new chemical hazard. New Scientist 32: 612
[11] Farman, JC; Gardiner, BG and Shanklin, JD (1985) Large losses of total ozone in Antarctica reveal seasonal ClO and NOx interaction. Nature (London) 315: 207-210
[12] Rasche, ME; Hicks, RE; Hyman, MR and Arp, DJ (1990) Oxidation of monohalogenated ethanes and n-chlorinated alkanes by whole cells of *Nitrosomonas europaea*. J Bacteriol 172: 5368-5373
[13] Wackett, LP (2003) Evolution of new enzymes and pathways: Soil microbes adapt to s-triazine herbicides. 37-50, in MA Boston, JJ Gan, Eds, Pesticide Decontamination and Detoxification; American Chemical Society, Washington DC
[14] Vandermeer, JR; Devos, WM; Harayama, S and Zehnder, AJB (1992) Molecular mechanisms of genetic adaptation to xenobiotic compounds. Microbiol Rev 56: 677-694
[15] Vandermeer, JR (1994) Genetic adaptation of bacteria to chlorinated aromatic-compounds. FEMS Microbiol Rev 15: 239-249
[16] Liu, S and Suflita, JM (1993) Ecology and evolution of microbial populations for bioremediation. Trends Biotechnol 11: 344-352
[17] Rausher, MD (2001) Co-evolution and plant resistance to natural enemies. Nature (London) 411: 857-864
[18] Singer, AC; Crowley, DE and Thompson, IP (2003) Secondary plant metabolites in phytoremediation and biotransformation. Trends Biotechnol 21: 123-130
[19] Singer, AC; Thompson, IP and Bailey, MJ (2004) The tritrophic trinity: A source of pollutant-degrading enzymes and its implications for phytoremediation. Curr Opin Microbiol 7: 1-6
[20] Mandelbrot, BB (1982) The Fractal Geometry of Nature. WH Freeman. San Francisco, CA, USA
[21] Firn, RD and Jones, CG (2000) The evolution of secondary metabolism - a unifying model. Mol Microbiol 37: 989-994
[22] Pazos, F; Valencia, A and De Lorenzo, V (2003) The organization of the microbial biodegradation network from a systems-biology perspective. EMBO Reports 4: 994-999
[23] Pazos, F; Guijas, D; Valencia, A and De Lorenzo, V (2005) MetaRouter: Bioinformatics for bioremediation. Nucleic Acids Res 33: D588-D592
[24] Ellis, LBM; Hou, BK; Kang, WJ and Wackett, LP (2003) The University of Minnesota Biocatalysis/Biodegradation Database: Post-genomic data mining. Nucleic Acids Res 31: 262-265
[25] Adams, R; Huang, C; Higson, F; Brenner, V and Focht, D (1992) Construction of a 3-chlorobiphenyl-utilizing recombinant from an intergeneric mating. Appl Environ Microbiol 58: 647-654
[26] Furukawa, K (1994) Molecular genetics and evolutionary relationship of PCB-degrading bacteria. Biodegradation 5: 289-300
[27] Maeda, M; Chung, S-Y; Song, E and Kudo, T (1995) Multiple genes encoding 2,3-dihydoroxybiphenyl 1,2-dioxygenase in the Gram-positive polychlorinated biphenyl-degrading bacterium *Rhodococcus erythropolis* TA421, isolated from a termite ecosystem. Appl Environ Microbiol 61: 549-555
[28] Chung, S-Y; Maeda, M; Song, E; Horikoshi, K and Kudo, T (1994) Isolation and characterization of a Gram-positive polychlorinated biphenyl-degrading bacterium, *Rhodococcus erythropolis* strain TA421, from a termite ecosystem. Biosci Biotechnol Biochem 58: 2111-2113

[29] Burdon, J (2001) Are the traditional concepts of the structures of humic substances realistic. Soil Sci 166: 752-769
[30] Donnelly, PK; Hegde, RS and Fletcher, JS (1994) Growth of PCB-degrading bacteria on compounds from photosynthetic plants. Chemosphere 28: 981-988
[31] Bedard, DL; Unterman, R; Bopp, LH; Brennan, MJ; Haberl, ML and Johnson, C (1986) Rapid assay for screening and characterizing microorganisms for the ability to degrade polychlorinated biphenyls. Appl Environ Microbiol 51: 761-768
[32] Focht, DD (1995) Strategies for the improvement of aerobic metabolism of polychlorinated biphenyls. Curr Opin Biotechnol 6: 341-346
[33] Hernandez, BS; Koh, SC; Chial, M and Focht, DD (1997) Terpene-utilizing isolates and their relevance to enhanced biotransformation of polychlorinated biphenyls in soil. Biodegradation 8: 153-158
[34] Dzantor, EK and Woolston, JE (2001) Enhancing dissipation of Aroclor 1248 (PCB) using substrate amendment in rhizosphere soil. J Environ Sci Health 36: 1861-1871
[35] Narasimhan, K; Basheer, C; Bajic, VB and Swarup, S (2003) Enhancement of plant-microbe interactions using a rhizosphere metabolomics-driven approach and its application in the removal of polychlorinated biphenyls. Plant Physiol 132: 146-153
[36] Kuiper, I; Bloemberg, GV and Lugtenberg, BJJ (2001) Selection of a plant-bacterium pair as a novel tool for rhizostimulation of polycyclic aromatic hydrocarbon-degrading bacteria. Mol Plant Microbe Interact 14: 1197-1205
[37] Kuiper, I; Kravchenko, LV; Bloemberg, GV and Lugtenberg, BJJ (2002) *Pseudomonas putida* strain PCL 1444, selected for efficient root colonization and naphthalene degradation, effectively utilizes root exudate components. Mol Plant Microbe Interact 15: 734-741
[38] Gilbert, ES and Crowley, DE (1997) Plant compounds that induce polychlorinated biphenyl biodegradation by *Arthrobacter* sp. strain B1B. Appl Environ Microbiol. 63: 1933-1938
[39] England, PA; Harford-Cross, CF; Stevenson, JA; Rouch, DA and Wong, LL (1998) The oxidation of naphthalene and pyrene by cytochrome P450(cam). FEBS Lett 424: 271-274
[40] Jones, JP; O'Hare, EJ and Wong, LL (2001) Oxidation of polychlorinated benzenes by genetically engineered CYP101 (cytochrome P450(cam)). Eur J Biochem 268: 1460-1467
[41] Wolkers, J; Burkow, IC; Lydersen, C; Dahle, S; Monshouwer, M and Witkamp, RF (1998) Congener specific PCB and polychlorinated camphene (toxaphene) levels in Svalbard ringed seals (*Phoca hispida*) in relation to sex, age, condition and cytochrome P450 enzyme activity. Sci Total Environ 216: 1-11
[42] Harms, G; Rabus, R and Widdel, F (1999) Anaerobic oxidation of the aromatic plant hydrocarbon *p*-cymene by newly isolated denitrifying bacteria. Arch Microbiol 172: 303-312
[43] Eaton, R (1996) *p*-Cumate catabolism pathway in *Pseudomonas putida* F1: Cloning and characterization of DNA carrrying the *cmt* operon. J Bacteriol 178: 1341-1362
[44] Park, YI; So, JS and Koh, SC (1999) Induction by carvone of the polychlorinated biphenyl (PCB)-degradative pathway in *Alcaligenes eutrophus* H850 and its molecular monitoring. J Microbiol Biotechnol 9: 804-810
[45] Jung, KJ; Kim, BH; Kim, E; So, JS and Koh, SC (2002) Monitoring expression of *bphC* gene from *Ralstonia eutropha* H850 induced by plant terpenes in soil. J Microbiol 40: 340-343
[46] Oh, ET; Koh, SC; Kim, E; Ahn, YH and So, JS (2003) Plant terpenes enhance survivability of polychlorinated biphenyl (PCB) degrading *Pseudomonas pseudoalcaligenes* KF707 labeled with gfp in microcosms contaminated with PCB. J Microbiol Biotechnol 13: 463-468
[47] Choi, KY; Kim, D; Koh, SC; So, JS; Kim, JS and Kim, E (2004) Molecular cloning and identification of a novel oxygenase gene specifically induced during the growth of *Rhodococcus* sp. strain T104 on limonene. J Microbiol 42: 160-162
[48] Kim, D; Park, MJ; Koh, SC; So, JS and Kim, E (2002) Three separate pathways for the initial oxidation of limonene, biphenyl, and phenol by *Rhodococcus* sp. strain T104. J Microbiol 40: 86-89
[49] Kim, BH; Oh, ET; So, JS; Ahn, Y and Koh, SC (2003) Plant terpene-induced expression of multiple aromatic ring hydroxylation oxygenase genes in *Rhodococcus* sp. strain T104. J Microbiol 41: 349-352
[50] Warren, R; Hsiao, WWL; Kudo, H; Myhre, M; Dosanjh, M; Petrescu, A; Kobayashi, H; Shimizu, S; Miyauchi, K; Masai, E, *et al.* (2004) Functional characterization of a catabolic plasmid from polychlorinated-biphenyl-degrading *Rhodococcus* sp. Strain RHA1. J Bacteriol 186: 7783-7795
[51] Master, ER and Mohn, WW (2001) Induction of *bphA*, encoding biphenyl dioxygenase, in two polychlorinated biphenyl-degrading bacteria, psychrotolerant *Pseudomonas* strain Cam-1 and mesophilic *Burkholderia* strain LB400. Appl Environ Microbiol 67: 2669-2676

[52] Tandlich, R; Brezna, B and Dercova, K (2001) The effect of terpenes on the biodegradation of polychlorinated biphenyls by *Pseudomonas stutzeri*. Chemosphere 44: 1547-1555

[53] Nishio, T; Patel, A; Wang, Y and Lau, PCK (2001) Biotransformations catalyzed by cloned *p*-cymene monooxygenase from *Pseudomonas putida* F1. Appl Microbiol Biotechnol 55: 321-325

[54] Qiu, XJ; Reed, BE and Viadero, RC (2004) Effects of flavonoids on C-14 7,10 -benzo a pyrene degradation in root zone soil. Environ Eng Sci 21: 637-646

[55] Amaral, JA and Knowles, R (1997) Inhibition of methane consumption in forest soils and pure cultures of methanotrophs by aqueous forest soil extracts. Soil Biol Biochem 29: 1713-1720

[56] Amaral, JA; Ekins, A; Richards, SR and Knowles, R (1998) Effect of selected monoterpenes on methane oxidation, denitrification, and aerobic metabolism by bacteria in pure culture. Appl Environ Microbiol 64: 520-525

[57] Kohler, H-PE; Angst, W; Giger, W; Kanz, C; Muller, S and Suter, MJF (1997) Environmental fate of chiral pollutants-the necessity of considering stereochemistry. Chimia 51: 947-951

[58] Singer, AC; Wong, CS and Crowley, DE (2002) Differential enantioselective transformation of atropisomeric polychlorinated biphenyls by multiple bacterial strains with different inducing compounds. Appl Environ Microbiol 68: 5756-5759

[59] Goris, J; De Vos, P; Caballero-Mellado, J; Park, J; Falsen, E; Quensen, JF; Tiedje, JM and Vandamme, P (2004) Classification of the biphenyl- and polychlorinated biphenyl-degrading strain LB400(T) and relatives as *Burkholderia xenovorans* sp nov. Int J Syst Evol Microbiol 54: 1677-1681

[60] Yu, Z; Stewart, GR and Mohn, WW (2000) Apparent contradiction: Psychrotolerant bacteria from hydrocarbon-contaminated Arctic tundra soils that degrade diterpenoids synthesized by trees. Appl Environ Microbiol 66: 5148-5154

[61] Button, DK (1984) Evidence for a terpene-based food chain in the Gulf of Alaska. Appl Environ Microbiol 48: 1004-1011

[62] Leigh, MB; Fletcher, JS; Fu, XO and Schmitz, FJ (2002) Root turnover: An important source of microbial substrates in rhizosphere remediation of recalcitrant contaminants. Environ Sci Technol 36: 1579-1583

[63] Olson, PE; Wong, T; Leigh, MB and Fletcher, JS (2003) Allometric modeling of plant root growth and its application in rhizosphere remediation of soil contaminants. Environ Sci Technol 37:638-643

[64] Kim, SY; Jung, JY; Lim, YH; Ahn, JH; Kim, SI and Hur, HG (2003) Cis-2', 3'-dihydrodiol production on flavone B-ring by biphenyl dioxygenase from *Pseudomonas pseudoalcaligenes* KF707 expressed in *Escherichia coli*. Antonie Van Leeuwenhoek 84: 261-268

[65] Parrish, ZD; Banks, MK and Schwab, AP (2005) Effect of root death and decay on dissipation of polycyclic aromatic hydrocarbons in the rhizosphere of yellow sweet clover and tall *Fescue*. J Environ Qual 34: 207-216

[66] Shaw, LJ and Burns, RG (2005) Rhizodeposition and the enhanced mineralization of 2,4-dichlorophenoxyacetic acid in soil from the *Trifolium pratense* rhizosphere. Environ Microbiol 7: 191-202

[67] Furukawa, K; Simon, JR and Chakrabarty, AM (1983) Common Induction and regulation of biphenyl, xylene toluene, and salicylate catabolism in *Pseudomonas paucimobilis*. J Bacteriol 154: 1356-1362

[68] Yen, K-M and Gunsalus, IC (1982) Plasmid gene organization: Naphthalene/salicylate oxidation. Proc Natl Acad Sci USA 79: 874-878

[69] Chen, S-H and Aitken, MD (1999) Salicylate stimulates the degradation of high-molecular weight polycyclic aromatic hydrocarbons by *Pseudomonas saccharophilia* P15. Environ Sci Technol 33: 435-439

[70] Rentz, JA; Alvarez, PJJ and Schnoor, JL (2004) Repression of *Pseudomonas putida* phenanthrene-degrading activity by plant root extracts and exudates. Environ Microbiol 6: 574-583

[71] Filonov, AE; Karpov, AV; Kosheleva, IA; Puntus, IF; Balashova, NV and Boronin, AM (2000) The efficiency of salicylate utilization by *Pseudomonas putida* strains catabolizing naphthalene via different biochemical pathways. Process Biochem 35: 983-987

[72] Bogan, BW; Lahner, LM and Paterek, JR (2001) Limited roles of salicylate and phthalate in bacterial PAH bioremediation. Bioremed J 5: 93-100

[73] Uz, I; Duan, YP and Ogram, A (2000) Characterization of the naphthalene-degrading bacterium, *Rhodococcus opacus* M213. FEMS Microbiol Lett 185: 231-238

[74] Aitken, MD; Stringfellow, WT; Nagel, RD; Kazunga, C and Chen, SH (1998) Characteristics of phenanthrene-degrading bacteria isolated from soils contaminated with polycyclic aromatic hydrocarbons. Can J Microbiol 44: 743-752
[75] Kamath, R; Schnoor, JL and Alvarez, PJJ (2004) Effect of root-derived substrates on the expression of nah-lux genes in *Pseudomonas fluorescens* HK44: Implications for PAH biodegradation in the rhizosphere. Environ Sci Technol 38: 1740-1745
[76] Sze, CC and Shingler, V (1999) The alarmone (p)ppGpp mediates physiological-responsive control at the sigma54-dependent Po promoter. Mol Microbiol 31: 1217-1228
[77] Sze, CC; Moore, T and Shingler, V (1996) Growth phase-dependent transcription of the sigma(54)-dependent Po promoter controlling the *Pseudomonas*-derived (methyl)phenol *dmp* operon of pVI150. J Bacteriol 178: 3727-3735
[78] Hugouvieuxcottepattat, N; Kohler, T; Rekik, M and Harayama, S (1990) Growth-phase-dependent expression of the *Pseudomonas putida* Tol plasmid *Pww0* catabolic genes. J Bacteriol 172: 6651-6660
[79] Denef, VJ; Park, J; Tsoi, TV; Rouillard, J-M; Zhang, H; Wibbenmeyer, JA; Verstraete, W; Gulari, E; Hashsham, SA and Tiedje, JM (2004) Biphenyl and benzoate metabolism in a genomic context: outlining genome-wide metabolic networks in *Burkholderia xenovorans* LB400. Appl Environ Microbiol 70: 4961-4970
[80] Yoshitomi, KJ and Shann, JR (2001) Corn (*Zea mays* L.) root exudates and their impact on C-14-pyrene mineralization. Soil Biol Biochem 33: 1769-1776
[81] Cases, I and de Lorenzo, V (2005) Promoters in the environment: Transcriptional regulation in its natural context. Nat Rev Microbiol 3: 105-118
[82] Cases, I; deLorenzo, V and PerezMartin, J (1996) Involvement of sigma(54) in exponential silencing of the *Pseudomonas putida* TOL plasmid Pu promoter. Mol Microbiol 19: 7-17
[83] Huang, WE (Personal Communication). Oxford, UK; 2005

DENDROREMEDIATION: THE USE OF TREES IN CLEANING UP POLLUTED SOILS

TAMAS KOMIVES AND GABOR GULLNER
Plant Protection Institute of the Hungarian Academy of Sciences, Herman Otto ut 15, 1022 Budapest, Hungary, FAX +36-1-4877555, E-mail: tkom@nki.hu

1. Introduction

Forests have provided shelter and habitat for our ancestors for many millennia. During historical times, man grew trees for a number of uses, including energy, furniture, building material, production of paper, fruit and rubber, etc. Recently, trees were introduced for use in dendroremediation, i.e. to depollute contaminated soils. The word dendroremediation comes from the Ancient Greek *dendron* meaning "tree" and Latin *remediare* meaning "reuse" [1-3]. Dendroremediation is an emerging phytoremediation [4-7] technology for cleaning up environment contaminated with organic or inorganic pollutants by using living trees to remove, sequester, or chemically decompose the pollutant [1, 2].

From the point of view of dendroremediation a tree may be considered as a solar driven pump-and-treat system, which may contain a contaminant plume and prevent the spread of contamination by reducing the movement of contaminated water and the erosional transport of contaminated soil. The efficiency dendroremediation has been proven in cleaning up soils polluted with crude oil, explosives, landfill leachates, metals, pesticides, polycyclic aromatic hydrocarbons, and solvents [1-7].

Trees are woody plants characterized with a large biomass, a permanent central self-supporting stem, a stable root system, and a long lifespan. Trees are highly efficient competitors for light, nutrients, and water and tend dominate the vegetation wherever conditions are favourable for plant growth. Since trees are exposed to highly variable biotic and abiotic stresses during their long lifespan they had evolved mechanisms to cope with them. For example, formation of wood can be viewed as an adaptive mechanism that enables trees to secure a dominant position in ecosystems. Wood has many functions that may be important for efficient dendroremediation, e.g. water and nutrient transport and storage of organic compounds and gases [8].

Dendroremediation considers the tree with its physical and biological environment including the soil and the associated microflora [2]. Tree roots are know to produce and release organic chemicals and create a rhizosphere zone more amenable to the microbes that degrade the contaminant. Root exudates such as organic acids and ketones may

M. Mackova et al. (eds.), Phytoremediation Rhizoremediation, 23–31.

promote microbial growth, as may the increase in soil organic matter caused by the roots. Microorganisms fostered by trees in their root zone may contribute significantly to the success of dendroremediation by enhancing the availability of the pollutant for uptake by the plant root system, as well as by degrading some organic pollutants [9, 10].

Much is expected from a plant to be successful in dendroremediation. For efficient uptake of the pollutant a large and deep penetrating root system and a high transpiration rate is important. Large biomass producing, fast-growing, stress-tolerant trees are preferred that are characterised with low nutrient and soil-quality requirement and are capable to survive in a hostile environment and tolerate the phytotoxic effects of the pollutants. In addition, feasible reproduction, propagation and production of the trees are also highly important. Recently, the genera *Salix* (willows and osiers) and *Populus* (i.e. poplars, including aspens and cottonwoods) have emerged as the most efficient systems for dendroremediation [1-7]. Very importantly, the power of poplar as a model system among tree species has been dramatically enhanced by the recent sequencing of *P. trichocarpa* (black cottonwood) [11].

2. Uptake and translocation of the pollutants in trees

Although the binding of pollutant molecules to soil particles can be irreversible, usually desorption occurs: pollutants may move with the soil solution and ultimately reach the groundwater. Efficacy of dendroremediation strongly depends on the bioavailability of the pollutant. Bioavailability is determined by the physical and chemical properties of the pollutant, as well as those of the soil. Uptake of aqueous solutions of inorganic and organic pollutants and their translocation within tree tissues are usually passive processes, regulated by the water transport into the cells. Alternatively, they may also be mediated by membrane-bound transporter systems [6]. Thus, uptake and translocation of a pollutant in trees depends on the pollutant's concentration in the soil solution, its efficiency to enter the root system, and the rate of transpiration in the tree. Trees are known to take up large amounts of water lost from the leaf surface in the transpiration stream. For example, mature poplar trees can transpire 200–1000 liters of water per day [12].

2.1. UPTAKE AND TRANSLOCATION OF INORGANIC POLLUTANTS

In soils metal ions are usually strongly bound to soil particles. To improve the bioavailability of metal micronutrients trees have evolved several strategies [6], e.g. producing and secreting metal-chelating chemicals which, by chelation, mobilize iron, copper and zinc, as well as exuding protons in order to change the pH of the soil in the root zone, thereby solubilising the soil-bound metal ions [13]. The physiological and biochemical mechanisms that explain differences in metal mobility in trees are not well understood [8]. Since in trees metals are transported through the xylem their mobility towards the shoots may be strongly retarded by the high cation exchange capacity of the xylem cell walls. As a result, anionic metal-chelate complexes are more efficiently transported in the transpiration stream. Thus, in the practice of dendroremediation, uptake and accumulation of metals in aerial tissues of plants can be enhanced through the application synthetic and/or natural chelating amendments, such as EDTA and citric acid to the soil [14, 15].

2.2. UPTAKE AND TRANSLOCATION OF ORGANIC POLLUTANTS

Physicochemical and structural properties determine the uptake of organic chemicals by plant roots from the soil [16]. Aqueous solutions of moderately hydrophobic organic chemicals (characterized with an octanol-water partition coefficient [log K_{ow}] of 1.0-3.5), such as low molecular weight aliphatics and aromatics, and chlorinated solvents dissolved in water are readily taken up by roots of trees and translocated to the aerial parts of the plant. Uptake of hydrophilic (log $K_{ow} < 1.0$) and strongly hydrophobic (log $K_{ow} > 3.5$) compounds is much slower and they may be practically unavailable for uptake because of their strong bonding to soil particles or to the roots of the tree. It is interesting to note that efficiencies of uptake of organic pesticides into the crop plant barley [17] and organic pollutants into poplars [16] are closely correlated. Although bioavailability of organic contaminants is typically low when compared to water-soluble inorganics, much less is known about the roles of amendments in the dendroremediation of soils polluted with organic compounds. Thus, contamination by benzene and its alkyl-derivatives (toluene, ethylbenzene, and xylenes) seems to be ideally suited for dendroremediation. However, removal of these aromatics from soil is possible only by increasing their apparent water-solubility. A new approach takes advantage of the ability of cyclodextrins to increase the elution of organic compounds from soils. Cyclodextrins have dual solubilising potency: they may act as surfactants as well as complexing agents that form inclusion complexes with hydrophobic compounds [18].

3. Biotransformation of pollutants in trees

Plant tissues are capable of transforming pollutants by a wide variety of chemical/biochemical metabolic reactions. Rate of metabolism of a pollutant is the main factor in determining sensitivity/tolerance between plant species and has been found to play an important role in the development of stress-resistant plants. Biotransformation reactions of xenobiotics are generally referred to as *Phases I* and *II*, where *Phase I* includes oxidation of xenobiotics and *Phase II* deals with the conjugation of *Phase I* products.

In trees, the oxidative metabolism in the *Phase I* system is usually mediated by cytochrome P-450-containing mixed function oxygenases (CYP, E.C.1.14.-.-) [19, 20]. These enzymes support the oxidative, peroxidative and reductive metabolism of both endogenous and xenobiotic substrates. They comprise a superfamily of heme-thiolate proteins present in every class of organism, including Archaea, and in humans they are responsible for 70-80% of all *Phase I* dependent metabolism of clinically used drugs [21]. In plants there are a surprisingly high number of CYP genes: 246 in *Arabidopsis* (representing approximately 1% of the plant's gene complement) compared with less than a 100 in humans [22]. CYP enzymes are characterized by the high diversity of reactions that they catalyze and the high range of their chemically divergent substrates.

Increasing emphasis on functional genomic approaches to CYP research recently has greatly advanced our understanding of CYP-mediated reactions in plants [22].

In the *Phase II* systems hydrophobic xenobiotics functionalized by the *Phase I* system are converted to more hydrophilic forms *via* conjugation with sugars or sulfhydryl (-SH) group-containing tripeptides, such as glutathione (γ-L-glutamyl-L-cysteinyl-glycine, GSH) [23, 24]. Since endogenous sulphhydryl group-containing chemicals give protection against toxic metal ions as well as against alkylating organic compounds, it is not surprising that GSH-homeostasis in trees is powerfully regulated by pollutants such as the heavy metal cadmium, or the -SH reactive chloroacetanilide herbicides [23].

Xenobiotic (hydrophobic)

↓ ***Phase I***

Functionalized derivative (less hydrophobic)

↓ ***Phase II***

Conjugate (hydrophilic)

↓ ***Phase III***

Terminal metabolite (compartmentised)

Figure 1. Metabolic scheme of xenobiotics in plants.

3.1. TRANSFORMATION PRODUCTS

Detailed information on the chemistry of transformation products of inorganic and organic pollutants and metabolic pathways in susceptible and tolerant trees are scarcely available. In tolerant trees heavy-metal ions may be detoxified via chemical transformation into insoluble forms or chelated with cellular thiols or carboxylic acids and are eventually sequestered into the cell vacuole as described in herbaceous plants [8]. Of the different Phase II reactions that are most commonly involved in pollutant metabolism in trees, conjugation with GSH is one of the most important reactions and often the rate limiting step in the detoxification of an organic compound [23]. GSH transferases (GST, EC. 2.5.1.18) mediate the GSH-conjugation of chloroacetanilide herbicides in poplar trees according to the reaction:

$$\text{GSH} + \text{X-R} \xrightarrow{\textit{GST}} \text{GS-R} + \text{XH}$$

X-R = pollutant
GS-R = glutathione conjugate of the pollutant

Figure 2. Glutathione-conjugation reaction of a pollutant containing an X^- *leaving group.*

GSTs represent a family of enzymes with usually broad and overlapping substrate specificities, which facilitate the above reactions of hydrophobic, electrophilic substrates. Our knowledge on plant GSTs in trees has expanded greatly in recent years. Evidence is accumulating on the regulation of gene expression, molecular characteristics, and specific catalytic action of the multiple forms of these enzymes. The majority of the information on plant GSTs concerns enzymes which are involved in the detoxification of a number of herbicides [23, 24], but evidence is gathering that plant GSTs have a much wider role, and may be involved in general plant stress phenomena [25]. Although the *Phase II* conjugation system is regarded as a detoxification process of xenobiotics, GSH conjugates are not devoid of biological activity. Therefore, processes reducing the concentration of GSH conjugates in the cytosol are important detoxification steps [23].

3.2. BIOTRANSFORMATION OF INORGANIC POLLUTANTS

Because of their active metal-uptake systems plants are vulnerable to toxic levels of metals in their rhizosphere. Plants have evolved several mechanisms to reduce high concentrations of free metal ions in their tissues. These involve metal-chelate formation with amino acids and carboxylic acids, as well as with thiol-rich proteins (metallothioneins, MTs) [26] and oligopeptides (phytochelatins, PCs) [27] followed by a transport of the metal-ligand complex to a subcellular location or to a specific tissue of the plant. MTs are gene-encoded, low-molecular-weight proteins rich in cysteine. They are induced by copper and have high affinity for this metal [26]. PCs, on the other hand, are cysteine-rich oligopeptides that are synthesized from GSH in an enzyme (phytochelatin synthase, EC. 1.2.3.4) catalyzed reaction which is powerfully upregulated by traces of heavy metal ions in the cytosol [27]. Plants can detoxify such metals and metalloids as arsenic, chromium, mercury, and selenium by chemically reducing the element and/or incorporation it into organic compounds [4].

3.3. BIOTRANSFORMATION OF ORGANIC POLLUTANTS

Petroleum products (such as o-xylene, Figure 3) and chlorinated organic solvents, such as carbon tetrachloride, chloroform, and trichloroethylene (TCE, Figure 3) are used in large quantities and are among the most common of the toxic substances found at hazardous waste sites. Leaking underground gasoline storage tanks are the most probable sources of groundwater contamination at numerous polluted sites. One major component of gasoline is methyl-*t*-butyl ether or MTBE (Figure 3), a volatile, toxic chemical. Poplars were successfully used in the dendroremediation of these low-molecular weight compounds [28, 29].

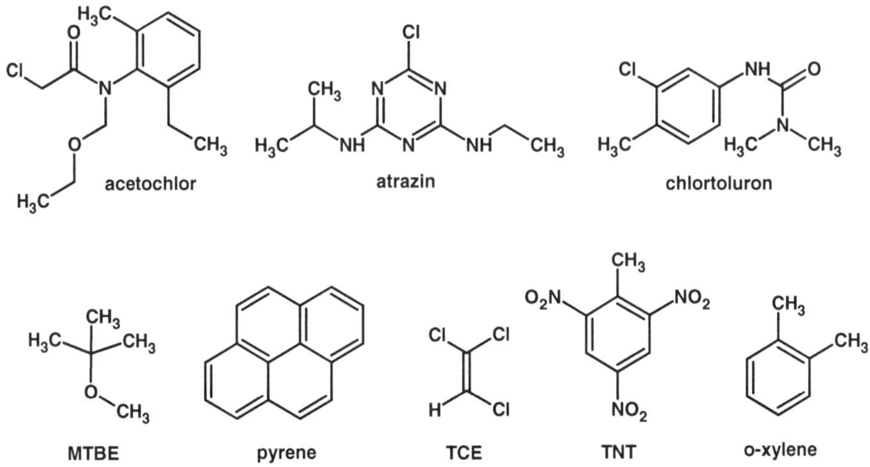

Figure 3. Chemical structures of acetochlor, atrazin, chlortoluron, MTB, pyrene, TCE, TNT, and o-xylene.

Soil and groundwater contamination due to explosives such as glycerol trinitrate, hexahydro-1,3,5-trinitro-1,3,5-triazine, and 2,4,6-trinitrotoluene (TNT, Figure 3) is a problem at many ammunition plants. The presence of trees did enhance removal of these explosives from groundwater [25-27].

Contamination of some soils with herbicides has become a serious environmental problem. Chloroacetanilides, especially acetochlor (Figure 3), alachlor and metolachlor, the chlorotriazine herbicide atrazin (Figure 3) and the urea-derivative chlortoluron (Figure 3) are common contaminants in agricultural settings and at herbicide manufacturing sites [28-30]. Since poplar trees tolerate low concentrations of these herbicides, dendroremediation is an attractive option to reduce their levels in the soil and in the roundwater [28, 29].

Dendroremediation of pollutants that are highly phytotoxic (such as herbicides) or highly persistent (such as polyaromatic hydrocarbons [PAHs], polychlorinated biphenyls [PCBs], and dioxins) are most difficult. [31-34]. PAHs (for example pyrene [Figure 3] and POPs are very hydrophobic. Therefore, their uptake by roots is very low. In addition, chemically they are very stable: plant and microbial enzymes can degrade them only very slowly. In spite of all the difficulties, introduction of xenobiotics-detoxifying enzymes from other plants or from heterologous species (animals and bacteria) to trees has been shown to enhance metabolism of organic pollutants. For example, poplar trees expressing the mammalian cytochrome P450 2E1 also metabolized TCE at an elevated rate [35]. Poplars, containing the bacterial γ-EC synthase had increased ability to tolerate and detoxify chloroacetanilide herbicides [23].

Plants lack the excretion system of animals. In plant cells toxic metabolites and pollutants are sequestered into the vacuole. This *Phase III* type process is an active one and is catalyzed by membrane-bound ATP-driven pumps. A recent study indicated the existence of a *Phase III* system in poplar [36].

4. Detoxification of the active oxygen species generated by the pollutant in trees

Activated oxygen species (hydrogen peroxide, superoxide anion, and hydroxyl radical) are produced at various electron transfer sites or *via* different oxidation reactions in tree tissues [37]. Under chemical stress conditions production of these radicals is powerfully enhanced. For example, cadmium-induced inhibition of ascorbate peroxidase (E.C. 1.11.1.11) and catalase (E.C. 1.11.1.6) was also associated with H_2O_2 accumulation and growth retardation in the poplar roots [38].

Trees contain a variety of defences to protect against the damaging effects of oxygen radicals that are produced. It has been shown that a critical balance exists between oxyradical-generating factors and the activity of the systems that protect the cell from their harmful effects. Antioxidant defences belong to three general classes including:

- water-soluble reductants, e.g. compounds that contain thiol-groups (cysteine, GSH, etc.), ascorbate and catechols;
- lipid-soluble compounds, e.g. γ-tocopherol and β-carotene; and
- enzymatic antioxidants, e.g. GSH peroxidase (GP, E.C. 6.4.11.6), ascorbate peroxidase, catalase, and superoxide dismutase (E.C. 1.15. 1.1) [39].

Microsomal and cytosolic GST enzymes in mammals may act as GP by catalyzing the reaction between GSH and lipophilic hydroperoxides (Figure 4) thereby protecting cell membrane polyunsaturated fatty acid moieties against lipid peroxidation. Poplar trees were recently shown to contain a number of GST isozymes [36], including those capable of detoxifying lipid hydroperoxides.

$$\mathrm{GSH} + \mathrm{ROOH} \xrightarrow{GST} \mathrm{GS\text{-}SG} + \mathrm{ROH} + \mathrm{H_2O}$$

ROOH = lipid hydroperoxide
ROH = lipid alcohol

Figure 4. Glutathione-peroxidase activity of plant GST enzymes.

4. Conclusions

Trees are well suited to be the key plants for phytoremediation – a vehicle with which the growing needs to depollute agricultural, industrial, military and communal sites can be fulfilled. Poplars seem to have many advantages as the first choice for the above purpose but continued research is necessary to fully exploit their potential. This can be achieved if simple, standardized methods for characterizing clone properties useful for the dendroremediation of a range of inorganic and organic pollutants alone or in mixtures are developed and a functional analysis of the genes that are most important for tree growth, development, and stress tolerance will be available. It is also critical to identify and characterize the genes that are involved in the uptake, translocation and detoxification (accumulation) of pollutants in plants to make dendroremediation applicable in under various environmental conditions. Finally, further research is necessary to clarify the roles woody tissues play in adaptation of trees to environmental stress.

References

[1] Dickmann, DI; Nguyen, PV and Pregitzer, KS (1996) Effects of irrigation and coppicing on above-ground growth, physiology, and fine-root dynamics of two field-grown hybrid poplar clones. J Forest Ecology and Management 80: 163-174
[2] Schoenmuth, BW and Pestemer, W (2004) Dendroremediation of trinitrotoluene (TNT). Part 1: Literature overview and research concept. Environ Sci Pollut Res 11: 273-278
[3] Schoenmuth, BW and Pestemer, W (2004) Dendroremediation of trinitrotoluene (TNT). Part 2: Fate of TNT in morphological compartments of trees. Environ Sci Pollut Res 11: 331-339
[4] Komives, T and Gullner, G (2000) Phytoremediation, in: RE Wilkinson, Ed., Plant-Environment Interactions, pp. 437-452. Marcel Dekker Inc., New York, U.S.A.
[5] Macek, T; Mackova, M and Kas, J (2000) Exploitation of plants for the removal of organics in environmental remediation. Biotechnol Advances 18: 23-35
[6] Pilon-Smits, EAH (2005) Phytoremediation. Ann Rev Plant Biol 56: 15-39
[7] Dietz, AC and Schnoor, JL (2001) Advances in phytoremediation. Environmental Health Perspectives, 109: 163-168
[8] Pulford, ID and Watson, C (2003) Phytoremediation of heavy metal-contaminated land by trees - a review. Environ Int 29: 529-540
[9] Tesar, M; Reichenbauer, TG and Sessitsch, A (2002) Bacterial rhizosphere populations of black poplar and herbal plants to be used for phytoremediation of diesel fuel. Soil Biol Biochem 34: 1883-1892
[10] Jordahl, JL; Foster, L; Schnoor, JL and Alvarez, PJJ (1997) Effect of hybrid poplar trees on microbial populations important to hazardous waste bioremediation. Environ Toxicol Chem16: 1318-1321
[11] Brunner, AM; Busov, VB and Strauss, SH (2004) Poplar genome sequence: functional genomics in an ecologically dominant plant species. Trends in Plant Science 9: 49-55
[12] Wullschleger, S; Meinzer, F and Vertessy, RA (1998) A review of whole-plant water use studies in trees. Tree Physiol 18: 499-512
[13] Robinson, BH; Mills, TM; Petit, D; Fung, LE; Green, SR and Clothier, BE (2000) Natural and induced cadmium-accumulation in poplar and willow: Implications for phytoremediation, Plant and Soil 227: 301-306
[14] Fodor, F; Gaspar, L; Morales, F; Gogorcena, Y; Lucena, JJ; Cseh, E; Kropfl, K; Abadia, J and Sárvári, E (2005) Effects of two iron sources on iron and cadmium allocation in poplar (*Populus alba*) plants exposed to cadmium. Tree Physiology 25: 1173-1180
[15] Liphadzi, MS; Kirkham, MB; Mankin, KR and Paulsen, GM (2003) EDTA-assisted heavy-metal uptake by poplar and sunflower grown at a long-term sewage-sludge farm. Plant and Soil 257: 171-182
[16] Burken, JG and Schnoor, JL (1998) Predictive relationships for uptake of organic contaminants by hybrid poplar trees. Environ Sci Technol 32: 3379-3385
[17] Briggs, GG; Bromilow, RH and Evans, AA (1982) Relationship between lipophilicity and root uptake and translocation of nonionized chemicals by barley. Pestic Sci 13: 495-504
[18] Tanada, S; Nakamura, T; Kawasaki, N; Torii, Y and Kitayama, S (1999) Removal of aromatic hydrocarbon compounds by hydroxypropyl-cyclodextrin. J Colloid Interface Sci 217: 417-419
[19] Komives, T and Gullner, G (2005) Phase I xenobiotic metabolic systems in plants. Z Naturforsch C 60: 179-185
[20] Coleman, JOD; Frova, C; Schroder, P and Tissut, M (2002) Exploiting plant metabolism for the phytoremediation of persistent herbicides. Environ Sci Pollut Res Int 9: 18-28

[21] Danielson, PB (2002) The cytochrome P450 superfamily: biochemistry, evolution and drug metabolism in humans. Current Drug Metab 3: 561-97
[22] Nelson, DR; Schuler, MA; Paquette SM, Werck-Reichhart, D and Bak, S (2004), Comparative genomics of rice and arabidopsis. Analysis of 727 cytochrome P450 genes and pseudogenes from a monocot and a dicot. Plant Physiol 135: 756-772
[23] Gullner, G; Komives, T and Rennenberg, H (2001) Enhanced tolerance of transgenic poplar plants overexpressing γ-glutamylcysteine synthase towards chloroacetanilide herbicides. J Exp Bot 52: 971-979
[24] Schroeder, P and Collins, C (2002) Conjugating enzymes involved in xenobiotic metabolism of organic xenobiotics in plants. Int J Phytoremediation 4: 247-265
[25] Tausz, M; Gullner, G; Komives, T and Gril, D (2003) The role of thiols in plant adaptation to environmental stress, in Sulphur in Plants, Abrol,YP and Ahmad, A, Eds., Kluwer Academic Publishers, Dordrecht, The Netherlands. pp. 221-244
[26] Kohler, A; Blaudez, D; Chalot, M and Martin, F (2004) Cloning and expression of multiple metallothioneins from hybrid poplar. New Phytologist 164: 83-93
[27] Cobbett, CS (2000). Phytochelatins and their roles in heavy metal detoxification. Plant Physiol 123: 825-832
[28] Hong, MS; Farmayan, WF; Dortch, IJ; Chiang, CY; McMillan, S. and Schnoor, JL (2001) Phytoremediation of MTBE from a groundwater plume. Environ Sci Technol 35: 1231-1239
[29] Gordon, M; Choe, N; Duffy, J; Ekuan, G; Heilman, P; Muiznieks, I; Ruszaj, M; Shurtleff, BB; Strand, S; Wilmoth, J and Newman, LA (1998) Phytoremediation of trichloroethylene with hybrid poplars. Environ Health Perspect 106: 1001-1004
[25] Thompson, PL; Ramer, LA and Schnoor, JL (1998) Uptake and transformation of TNT by hybrid poplar trees. Environ Sci Technol 32: 975-980
[26] Thompson, PL; Ramer, LA and Schnoor, JL (1999) Hexahydro-1,3,5-trinitro-1,3,5-triazine translocation in hybrid poplar trees. Environ Toxicol Chem 18: 279-284
[27] Yoon, JM; Oh, BT; Just, CL and Schnoor, JL (2002) Uptake and leaching of octahydro-1,3,5,7-tetranitro-1,3,5,7-tetrazocine by hybrid poplar trees. Environ Sci Technol 36: 4649-55
[28] Komives, T; Gullner, G; Rennenberg, H and Casida, JEC (2003) Ability of poplar (*Populus* spp.) to detoxify chloroacetanilide herbicides. Water, Air and Soil Pollution: Focus. 3: 277-283
[29] Burken, JG and Schnoor, JL (1997) Uptake and metabolism of atrazine by poplar trees. Environ Sci Technol 31: 1399-1406
[30] Mezzari, MP; Walters, K; Jelinkova, M, Shih, M-C; Just, CL and Schnoor JL (2005) Gene expression and microscopic analysis of Arabidopsis exposed to chloroacetanilide herbicides and explosive compounds. A phytoremediation approach. Plant Physiol 138: 858-869
[31] Liste, HH and Alexander, M (2000) Plant-promoted pyrene degradation in soil. Chemosphere 40: 7-10
[32] Francova, K; Mackova, M; Macek, T and Sylvestre, M (2004) Ability of bacterial biphenyl dioxygenases from *Burkholderia* sp. LB400 and *Comamonas testosteroni* B-356 to catalyse oxygenation of ortho-hydroxychlorobiphenyls formed from PCBs by plants. Environmental Pollution, 127: 41-48
[33] Campanella, B; Bock, C and Schroder, P (2002) Phytoremediation to increase the degradation of PCBs and PCDD/Fs. Environ Sci Pollut Res 9: 73-85
[34] Macek, T; Sura, M; Pavlikova, D; Francova, K; Scouten, WH; Szekeres, M; Sylvestre, M and Mackova, M (2005) Can tobacco have a potentially beneficial effect to our health? Z Naturforsch C 60: 292-299
[35] Strand, SE; Dossett, M; Harris, C; Wang, X and Doty, SL (2005) Mass balance studies of volatile chlorinated hydrocarbon phytoremediation Z Naturforsch C 60: 325-330
[36] Rishi, AS; Munir, S; Kapur, V; Nelson, ND and Goyal, A (2004) Identification and analysis of safener-inducible expressed sequence tags in *Populus* using a cDNA microarray. Planta 220: 296-306
[37] Noctor, G; Arisi, A C M., Jouanin, L; Kunert, KJ; Rennenberg, H and Foyer, CH (1998) Glutathione: biosynthesis, metabolism and relationship to stress tolerance explored in transformed plants. J Exp Botany 49: 623-47
[38] Schutzendubel. A; Nikolova, P; Rudolf, C and Polle, A (2002) Cadmium and H_2O_2-induced oxidative stress in *Populus x canescens* roots. Plant Physiol Biochem 40: 577-584
[39] Komives, T; Gullner, G and Kiraly, Z (1997) The ascorbate-glutathione cycle and oxidative stresses in plants, in KK Hatzios, Ed., Regulation of Enzymatic Systems Detoxifying Xenobiotics in Plants, Dordrecht: Kluwer Academic Publishers, pp. 85-96

METHODS FOR RHIZOREMEDIATION RESEARCH

Approaches to experimental design and microbial analysis

MARY BETH LEIGH
Center for Microbial Ecology, Michigan State University, 540 Plant and Soil Sciences Building, East Lansing, MI 48824-1325 USA, Fax:(517)353-2917, E-mail: leigh@msu.edu

1. Introduction

Rhizoremediation is an elegant form of bioremediation that seeks to harness light energy via plants to biostimulate pollutant degradation by the indigenous soil microbial community. In pursuit of this goal, rhizoremediation draws upon the fields of rhizosphere ecology and microbial biodegradation, and in doing so confronts many of the same questions that have challenged these fields for decades, including: What mechanisms dictate microbial community structure and activity in the rhizosphere? Which microorganisms and degradative genes are actively involved in biodegradation, and how does their activity respond to different treatments? One major obstacle to answering these fundamental mechanistic questions over the years, and subsequently developing successful rhizoremediation technology, has been our limited methodological capabilities to quantify the degradative potential of microbial communities and to demonstrate and understand their response to the unique and multi-faceted environment of the plant-soil interface.

Microbiological research methods are the windows through we view the hidden world of microorganisms, with each being of different size and dimension, yet none affording a complete view of the diversity, function and abundance of microbial populations within a community. Cultivation methods are notoriously limited in their ability to detect the uncultivable majority of organisms in the environment, yet remain informative and indispensable for studying the metabolic capabilities of particular strains. Molecular biological methods like functional gene detection and stable isotope probing open much larger windows by circumventing culture bias and allowing the direct examination of microbial populations important to bioremediation, although they are limited somewhat by available sequence data. One aim of this chapter is to provide an overview of microbial methodologies, both cultivation-dependent and independent, which have already or are likely to provide new insight into the multitude of mechanistic and applied rhizoremediation questions to be answered.

M. Mackova et al. (eds.), Phytoremediation Rhizoremediation, 33–55.

Microbial analyses are most valuable to rhizoremediation when conducted within the context of comprehensive, carefully designed experiments to evaluate the effectiveness of different plant species and to uncover the mechanisms that drive rhizodegradation. Accomplishing conclusive, informative experiments can be a challenging task in light of the extraordinary heterogeneity of soil and rhizosphere systems and the long-term nature of rhizostimulation. This chapter considers a variety of different experimental approaches, ranging from controlled greenhouse pot experiments to forensic field studies, and discusses the relationship between the hypothesized mechanism of rhizostimulation and selection of the appropriate experimental system.

While this overview is primarily drawn from rhizoremediation studies of organic aromatic pollutants like polychlorinated biphenyls (PCBs) and polyaromatic hydrocarbons (PAHs), the guidelines are easily adaptable to many other organic contaminants. Together, the experimental design strategies and microbial analytical methods discussed are aimed at assisting researchers in the creation of studies that will successfully answer an array of rhizoremediation questions, both mechanistic and applied.

2. Experimental approaches and design

When planning a study to assess the influence of plants on pollutant degradation, selecting the appropriate experimental approach is critical. Will conclusive results best be achieved by pot studies, field plots or less conventional approaches like forensic field studies of long-term sites? The answer lies in part in the mechanisms and time scale by which detectable rhizostimulation and/or contaminant disappearance is hypothesized to occur. In this section, the benefits and drawbacks of different experimental approaches are discussed and guidelines are provided to assist in their successful implementation.

2.1. FORENSIC FIELD STUDIES

While conventional wisdom may dictate a progression from bench to greenhouse to field-testing of remedial technologies, there are distinct advantages to starting directly in the field to examine the rhizoremediation capabilities of natural vegetation. This approach affords a view of the long-term impacts of a diverse array of plant species, including mature trees, on microbial communities and/or contaminant disappearance otherwise inaccessible due to the practical time limitations of most planted studies. Forensic studies are also invaluable as a means of screening species for their ability to prosper in contaminated soil and under environmental stresses (i.e. drought and nutrient stress) without active cultivation. However, the numerous abiotic heterogeneities and lack of controlled conditions at natural field sites can make determinations of true rhizoremediation effects challenging.

The forensic approach was first applied by Olson *et al.* [1-4] at a naturally-vegetated PAH-contaminated sludge disposal basin in Texas City, Texas. Historical aerial photographs and tree ring analyses revealed that plants colonized the basin over the course of 16 years following the drainage of standing water. Through plant invasion and succession, a diverse plant community (51 species and 22 families) developed that was dominated by mulberry trees (*Morus* spp.), Bermuda grass (*Cynodon dactylon*) and common sunflowers (*Helianthus annuus*). In the absence of baseline data regarding original contaminant levels, PAH concentrations beneath the root zone were considered representative of the parent sludge which averaged 16,854 mg/kg total PAHs. Coring, excavation and chemical analyses throughout the site indicated that PAH concentrations in the root zones of plants were reduced to 10-50% of that in the original sludge. Microbial analyses indicated that cultivable PAH utilizing bacteria were five-fold more numerous ($p < 0.01$) in the mulberry root zone than in non-rooted samples of the same depth, and these results were confirmed with real-time PCR quantitation of degradative genes [5-8]. The close correspondence between root zone depth, sludge PAH concentration and abundance of PAH-degrading bacteria suggests that plants and associated microflora facilitated pollutant degradation.

Areas with deep and relatively homogeneous contamination like the sludge basin in Texas City are fortuitous opportunities for forensic rhizoremediation studies based on chemical disappearance. However in many cases contaminated field sites are a result of heterogeneous surface spillage of chemicals, making a forensic determination of chemical disappearance nearly impossible. Even in the absence of chemical disappearance data, naturally vegetated contaminated sites afford a good opportunity to screen various plant species for their tolerance to contaminants and their long-term influence on the degradative potential of the associated microbial community.

A forensic field study focusing on rhizosphere microbial populations was performed in a PCB contaminated site near Uherské Hradiste in South Moravia, Czech Republic [9-11]. Accidental spillage of PCBs onto the soil occurred from the mid 1950s to the mid 1980s, resulting in a heterogeneously contaminated area with soil PCB concentrations ranging from 1-500 mg/kg. The site was naturally vegetated with 25 different plant species, including 5 tree species, all rooted directly in contaminated soil of varying concentrations. The abundance and identity of cultivable PCB-metabolising bacteria were determined in soil and rhizosphere samples collected at various depths beneath the trees, grasses and forbs on four different dates during one year. In order to distinguish effects of plants from abiotic factors on microbial populations, PCB concentration and soil moisture content were also analysed but were not found to significantly correlate with numbers of PCB-degrading bacteria. Austrian pine (*Pinus nigra*) and goat willow (*Salix caprea*) trees fostered significantly higher numbers of cultivable PCB-metabolising bacteria in their root zones than other plant species or non-rooted soil of equivalent depth (collected beneath shallow-rooted grasses). The results imply that long term growth of certain plant species can increase numbers of contaminant degraders in the bulk soil.

Forensic field studies, by virtue of their scale and scope, are large interdisciplinary undertakings that incorporate plant taxonomy and ecology, soil biochemistry, soil/rhizosphere microbiology and complex statistical analyses. If planning to embark on a forensic field study, is it recommended that the assistance of collaborators with expertise in these fields be sought early in the process. As much historical information as possible should be collected regarding the time of contamination, previous measurements of concentration, presence of co-contaminants and vegetation history. A comprehensive sampling strategy should be planned (see Sampling Methods section) and discussed in depth with an ecological statistician. A particular challenge in forensic studies is distinguishing true plant-mediated effects from those created by numerous abiotic heterogeneities. Multiple measurements (i.e. contaminant concentration, soil moisture, pH, nutrients, depth, etc.) should be made of each sample and subjected to statistical analyses along with microbial population data. Multivariate statistical methods are particularly well-suited to this purpose, and a number of useful books on ecological multivariate statistics are available to provide guidance [12-15].

2.2. PLANTED STUDIES

Rhizoremediation experiments conducted by planting and monitoring microbial populations and/or contaminant levels over time are powerful approaches since they can be constructed and operated with replication, randomisation and controlled conditions. Perhaps the major limitation of planted studies is time, since for reasons of practicality planted studies are typically only run for 1-3 years. Because roots are estimated to occupy less than 1% of soil [16], it is reasonable to anticipate that many years of root exploration would be required to thoroughly treat a contaminated area. Rates of contaminant disappearance also indicate that rhizoremediation should be regarded as a long-term process [17]. Nonetheless, losses in soil pollutant concentration and/or increased populations of degraders have been observed within several years or less [17-21]. In the event that pollutant disappearance occurs too slowly to detect, a comprehensive planted study still affords an excellent opportunity to investigate mechanisms of remediation, such as microbial population shifts at the root-soil interface.

When designing planted studies, as with any controlled experiment, careful and statistically conscious design is critical to ensure conclusive results. Replicates of each treatment including unplanted soil as a control should be included. Numbers of replicates should be as high as reasonably achievable without compromising the quality of the data since higher numbers of replicates produce more accurate results and a lower likelihood of producing false positive/negative results. Randomised arrangements of treatments, whether in pots or field plots, should be employed to ensure that differences observed are due to treatments rather than variations in light, soil composition, moisture, etc. associated with different locations in a greenhouse or outdoor area. Additional guidance may be found in books focusing on basic statistics [22] and experimental design [23]. The following sections discuss additional design considerations specific to pot or field plot studies.

2.2.1 Pot studies

Performing studies in containers is an immediately attractive option because it affords greater homogeneity than can be achieved in an intact field site, keeps plant roots contained and therefore maintains discrete species-specific influences on soil. However, pot studies are subject to criticism regarding their ability to realistically replicate natural conditions that may impact rhizoremediation processes, especially when conducted in a greenhouse.

Pot experiments performed in a greenhouse are convenient in that they provide independence from seasonal limitations on outdoor research. Unfortunately, plants are shielded from environmental cues that influence plant physiological responses important to some rhizoremediation mechanisms. One issue is that greenhouse glass filters out a portion of ultraviolet light from solar radiation. Ultraviolet light is reported to promote the production of plant secondary compounds [24], which are thought to play a major role in biostimulation and/or induction of aromatic pollutant degrading bacteria [25] (detailed in the following chapter by Andrew C. Singer). If studies are conducted in the greenhouse, supplementation with UV light is recommended since it has been demonstrated to increase the concentrations of phenolic compounds produced by plants to levels similar to those obtained under outdoor growth conditions [24].

Greenhouse studies also shield perennial plants from seasonal signals that dictate cycles of senescence and dormancy, including tree root turnover events [26]. Root turnover may be an important mechanism in the delivery of secondary plant metabolites that support the growth of PCB-degrading bacteria [27]. If a concern, this can be remedied by running the pot study outdoors where plants can be exposed to natural environmental conditions. For additional realism, pots may be buried in the soil to help maintain ambient soil temperature. Whether indoors or outdoors, root growth in pots is often very different from bulk soil. Roots tend to grow along the pot edges, coil along the bottom, and may be significantly less branched than field-grown roots, which could deleteriously effect rhizoremediation results (John Fletcher, 2005, personal communication).

Pot studies are advantageous over field studies since they afford the opportunity to minimize heterogeneity of contaminant concentrations. To take advantage of this, care should be taken to homogenize soil thoroughly in a large soil mixer or similar equipment before filling pots, since soil either collected from a contaminated site or manually spiked with pollutants tends to be heterogeneous. At or prior to the initiation of the study, chemical analyses of replicate samples should be performed to assess the variability of pollutant concentrations within and among pots and help determine adequate sample replication. When natural environmental soils are used without additives to provide loose texture (i.e. vermiculite), they can become very compacted after continued watering. If considering the addition of organic material to the potting mix, note that some materials (i.e. bark mulch) contain high concentrations of plant aromatic compounds that can interfere with gas chromatographic analyses of polyaromatic hydrocarbons in soil as well as impact microbial communities.

Another consideration with pot studies is that contaminants may leach out of drainage holes posing not only a chemical hygiene problem but also creating a false positive response when chemical disappearance data are analyzed. Care should be taken to avoid excessive watering, and leachates should be collected for chemical analyses especially if working with water-soluble contaminants. Unplanted controls introduce unique complications with watering regime. If watered uniformly with planted treatments, soil in unplanted pots may remain saturated, generating hypoxic conditions distinctly different from soil conditions in planted pots. This compromises important control data for background levels of microbial activity as well as for volatilization of pollutants through soil pores. A solution to this problem is to maintain soil moisture at similar levels rather than to use equal water volumes, and to monitor outflow for escaping contaminants.

2.2.2 Field plots

Planted field plots afford the opportunity to evaluate the rhizoremediation potential of a variety of plant species under more realistic environmental conditions than pot studies. While heterogeneity of soil contamination is often greater, it can generally be overcome with a randomised layout such as Latin square or randomised block design and careful statistical analyses [22, 23]. Many precautions mentioned in the previous section also apply, including attention to the watering regime and chemical analysis of preliminary samples to determine contaminant variability.

A unified experimental design strategy for field investigations of rhizoremediation has been developed by the U.S. Environmental Protection Agency's Remediation Technologies Development Forum (RTDF) Phytoremediation Action Team. The strategy was designed to evaluate the efficacy of various agricultural and non-crop herbaceous plants for the rhizoremediation of weathered petroleum hydrocarbons. Use of the unified protocol permits direct comparisons of data by researchers in a wide range of geographic locations and climates. Having been carefully developed and debated by numerous researchers, the RTDF experimental design provides a useful general framework for field plot studies, and may be easily adapted to study other plants or contaminants. The experimental method involves three different planted treatments in a randomised block design with each block being a minimum of 6.1 m square in size. The three treatments are 1) a mixture of species optimised for local conditions which may include grasses and trees 2) an unplanted control and 3) a standard seed mixture, including some flexibility in plant species to accommodate local conditions, of 10-15% rye, 20-25% legume and 60-70% fescue. Additional blocks may be added with plants of particular interest to the investigator. Plots are planted and monitored over the course of 3 growing seasons for chemical disappearance and microbial populations. Detailed sampling methods, chemical, microbial and plant analyses are provided in the protocol, which is available at http://www.engg.ksu.edu/HSRC/appa.html.

3. Sampling methods

Strategies for sampling forensic or planted rhizoremediation studies are an integral component of experimental design, and ultimately will impact how accurately the data reflect true conditions. The first step in developing a sampling strategy is to define the compartments of interest within the complex root-soil interface (rhizosphere, rhizoplane, bulk soil, etc.) and select a sampling method that is both appropriate and feasible. Multiple samples must be collected from each replicate treatment in a representative manner, with consideration of temporal and spatial issues. Lastly, samples must be transported and stored for microbial or chemical analyses in ways that will minimize the introduction of artefacts.

3.1. DEFINING AND SAMPLING THE ROOT-SOIL INTERFACE

Before sampling the root-soil interface, one has first to define it. When the term "rhizosphere" was first coined by Lorenz Hiltner in 1904, it was defined simply as the zone of soil in which the microflora is influenced by plant roots. The zone of influence has since been subdivided into several subgroups to distinguish microbial consortia based on their physical location relative to the root. Although the terminologies and definitions of each subgroup vary somewhat, in general the root-influenced microbial community is divided into those residing in the interior of roots (endophytes or endorhizosphere), on the root surface (rhizoplane), in the soil immediately surrounding the root (rhizosphere), and beyond the root into the root zone or bulk soil. Dead roots are rarely included in rhizosphere studies, however they may harbour distinct microbial populations including bacteria relevant to rhizoremediation [27, 28].

In practice, the rhizosphere is defined operationally. The traditional method for sampling the rhizosphere is to excavate roots carefully and then to shake off loose soil, keeping roots with the small amount of soil that remains adhered for analysis [29]. Although this rhizosphere sampling method may be imprecise since the amount of soil that adheres varies with soil moisture and composition, it remains the most practical, widely used and accepted technique. Following transport to the laboratory, the rhizosphere soil is separated from the root by washing with a diluent, and the resultant suspension is subjected to microbial analyses. The remaining root may then be further processed to recover rhizoplane and endorhizosphere organisms, as detailed later in the microbial analysis section.

The soil beneath a plant that is not defined operationally as the rhizosphere is often referred to as the bulk soil or the root zone. Plant species-specific effects on microbial communities are frequently observed in this soil fraction [9, 30-32], including effects on numbers of pollutant degraders in rhizoremediation studies [9, 30]. These findings indicate that the plant's influence extends beyond the immediate vicinity of roots, which is not surprising considering that fine roots are continually exploring new regions of soil and dying back, impacting the soil both with exudates from living roots and lysates from detritus.

In the field, sampling the root zone of trees and shrubs is achieved by excavation in the densely rooted region of soil, which generally occupies the area beneath the canopy. Root zone soil samples can be accessed by excavation or coring, although the former is often most practical, especially when root and rhizosphere samples are targeted as well. It is recommended that sampling tools (shovels, augers) be disinfected with alcohol between samples to minimize cross contamination.

For pot studies, soil and root sampling may easily be performed by removal of the soil and root complex from the pot followed by dissection. If plants are to be repeatedly sampled, coring may also be used to collect soil samples, although it is generally difficult to obtain sufficient root material for analysis from small cores.

3.2. SPATIAL SAMPLING STRATEGY

Soil is notoriously heterogeneous at both the macro and the micro scale, and plant roots introduce further variation. The main focus of this section is to provide guidelines on sampling schemes that can provide an accurate and statistically valid representation of the treatment in question. As discussed previously, the appropriate sample size can be determined by careful preliminary studies to determine the variability within the experimental system.

When sampling a field plot study with a randomised block design, multiple replicate samples should be collected within each block. The arrangement of the sampling locations within the block may be either random or systematic. There are a number of ways to design systematic sampling schemes, such as along a grid pattern or several transects [33, 34]. Composing of samples from several different locations is not recommended because it reduces sample size and precludes determinations of variability that are important to determining statistical differences among treatments.

In pot studies, multiple samples should be collected from each pot. If the whole plant root system is harvested, different zones of soil may be separated, such as into quadrants and by depth. Analysing multiple separate samples from each pot helps provide the statistical power needed to discern relatively small but significant effects with confidence, as well as to reveal the heterogeneity of the system.

In forensic studies, the distribution of plants will be random and hence rhizosphere sampling will be dictated primarily by the location of the plants. Multiple replicate samples should be collected from each plant, such as in a circumference around a tree. For purposes of site characterization and mapping of contaminant concentration and other factors, it is recommended that a comprehensive set of samples also be collected throughout the entire site on at least one time point using either a randomised or systematic pattern.

3.3. SEASONAL SAMPLING CONSIDERATIONS

Seasonal fluxes in environmental conditions are unavoidable, and not necessarily detrimental, in real-world applications of rhizoremediation technology. Since both plant root physiology (i.e. root turnover) [26] and soil microbial communities [35, 36] respond to seasonal changes, it is anticipated that degradative potential of the microbial community will also fluctuate. For this reason, research efforts should seek whenever possible to understand rhizoremediation process not just in the active summer growing season but throughout the seasonal cycle and under varying moisture conditions. Recognizing these influences on microbial populations important to rhizoremediation may help to formulate site management practices and provide new insights into the mechanisms of rhizoremediation.

At minimum, it is recommended that samples be collected from outdoor rhizoremediation studies at 3-4 times per year to reflect early spring, summer and late autumn conditions. Because major root turnover events typically occur in the autumn in rough synchrony with leaf senescence [26], samples should be collected after occurrence of complete leaf fall for deciduous plants or visible shoot dieback for annuals or biennials. While more frequent sampling events are desirable, care should be taken to ensure that analysing the large number of samples resulting from aggressive sampling is feasible.

At a PCB-contaminated site in the Czech Republic, populations of PCB-degrading bacteria were enumerated in June, August, November and the following May. Significant seasonal differences in population size were detected beneath the Austrian pine tree, in which numbers of degraders increased significantly between August and November [9], coinciding with expected root turnover events. Although the cause of the increase remains unknown, it exemplifies the seasonal flux in degradative populations that can occur in a rhizoremediation setting.

3.4. SAMPLE STORAGE

Root and soil samples collected in the field can be conveniently stored in self-sealing thin-walled plastic bags, which permit slow gas exchange (O_2 and CO_2) while preventing soil drying [33]. If samples are destined for chemical analyses of contaminants, exposure to common flexible plastics should be avoided because they can introduce phthalate esters into the sample that interfere with analyses of many pollutants including pesticides and PCBs. Instead, inert sample vessels such as glass bottles should be used. For more detailed recommendations regarding sample storage for chemical analyses for particular pollutants, consult U.S. EPA Methods such as Method 8081.

Samples are commonly transported to the laboratory as quickly as possible on ice to reduce microbial activity, and then they are stored at 4°C until analysis. Cultivable or direct microbial analyses should be performed as soon as possible, since sample storage results in changes in microbial properties [33]. For molecular-based analyses of microbial communities, subsamples should be frozen at –20°C or –70°C for later extraction and analyses of DNA or RNA, respectively. Commercially available RNA protectants may be added before freezing to inhibit RNase activity when samples thaw. However, nucleic acid extraction recovery in the presence of the protectant should be tested in advance since these protectants can interfere with some protocols.

If microbial, molecular or pollutant analyses are to be reported per unit of dry weight of soil or root material, then aliquots of each samples should be weighed and dried to obtain ratios of fresh weight to dry weight.

4. Microbial analyses

Once samples have been collected, one turns to the task of investigating the microbial populations contained therein. What is the best approach to identify, quantify or characterize pollutant degraders and to compare them among samples? There are a host of different microbiological methods for these purposes, each with its own set of advantages and limitations. Cultivation-based methods present a notoriously incomplete picture of the microbial community due to the "great plate count anomaly", in which only 0.1 - 10% of bacteria in the environment are cultivable in the laboratory. However, cultivation methods are widely used for their simplicity and cost-effectiveness, and because they remain the primary way to demonstrate metabolic capabilities of individual strains. Molecular tools including stable isotope probing, quantitative real-time PCR, high-throughput sequencing and microarrays provide unprecedented access to data not subject to issues of culture bias, yet are somewhat constrained by our existing knowledge of target genes and sequence diversity. Because of the unique insights provided by both culture-based and direct molecular approaches, their combined application can provide a complementary view of the microbial community and its function. This section aspires to acquaint investigators embarking on rhizoremediation studies with an array of both common and innovative approaches to studying the diversity, abundance and activity of microorganisms important to biodegradation.

4.1. CULTIVATION-BASED METHODS

4.1.1 Microbial recovery from the soil-root interface

In preparation for cultivation, it is often necessary to extract microorganisms from the soil, rhizosphere, rhizoplane or endorhizosphere. Typically a bacterial suspension is created that may then be serial diluted or used directly as an inoculum using the following methods.

Soil bacteria are suspended by shaking or vortexing soil in liquid along with sterile glass beads [37], sometimes followed by standing for 30 min or short low-speed centrifugation (500-1000 × g) [33] to separate soil particles from suspended bacteria. Sodium pyrophosphate solution is an effective soil aggregate dispersal agent [38] and so is commonly used as a suspension medium [33, 37]. Alternatively, other buffers, saline solution or simply sterile water may be used. Media containing carbon sources should be avoided since growth may occur that would skew populations during processing. A side-by-side study evaluating cell recovery from soil using water, sodium pyrophosphate and several buffers showed no significant difference (Terence Marsh, 2003, personal communication).

Rhizosphere soil bacteria can be suspended as described for soil, although glass beads may be omitted. Some rhizoplane organisms will invariably be recovered along with the rhizosphere, however gentle washing is thought to leave the rhizoplane largely intact since it reportedly removes only 10% of the bacteria removed by vigorous shaking with glass beads [39]. After shaking a defined quantity (by fresh weight or length) of root with adhering rhizosphere soil in liquid to suspend cells, the roots may be recovered, dried and weighed so that rhizosphere bacterial numbers can be based on root dry weight.

Following removal of the rhizosphere soil, the rhizoplane and endorhizosphere microflora may be recovered separately or together for analyses. Rhizoplane organisms remaining on the surface of the root after rhizosphere washing may be extracted by vigorous treatments such as shaking with glass beads, vortexing, or using a Stomacher blender [40, 41]. For recovery of endorhizosphere organisms, roots may be ground with a small amount of water or buffer in a Warring blender [42] or mortar and pestle [43] to create a suspension. To ensure that truly endophytic organisms are recovered in the absence of rhizoplane contaminants, the roots may be surface disinfected prior to grinding [43].

4.1.2 Liquid enrichment cultures

Because pollutant-degrading bacteria typically comprise a small proportion of the total microbial community, enrichment methods are useful to generate a mixed culture in which degradative bacteria are predominant. Many of the earliest isolates of bacteria capable of degrading pesticides and pollutants were isolated using enrichment methods [44, 45]. Constructing enrichment cultures is a simple matter of inoculating a flask, tube or bottle containing sterile minimal salts medium with an environmental sample, taking care to minimize introduction of foreign substrates, providing the compound of interest as the sole carbon source, and then incubating with shaking for a period of days to weeks until turbid. Often the enrichment is subjected to several passages to achieve a stable consortium.

Although enrichment cultures are valuable for producing isolates for study, they are not quantitative reflections of the diversity or relative abundance of contaminant-degrading populations within a sample. Studies of environmental samples have demonstrated dramatically reduced diversity of 2,4-D and PAH degrading organisms as well as degradative genes when cultivated using enrichment cultures in comparison to direct agar plates or biofilm culture methods [46, 47]. The difference is explained by competitive interactions in the enrichment culture in which slower-growing organisms are outcompeted by others with higher maximum specific growth rates. Thus, enrichment methods are not recommended for making comparisons of diversity or relative abundance among different samples.

4.1.3 Direct agar plate methods

Direct agar plating methods are a useful means to both enumerate and investigate the diversity of the cultivable fraction of bacteria capable of utilizing pollutants. As opposed to enrichment cultures, direct plates afford the opportunity for organisms of the same functional group but with different growth rates to form colonies and be detected [48]. Direct plates are also relatively simple to perform, require little specialized equipment and can easily be appended to existing protocols for plate counts of total cultivable bacteria. For these reasons, direct plate methods are the most widely used technique for studying abundance and diversity of microbial populations important to pollutant degradation.

Direct plating of soil or rhizosphere samples is performed by suspending bacterial cells from a defined quantity of soil or roots, spread-plating a dilution series onto a minimal medium with the contaminant (or an analogue) provided as a sole carbon source, incubating and colony counting. A secondary screening procedure, such as a clearing-zone test, may be employed to help verify that colonies are truly utilizing the substrate rather than impurities in the medium or carryover from the soil suspension. Details of these procedures are discussed below. For accurate plate counts, 2-3 replicate plates should be inoculated from each dilution generated from a sample and counts averaged. Likewise, multiple subsamples from each soil/root sample should be plated for statistically valid enumeration.

Growth conditions In order to selectively cultivate organisms that utilize a sole carbon source, care should be taken in the design and preparation of agar media to avoid introduction of unwanted growth substrates. A variety of different defined minimal media recipes may be used, taking care to avoid inclusion of any potential energy sources. Addition of yeast extract or vitamins should be avoided unless absolutely necessary for growth since they also function as carbon sources, even in low concentrations. Although expensive, highly purified Noble Agar is preferable over common agars, since most agars contain impurities that act as non-specific carbon sources. However, secondary screening methods discussed later can be used to distinguish true contaminant utilisers from non-degradative colonies growing on agar plates.

The contaminant or an analogue can be provided as a growth substrate in a number of ways depending on the nature of the compound. Water-soluble compounds may be added directly to the agar medium, preferably when relatively cool following autoclaving to prevent thermal decomposition. Practically insoluble but highly volatile compounds like biphenyl or naphthalene are easily provided in the vapour phase. Biphenyl is commonly provided by sprinkling a few crystals in the lid of the Petri plate [49]. For chemical hygiene, plates are then sealed with Parafilm or enclosing in a plastic sleeve or chamber to minimize volatilization into the ambient air. Naphthalene can be provided to plates all together in a sealed container [37]. Less volatile compounds may be added to the surface of agar plates by dissolving in water, ethanol or acetone and spreading, followed by removal of organic solvents by evaporation from open plates in a laminar flow hood [37]. Alternatively, compounds may be suspended in a small amount of agar or agarose and poured over the basal mineral agar layer [50].

Plates are commonly incubated or below 25°C, which presumably simulates environmental conditions better than the higher temperatures commonly used in microbiological research. The length of incubation required for colony formation of pollutant degraders varies depending on the nature of the population and the compound. It is recommended that when embarking on a new study that a pilot test be performed in which colony forming units are counted repeatedly over a period of up to 1 month in order to identify appropriate incubation times. Brief incubation periods can result in erroneously low counts and bias data toward the faster growing fraction of the population. A biphenyl-utilizing bacterial population comprised predominantly of relatively slow-growing Gram-positive rhodococci required 3 weeks for new colony formation to cease [9].

Colony screening methods Screening colonies pre-grown on agar plates for degradative abilities is often desirable since non-degradative colonies can grow on substrates other than the target compound such as impurities in agar, carryover from the soil suspension, or when cofactors in media are required. Several clearing-zone and colorimetric techniques for secondary screening are described below that that can help confirm that colonies are truly degrading the target substrate. Some of these methods may also be applied to colonies grown intentionally on a rich medium to detect degraders among the total cultivable community.

Following colony formation, clearing zone assays can be performed by depositing a thin cloudy layer of the target compound over the agar plate surface and then incubating until zones of clearing form. A 5 - 10% w/v solution of the target compound prepared in ether or acetone solution may be sprayed onto the agar surface using thin layer chromatography plate spraying apparatus in a fume hood. Alternatively, compounds may be overlayed by sublimation [51]. Spray methods have been used for a variety of aromatic compounds including biphenyl and chlorobiphenyl [52, 53], as well as phenanthrene, fluoranthene, pyrene, naphthalene and anthracene [53, 54]. Agar plates are then sealed in plastic sleeves to minimize volatilization of the compound and incubated for days to several weeks and checked periodically for the formation of zones of clearing around colonies.

Clearing zone assays may also be performed simultaneously with initial colony formation for compounds such as phenanthrene. The inoculum is mixed in an agarose suspension containing phenanthrene crystals and then spread onto a minimal medium agar plate. As phenanthrene-utilizing colonies form they create clearing zones [50]. This approach is advantageous over spraying methods by reducing the contamination of laboratory fume hoods with chemicals and preventing exposure of organisms to organic solvents, and also provides an effective means for efficient delivery of non-volatile substrates to organisms.

Colonies with aromatic ring dioxygenase activity can be screened using a simple colour indication assay that capitalizes on the ability of naphthalene enzyme to produce indigo from indole [54, 55]. Crystals of indole placed in the lid of the Petri dish will result in rapid (1 day or less) appearance of the blue indigo pigment in colonies in which aromatic dioxygenases are active. However, the assay only detects activity of dioxygenases enzymes that form cis-dihydrodiols from aromatic hydrocarbons [55].

4.1.4 Most probable number (MPN) method

Another useful approach for enumeration of pollutant degraders is the most probable number (MPN) method, which has been adapted for efficient use in 96-well microtiter plates to quantify bacteria that degrade polyaromatic hydrocarbon and alkanes [56] and crude oil [57]. In essence, a suspension of bacterial cells is diluted to extinction in a microtiter plate containing specific substrates as carbon sources. For PAH-degrading organisms, growth is detected in the wells by the presence of a yellow to green-brown colour generated by PAH oxidation products. For substrates without pigmented products such as alkanes, detection using the colour change of an iodonitrotetrazolium violet dye can be used. Empirical testing demonstrated that microtiter MPN produced the same result as enumeration by direct plate count for PAH-degraders [56]. The MPN method may be easily adapted for use with other compounds. When degrader population sizes are low and growth is only detected in wells with very low dilution factors, care should be taken to ensure that the signal is due to utilization of the substrate and not compounds carried over from the soil solution. To minimize carryover, cell suspensions may be washed by repeated centrifugation and resuspension to remove unwanted substrates prior to inoculation.

4.2. MOLECULAR ECOLOGICAL METHODS

Molecular microbial ecology methods bypass culture-bias to provide direct measures of a variety of microbial community parameters. This section highlights some methods that are of particular value to bioremediation studies because of their ability to directly detect the presence, activity, diversity and abundance of degradative organisms.

4.2.1 DNA and RNA extraction methods

Many molecular ecological analyses require bacterial DNA and/or RNA to be extracted from soil. Depending on both the soil itself and the protocol applied, results vary widely in terms of the quantity, size and purity of nucleic acids yielded. Numerous extraction methods exist in the published literature and in the form of commercial kits. Since the effectiveness of extraction protocols varies among soil types, it is recommended that several methods be tested, compared or sometimes adapted to identify a successful approach for a particular soil.

For the simultaneous extraction of DNA and RNA, a crude nucleic acid extract is first obtained using methods such as the rapid bead-beating method developed by Griffiths *et al.* [58] or a freeze grinding approach by Hurt *et al.* [59]. Bead beating tends to shear nucleic acids, so if high molecular weight extracts are needed (i.e. for metagenomics) then freeze grinding may be preferable. Precipitation of nucleic acids with polyethylene glycol (PEG) [58] is particularly effective in removal of humic materials that interfere with PCR, and may be used with other protocols when needed. RNA and DNA can then be separated from each other by gravity column systems such as the Qiagen DNA/RNA extraction kit (Qiagen Inc., Carlsbad, CA), followed by RNase or DNase treatment to remove residual amounts of undesirable nucleic acids. Alternatively, the crude extract may be split and then each fraction directly subjected to RNase or DNase treatment.

To extract DNA only, simple and effective commercial bead beating kits are commonly used such as those available from QBiogene, Inc. (Carlsbad, CA) or MoBio (Carlsbad, CA) or using a freeze-grinding method [60]. RNA extraction techniques are similar to DNA methods, however since RNA is very labile, extracting it from soil is much more challenging and requires conditions that inhibit RNase activity. Soil RNA extraction protocols are available in the literature [61], and commercial kits have recently arrived on the market (MoBio, Carlsbad, CA and QBiogene Inc., Carlsbad, CA).

To simultaneously extract endophytes, rhizoplane and rhizosphere soil, methods above may be adapted for use with whole roots and adhering soil. For extraction of endophytic root organisms only, other methods in the literature can be consulted [62-64].

4.2.2 Functional gene detection methods

Direct detection of genes that function in degradative processes may be performed on bacterial community DNA extracts either by PCR amplification or using microarrays. The primary limitations of these approaches lie in effective primer or probe design and detection limits. Primers and probes can be designed using bioinformatic methods to either broadly target groups of similar sequences or to be highly sequence specific. Inclusive primers or probes are designed by aligning multiple sequences and identifying conserved regions. Thus, they are limited by current knowledge of sequence diversity in the environment, which may prevent detection of unknown sequences. Detection limits for both PCR and microarray methods can be an obstacle since pollutant degraders often comprise a small proportion of the total community. Nonetheless, direct functional gene detection is a very powerful tool to investigate the presence, diversity and in some cases quantity of certain degradative genes in the microbial community.

PCR and quantitative real-time PCR PCR methods for detecting genes important to aromatic pollutant degradation have frequently targeted initial aromatic dioxygenase genes, which encode enzymes catalyzing the first, rate-limiting hydroxylation step of degradation. The aromatic dioxygenase gene family spans a wide substrate range, including PCBs, mono- and polyaromatic hydrocarbons, and are multimeric enzymes including a reductase subunit, a ferredoxin subunit, a large (alpha) and a small (beta) iron sulphur protein. Sequence variations in the large subunit are associated with substrate specificity [65]. For this reason, as well as the presence of a conserved region encoding the Rieske centre, the large subunit is an attractive target for PCR primer design. PCR primer sets have been successfully developed that can amplify groups of initial dioxygenase genes specific to certain substrates [66], while other primer sets are more organism-specific [66, 67]. Baldwin *et al.*, [66] developed multiple primer systems to differentially detect and enumerate subfamilies of aromatic oxygenase genes for biphenyl, naphthalene, toluene dioxygenases as well as monooxygenases involved in toluene/xylene and phenol degradation. Primers have been effectively designed targeting dioxygenases catalysing later steps in aromatic degradation as well [67, 68] For example, Erb and Wagner-Dobler successfully amplified *bphC* encoding the dioxygenase enzyme catalysing the meta-cleavage step in biphenyl degradation from the environment and achieved a detection limit of five copies per reaction mixture or 100 cells per g wet weight of sediment.

The genetic diversity of catabolic genes can be investigated when PCR products are subjected to sequence analysis. Following PCR amplification of degradative genes, clone libraries may be constructed for sequencing, or alternatively amplicons may be separated by methods such as denaturing gradient gel electrophoresis (DGGE) and excised bands sequenced. The relationship of sequences to previously known genes can be established by searching for similar sequences in public databases such as BLASTn searches of GenBank [69] and conducting phylogenetic analyses [70].

The degradative potential of a microbial community can be quantified using real-time PCR (RTm-PCR) with primers targeting functional genes. In RTm-PCR, a PCR reaction is performed in the presence of either fluorescent probes that hybridize to the specific target sequence (i.e. TaqMan ® probes, Applied Biosciences, Foster City, CA) or alternatively a non-specific fluorescent dye that binds to double stranded DNA (i.e. SYBR green). RTm-PCR thermocyclers are specially designed to measure fluorescence in the PCR reaction at each amplification cycle, permitting quantitation relative to a standard curve containing known copy numbers of the target sequence. When amplifying variable regions of functional genes, use of SYBR green is preferable over highly specific probes, however extra care must be taken to ensure that non-specific products are not being produced and measured. Running a melting curve analysis at the end of the reaction is an important quality control step for this purpose. For additional information regarding RTm-PCR methods, the reader is referred to books dedicated to the subject [71, 72]. RTm-PCR methods to enumerate aromatic dioxygenases have been developed, which used SYBR green and produced detection limits down to 2×10^2 copies per reaction using pure culture DNA [66].

Functional gene microarrays Using microarray techniques, it is now possible to probe community DNA samples for thousands of target sequences simultaneously, opening an enormous window into the functional gene diversity of microbial communities. Microarrays have already been designed that successfully detect a broad range of genes involved in the degradation of pollutants, including monoaromatic and polyaromatic hydrocarbons, PCBs and aliphatics [73-75]. Careful probe design, including the design of several probes per target when possible, followed by comprehensive testing is critical to help ensure the specificity of hybridisation and to prevent false positives or negatives. Unfortunately, as in the design of PCR primers, probe design is limited by current knowledge of the diversity of functional gene sequences so some important genes may not be detected.

Although not yet applied for quantitative purposes, strong linear relationships have been reported between signal intensity of hybridised probes and the number of gene copies in a sample with some functional gene arrays [73-75]. Thus, arrays may eventually be used to evaluate both the diversity and abundance of target genes in a sample.

A major challenge in applying arrays to bioremediation studies is their detection limits. As reported by Denef *et al.* [73], single-copy genes can only be detected from organisms that comprise 1% or more of the community with current technology. This is an issue for many environmental samples where the degradative population often comprises a small proportion of the total community. The detection limits of arrays could likely be overcome, however, if combined with stable isotope probing methods detailed in the following section.

4.2.3 Stable isotope probing

Stable isotope probing (SIP) methods now allow researchers to directly answer the previously evasive question: Which microorganisms are truly active in pollutant degradation? In addition to linking phylogeny with function, SIP also provides the opportunity to thoroughly investigate the genomic content of active degradative organisms. Although it does not provide quantitative measures of degrader abundance, SIP can provide insights into the identity and genetic diversity of organisms that are truly active in rhizoremediation.

SIP methodologies SIP involves provision of a substrate that is labelled with a heavy stable isotope, typically ^{13}C, to a microbial community. Organisms that actively incorporate the substrate are detected by virtue of the presence of ^{13}C in biomarkers such as phospholipid fatty acids (PLFAs), DNA or RNA. SIP based on nucleic acids provides higher resolution phylogenetic data than PLFAs [76], as well as access to functional genes of active species [77]. For DNA or RNA-SIP, nucleic acid extracts are subjected to density gradient centrifugation to separate heavier, ^{13}C-labelled nucleic acids from those of the inactive community. Heavy fractions may then be subject to PCR amplification, community profiling and/or sequence analyses. RNA-SIP is more sensitive than DNA-SIP, and is also advantageous in that growth (and potential community change) during the incubation is not necessary [78-80]. However, if functional genes are to be targeted, DNA-SIP is more attractive because of the difficulty of extracting mRNA from environmental samples. Recently, SIP methods have been developed to capitalize on the stable isotope, ^{15}N [81], which opens a window into the microorganisms that biodegrade N-containing contaminants.

To perform SIP of soil samples, microcosms are typically constructed in the laboratory, or using an *in situ* method that has also been developed [82] which may prove valuable in studying intact rhizospheres. Closed systems are desirable because they permit the sampling and monitoring of $^{13}CO_2$ evolution in the headspace as an indicator of substrate utilization. Substrates may be provided to soil in aqueous solution or, if volatile, as crystals enclosed in the microcosm. For SIP with ^{13}C-biphenyl, crystals on the interior wall of a sealed serum bottle were effective for enriching bacteria in soil spread loosely on the bottom [83, 84].

Care should be taken to avoid lengthy incubation times to minimize flow of ^{13}C to non-degraders via degradative intermediates or the biomass of primary degraders, and to avoid community change during the incubation. Thus, sampling over a time-course is recommended to ensure that primary utilisers are identified. In a timecourse SIP experiment with ^{13}C-biphenyl, significant $^{13}CO_2$ evolution into the headspace was detected within 24 hours and continued to be produced over the course of the 14-day incubation, indicating that biphenyl was being metabolised [83, 84].

Separation of ^{13}C-labeled DNA from unlabeled DNA can be performed by density gradient centrifugation in caesium chloride (CsCl) [79, 85] or in caesium trifluoroacetate (CsTFA) solutions [83, 84], while RNA is separated in CsTFA with the addition of formamide to reduce secondary structure [86]. Following ultracentrifugation, it is strongly recommended that density gradients be fractionated rather than extracting bands using a syringe needle since fractionation permits the detection of partially labelled nucleic acids. Fractionating samples along with appropriate controls is important to ensuring that organisms detected are truly labelled, since low levels of unlabeled nucleic acids are present throughout the density gradient. Control gradients should be constructed with unlabeled nucleic acids from the same community, such as a sample collected at the initiation of the experiment (t = 0). Control fractions should be subjected to the same fractionation and downstream molecular analyses. Then, any sequences detected in the background heavy control fractions can be subtracted from those identified in the ^{13}C-treated sample.

Following fractionation and nucleic acid recovery, PCR amplification of 16S rRNA genes (or RT-PCR of 16S rRNA) and community analyses are helpful to locate the ^{13}C-labeled fractions and to track changes in the labelled populations with increased incubation time. Denaturing gradient gel electrophoresis (DGGE) [78, 86] or terminal restriction fragment length polymorphism (T-RFLP) [79, 80] are common community profiling methods used in SIP.

Following the identification of the fractions containing ^{13}C-labeled nucleic acids, sequence analysis may be performed to phylogenetically identify organisms that derived carbon from the substrate. Clone libraries may be constructed from PCR amplicons, or if DGGE is used, bands of interest may be excised, reamplified and sequenced. Sequence analyses of ^{13}C-labeled DNA and RNA have focused primarily on the SSU rRNA for the purposes of obtaining phylogenetic information. However, a wealth of functional gene information is available in the labelled fraction that can prove informative in biodegradation studies. Functional gene analyses of SIP-derived ^{13}C-DNA have been reported for methylotrophs [77], but not yet for environmental pollutants.

SIP in bioremediation and the rhizosphere To date, SIP studies targeting aromatic pollutant-degrading bacteria have been achieved in a variety of settings, including phenol in activate sludge [78, 87] naphthalene and phenanthrene in a bioreactor treating soil [88], 2,2'-dichlorobiphenyl in a biofilm growing on PCB droplets [89], naphthalene in sediment [90], toluene in soil [91, 92], phenol, caffeine and naphthalene in soil [82]. Although as yet unexplored, SIP also provides an excellent means for identifying fungi that are significant in bioremediation by simply amplifying 18S rRNA genes in heavy nucleic acid fractions using universal fungal rRNA primers [93].

In the field of rhizosphere ecology, SIP has been applied in efforts to identify microbial communities that derive carbon from root exudates following provision of $^{13}CO_2$ to aboveground plant parts [94, 95]. A major challenge of this approach is the dilution of ^{13}C with the predominantly unlabeled carbon in the plant which can result in isotopic enrichment of the microbial biomarkers below levels needed for SIP [94]. If combined with functional gene detection, SIP of root exudates may prove valuable to demonstrate that pollutant-degrading bacteria are deriving carbon from the plant in the rhizosphere. Unfortunately this approach simultaneously probes for microbes using any of the plethoras of compounds exuded by plant roots, and thus does not permit differentiation of specific plant compound-utilizing bacteria. In order to determine which components of root exudates or lysates specifically feed pollutant degraders, SIP may be performed in a rhizosphere setting, or in root zone samples, with ^{13}C-labelled plant compounds provided individually.

5. Conclusions and research needs

Through the array of windows opened by powerful molecular ecological tools in combination with traditional microbial methods and carefully constructed experiments, new insights can be gained into bioremediation processes in the rhizosphere. What are the mechanisms for rhizostimulation of contaminant degraders? Which bacteria, fungi and degradative genes are most important? Which plant species, communities, or biosystems are most effective at rhizoremediation of various contaminants? Methods now available invite us to answer these questions and assemble the fundamental understanding of rhizoremediation ecology necessary for its successful implementation.

References

[1] Olson, PE; Wong, T; Leigh, MB and Fletcher, JS (2003) Allometric modeling of plant root growth andits application in rhizosphere remediation of soil contaminants. Environ Sci Technol 37(3): 638-643

[2] Olson, PE; Fletcher, JS and Philp, PR (2001) Natural attenuation/phytoremediation in the vadose zone of a former industrial sludge basin. Environmental Science and Pollution Research 8(4): 243-249

[3] Olson, PE and Fletcher, JS (1999) Field evaluation of mulberry root structure with regard to phytoremediation. Bioremediat J 3(1): 27-33

[4] Olson, PE and Fletcher, JS (2000) Ecological recovery of vegetation at a former industrial sludge basin and its implications to phytoremediation. Environ Sci Pollut R 7(4): 195-204

[5] Kyle, MD (2002) Rhizoremediation: Statistical and molecular analysis of PAH-utilizing bacteria enriched in the rhizozone of mulberry growing in a contaminated site, in American Society for Microbiology Missouri Valley Branch Joint Annual Meeting. 2002. Kansas City, MO, USA
[6] Kyle, MD; Fletcher, JS and Nagle, DP (2001) Apparent enrichment for PAH-degrading bacteria in the mulberry rhizozone of contaminated soil, in 8th International Petroleum Environmental Conference: Issues and Solutions in Exploration, Production and Refining. 2001. Houston, TX, USA
[7] Nagle, DP; Kyle, MD; Olson, PE and Fletcher, JS (1998) Bacterial microflora from a vegetated, PAH-contaminated industrial site, in IBC Third International Conference on Phytoremediation: Strategies and evaluation of phytoremediation's performance in the field. 1998. Houston, TX: International Business Communications, Inc.
[8] Nagle, DP; Kyle, MD; Olson, PE and Fletcher, JS (1998) Polyaromatic hydrocarbon-degrading microflora associated with vegetation at a phytoremediated site, in Fifth International Petroleum Environmental Conference. 1998. Albuquerque, NM.
[9] Leigh, MB (2003) Rhizosphere ecology of PCB-degrading bacteria at a contaminated site and implications for bioremediation, PhD Dissertation, Department of Botany and Microbiology, University of Oklahoma: pp. 88.
[10] Leigh, MB; Fletcher, JS; Nagle, DP; Mackova, M and Macek, T (2001) Vegetation and fungi at Czech PCB-contaminated sites as bioremediation candidates. International In Situ and On-Site Bioremediation Symposium, 6th, San Diego, CA, United States, June 4-7, 2001, 5: 61-68
[11] Demnerova, K; Stiborova, H; Leigh, MB; Pieper, D; Pazlarova, J; Brenner, V; Mackova, M and Macek, T (2001) Degradation of PCBs and CBs by Indigenous Bacteria Isolated from Contaminated Soil, in N. Kalogerakis, Ed., Proceedings of the First European Bioremediation Conference. 2001. Chania, Crete, Greece.
[12] Legendre, P and Legendre, L (1998) Numerical ecology. Amsterdam: Elsevier.
[13] Braak, CJF; Jongman, RHG; Ter-Braak, CJF and Van-Tongerer, OFR (1995) Data analysis in community and landscape ecology. Cambridge: Cambridge University. 299
[14] Ludwig, JA and Reynolds, JF (1988) Statistical ecology: a primer on methods and computing. Wiley: New York.
[15] Manly, BFJ (1994) Multivariate statistical methods: a primer. 2nd ed. London: Chapman & Hall.
[16] Jungk, AO (1991) Dynamics of nutrient movement at the soil-root interface, in Y Waisel, A Eshel and U Kafkafi, Eds., Plant Roots the Hidden Half, Marcel Dekker, New York. 455-481
[17] Banks, MK; Schwab, AP; Liu, B; Kulakow, PA; Smith, JS and Kim, R (2003) The effect of plants on the degradation and toxicity of petroleum contaminants in soil: a field assessment, in Phytoremediation, DT Tsao, Ed., Springer, Berlin. 206
[18] Demnerova, K; Mackova, M; Kucerova, P; Leigh, MB; Novakova, H; Burkhard, J and Macek, T (2000) Practical use of bioremediation for PCB removal from contaminated soil, in M. Hofman and J. Anné, Series Eds., Focus on Biotechnology, Vol. 3, Kluwer, Dordrecht
[19] Chekol, T; Vough, LR and Chaney, RL (2004) Phytoremediation of polychlorinated biphenyl-contaminated soils: the rhizosphere effect. Environ Int 30(6): 799-804
[20] Siciliano, SD and Germida, JJ (1998) Mechanisms of phytoremediation: biochemical and ecological interactions between plants and bacteria. Environ Rev 6: 65-79
[21] Miya, RK and Firestone, MK (2000) Phenanthrene-degrader community dynamics in rhizosphere soil from a common annual grass. Journal of Environmental Quality 29(2): 584-592
[22] Dytham, C (1999) Choosing and Using Statistics: A Biologists Guide. Oxford: Blackwell Science Ltd. 218
[23] Montgomery, DC (2005) Design and Analysis of Experiments. 6th ed. Hoboken, NJ: John Wiley & Sons. 643
[24] Rice, EL (1974) Allelopathy, Vol. 352. Academic Press: New York
[25] Singer, AC; Crowley, DE and Thompson, IP (2003) Secondary plant metabolites in phytoremediation and biotransformation. TRENDS in Biotechnology 21(3): 123-130
[26] Vogt, KA and Bloomfield, J (1991) Tree root turnover and senescence, in Y Waisel and U Kafkafi, Eds., Plant Roots the Hidden Half, Marcel Dekker: New York, 297-306
[27] Leigh, MB; Fletcher, JS; Fu, XO and Schmitz, FJ (2002) Root turnover: An important source of microbial substrates in rhizosphere remediation of recalcitrant contaminants. Environ Sci Technol 36(7): 1579-1583

[28] Miya, RK and Firestone, MK (2001) Enhanced phenanthrene biodegradation in soil by slender oat root exudates and root debris. J Environ Quality 30: 1911-1918
[29] Rovira, AD (1965) Interactions between plant roots and soil microorganisms. Annu Rev Microbiol 19: 241-266
[30] Siciliano, SD; Germida, JJ; Banks, K and Greer, CW (2003) Changes in microbial community composition and function during a polyaromatic hydrocarbon phytoremediation field trial. Appl Environ Microbiol 69(1): 483-489
[31] Nusslein, K and Tiedje, JM (1999) Soil bacterial community shift correlated with change from forest to pasture vegetation in a tropical soil. Appl Environ Microbiol 65(8): 3622-3626
[32] Saetre, P and Baath, E (2000) Spatial variation and patterns of soil microbial community structure in a mixed spruce-birch stand. Soil Biol Biochem 32(7): 909-917
[33] van Elsas, JD and Smalla, K (1997) Methods for sampling soil microbes, in CJ Hurst, GR Knudsen, MJ McInerney, LD Stetzenbach and MV Walter, Eds., Manual of Environmental Microbiology, ASM Press: Washington, DC, 893
[34] Tan, KH (1996) Soil Sampling, Preparation and Analysis. Marcel Dekker: New York, 408
[35] Smalla, K; Wieland, G; Buchner, A; Zock, A; Parzy, J; Kaiser, S; Roskot, N; Heuer, H and Berg, G (2001) Bulk and rhizosphere soil bacterial communities studied by denaturing gradient gel electrophoresis: Plant-dependent enrichment and seasonal shifts revealed. Appl Environ Microbiol 67(10): 4742-4751
[36] Buckley, DH and Schmidt, T (2003) Diversity and dynamics of microbial communities in soils from agro-ecosystems. Environ Microbiol 5(6): 441
[37] Kastner, M; Jammali, MB and Mahro, B (1994) Enumeration and characterization of the soil microflora from hydrocarbon-contaminated soil sites able to mineralize polycyclic aromatic hydrocarbons (PAH). Appl Microbiol Biotechnology 41: 267-273
[38] Strickland, TC; Sollins, P; Schimel, DS and Kerle, EA (1988) Aggregation and aggregate stability in forest and range soils. Soil Sci Soc Am J 52: 829-833
[39] Rovira, AD; Newman, EI; Bowen, HJ and Campbell, R (1974) Quantitative assessment of the rhizoplane microflora by direct microscopy. Soil Biol Biochem 6: 211-216
[40] Dandurand, L-MC and Knudsen, GR (1997) Sampling microbes from the rhizosphere and phyllosphere, in CJ Hurst, GR Knudsen, MJ McInerney, LD Stetzenbach and MV Walter, Eds., Manual of Environmental Microbiology, ASM Press, Washington, DC, 893
[41] Wieland, G; Neumann, R and Backhaus, H (2001) Variation of microbial communities in soil, rhizosphere, and rhizoplane in response to crop species, soil type and crop development. Appl Environ Microbiol 67(12): 5849-5854
[42] Rouatt, JW and Katznelson, H (1961) A study of the bacteria on the root surface and in the rhizosphere soil of crop plants. J Appl Bacteriol 24(2): 164-171
[43] Siciliano, SD and Germida, JJ (1999) Taxonomic diversity of bacteria associated with the roots of field-grown transgenic *Brassia napus* cv. Quest, compared to the non-transgenic *B. napus* cv. Excel and *B. rapa* cv. Parkland. FEMS Microbiol Ecol 29: 263-272
[44] Furukawa, K (1982) Microbial degradation of polychlorinated biphenyls (PCBs), in AM Chakrabarty, Ed., Biodegradation and detoxification of environmental pollutants, CRC Press: Boca Raton, FL. 147
[45] Bartha, R (1990) Isolation of microbes that metabolize xenobiotic compounds, in Isolation of Biotechnological Organisms from Nature, DP Labeda, Ed., McGraw Hill: New York. 283-305
[46] Dunbar, J; White, S and Forney, LJ (1997) Genetic diversity through the looking glass: effect of enrichment bias. Appl Environ Microbiol 63: 1326-1331
[47] Stach, JE and Burns, RG (2002) Enrichment versus biofilm culture: a functional and phylogenetic comparison of polycyclic aromatic hydrocarbon-degrading microbial communities. Environ Microbiol 4(3): 169-182
[48] Dunbar, J; Wong, DCL; Yarus, DC and Forney, LJ (1996) Autoradiography method for isolation of diverse bacterial species with unique catabolic traits. Appl Environ Microbiol 62: 4180-4185
[49] Gilbert, ES and Crowley, DE (1997) Plant compounds that induce polychlorinated biphenyl biodegradation by *Arthrobacter* sp. strain B1B. Appl Environ Microbiol 63(5): 1933-1938
[50] Bogardt, AH and Hemmingsen, BB (1992) Enumeration of phenanthrene-degrading bacteria by an overlayer technique and Its use in evaluation of petroleum-contaminated sites. Appl Environ Microbiol 58(8): 2579-2582

[51] Alley, JF and Brown, LR (2000) Use of sublimation to prepare solid microbial media with water-insoluble substrates. Appl Environ Microbiol 66(1): 439-442

[52] Sylvestre, M (1980) Isolation method for bacterial isolates capable of growth on p-chlorobiphenyl. Appl Environ Microbiol 39(6): 1223-1224

[53] Kiyohara, H; Nagao, K and Yana, K (1982) Rapid screen for bacteria degrading water-insoluble, solid hydrocarbons on agar plates. Appl Environ Microbiol 43(2): 454-457

[54] Ahn, Y; Sanseverino, J and Sayler, G (1999) Analyses of polycyclic aromatic hydrocarbon-degrading bacteria isolated from contaminated soils. Biodegradation 10(2): 149-157

[55] Ensley, BD; Ratzkin, BJ; Osslund., TD and Simon, MJ (1983) Expression of naphthalene oxidation genes in *Escherichia coli* results in the biosynthesis of indigo. Science 222: 167-169

[56] Wrenn, BA and Venosa, AD (1996) Selective enumeration of aromatic and aliphatic hydrocarbon degrading bacteia by a most probable number procedure. Can J Microbiol 42: 252-258

[57] Haines, JR; Wrenn, BA; Holder, EL; Strohmeier, KL; Herrington, RT and Venosa, AD (1996) Measurement of hydrocarbon-degrading microbial populations by a 96-well plate most-probable-number procedure. J Ind Microbiol 16: 36-41

[58] Griffiths, RI; Whiteley, AS; O'Donnell, AG and Bailey, MJ (2000) Rapid method for coextraction of DNA and RNA from natural environments for analysis of ribosomal DNA- and RNA-based microbial community composition. Appl Environ Microbiol 66(12): 5488-5491

[59] Hurt, RA; Qiu, X; Wu, L; Roh, Y; Palumbo, AV; Tiedje, JM and Zhou, JZ (2001) Simultaneous recovery of RNA and DNA from soils and sediments. Appl Environ Microbiol 67(10): 4495-4503

[60] Zhou, JZ; Bruns, MA and Tiedje, JM (1996) DNA recovery from soils of diverse composition. Appl Environ Microbiol 62(2): 316-322

[61] Sessitsch, A; Gyamfi, S; Stralis-Pavese, N; Weilharter, A and Pfeifer, U (2002) RNA isolation from soil for bacterial community and functional analysis:evaluation of different extraction and soil conservation protocols. J Microbiol Methods 51(2): 171-179

[62] Garbeva, P; Van Overbeek, LS; Van Vuurde, JWL and Van Elsas, JD (2001) Analysis of endophytic bacterial communities of potato by plating and denaturing gradient gel electrophoresis (DGGE) of 16S rDNA based PCR fragments. Microbial Ecology 41(4): 369-383

[63] Mullin, BC; Joshi, PA and An, CS (1983) The isolation and purification of endophyte DNA from *Alnus glutinosa* nodules. Can J Bot 61(11): 2855-2858

[64] Conn, V and Franco, CMM (2004) Analysis of the endophytic actinobacterial population in the roots of wheat (*Triticum aestivum* L.) by terminal restriction fragment length polymorphism and sequencing of 16S rRNA clones. Appl Environ Microbiol 70(3): 1787-1794

[65] Parales, RE; Lee, K; Resnick, SM; Jiang, H; Lessner, DJ and Gibson, DT (2000) Substrate specificity of naphthalene dioxygenase: effect of specific amino acids at the active site of the enzyme. Curr Opin Biotechnol 182(6): 1641-1649

[66] Baldwin, BR; Nakatsu, CH and Nies, L (2003) Detection and enumeration of aromatic oxygenase genes by multiplex and real-time PCR. Appl Environ Microbiol 69(6): 3350-3358

[67] Ringelberg, DB; Talley, JW; Perkins, EJ; Tucker, SG; Luthy, RG; Bouwer, EJ and Fredrickson, HL (2001) Successon of phenotypic, genotypic and metabolic community characteristics during *in vitro* bioslurry treatment of polycyclic aromatic hydrocarbon-contaminated sediments. Appl Environ Microbiol 67(1542-1550)

[68] Erb, RW and Wagner-Dobler, I (1993) Detection of polychlorinated biphenyl degradation genes in polluted sediments by direct DNA extraction and polymerase chain reaction. Appl Environ Microbiol 59(12): 4065-4073

[69] Altschul, SF; Madden, TL; Schäffer, AA; Zhang, J; Zhang, Z; Miller, W and Lipman, DJ (1997) Gapped BLAST and PSI-BLAST: a new generation of protein database search programs. Nucleic Acids Res 25: 3389-3402

[70] Hall, BG (2004) Phylogenetic Trees Made Easy, A How-To Manual. 2nd ed. Sunderland, Mass.: Sinauer Associates.

[71] Edwards, K; Logan J and Saunders, N Ed. (2004) Real-time PCR: an essential guide. Wymondham, Norfolk, UK: Horizon Bioscience.

[72] Meuer, S; Wittwer, C and Nakagawara, K Ed. (2001) Rapid cycle real-time PCR: methods and applications. Berlin: Springer.

[73] Denef, VJ; Park, J; Rodrigues, JLM; Tsoi, TV and Hashsham, S (2003) Validation of a more sensitive method for using spotted oligonucleotide DNA microarrays for functional genomics studies on bacterial communities. Environ Microbiol 5(10): 933-943

[74] Rhee, SK; Liu, XD; Wu, LY; Chong, SC; Wan, XF and Zhou, JZ (2004) Detection of genes involved in biodegradation and biotransformation in microbial communities by using 50-mer oligonucleotide microarrays. Appl Environ Microbiol 70(7): 4303-4317

[75] Wu, LY; Liu, XD; Schadt, CW; Tiedje, JM and Zhou, JZ (2005) Microarray-based detection of fento-to-nanogram levels of genomic DNA from microbial genomes and microbial communities using whole community genome amplification, in Proc. of American Society for Microbiology 105th General Meeting, Atlanta, GA.

[76] Manefield, M; Whiteley, AS and Bailey, MJ (2004) What can stable isotope probing do for bioremediation? International Biodeterioration and Biodegradation 54(2-3): 163-166

[77] McDonald, IR; Radajewski, S and Murrell, JC (2005) Stable isotope probing of nucleic acids in methanotrophs and methylotrophs: A review. Org Geochem 36(5): 779-787

[78] Manefield, M; Whiteley, AS; Griffiths, RI and Bailey, MJ (2002) RNA stable isotope probing, a novel means of linking microbial community function to phylogeny. Appl Environ Microbiol 68(11): 5367-5373

[79] Lueders, T; Manefield, M and Friedrich, MW (2004) Enhanced sensitivity of DNA- and rRNA-based stable isotope probing by fractionation and quantitative analysis of isopycnic centrifugation gradients. Environ Microbiol 6(1): 73-78

[80] Lueders, T; Wagner, B; Claus, P and Friedrich, MW (2004) Stable isotope probing of rRNA and DNA reveals a dynamic methylotroph community and trophic interactions with fungi and protozoa in oxic rice field soil. Environ Microbiol 6(1): 60-72

[81] Cadisch, G; Espana, M; Causey, R; Richter, M; Shaw, E; Morgan, JAW; Rahn, C and Bending, GD (2005) Technical considerations for the use of ^{15}N-DNA stable-isotope probing for functional microbial activity in soils. Rapid Commun Mass Spectrom 19(11): 1424-1428

[82] Padmanabhan, P; Padmanabhan, S; DeRito, C; Gray, A; Gannon, D; Snape, JR; Tsai, CS; Park, W; Jeon, C and Madsen, EL (2003) Respiration of ^{13}C-labeled substrates added to soil in the field and subsequent 16S rRNA gene analysis of ^{13}C-labeled soil DNA. Appl Environ Microbiol 69(3): 1614-1622

[83] Leigh, MB; Park, J; Tiedje, JM and Ostrom, N (2004) Stable isotope probing for (polychlorinated)-biphenyl degrading bacteria in soil, in Proc. of 10th International Symposium on Microbial Ecology (ISME), Cancun, Mexico

[84] Leigh, MB; Park, J; Bailey, MJ and Tiedje, JM (2005) Identification of (polychlorinated) biphenyl-degrading bacteria in contaminated soils, in Proc. of American Society for Microbiology (ASM) 105th General Meeting, Atlanta, GA, USA

[85] Radajewski, S; McDonald, IR and Murrell, JC (2003) Stable-isotope probing of nucleic acids: a window to the function of uncultured microorganisms. Curr Opin Biotechnol 14(3): 296-302

[86] Manefield, M; Whiteley, AS; Ostle, N; Ineson, P and Bailey, MJ (2002) Technical considerations for RNA-based stable isotope probing: An approach to associating microbial diversity with microbial community function. Rapid Commun Mass Spectrom 16(23): 2179-2183

[87] Manefield, M; Griffiths, RI; Leigh, MB; Fisher, R and Whiteley, AS (2005) Functional and compositional comparison of two activated sludge communities remediating coking effluent. Environ Microbiol 7(5): 715-722

[88] Singleton, DR; Powell, SN; Sangaiah, R; Gold, A; Ball, LM and Aitken, MD (2005) Stable-isotope probing of bacteria capable of degrading salicylate, naphthalene, or phenanthrene in a bioreactor treating contaminated soil. Appl Environ Microbiol 71(3): 1202-1209

[89] Tillmann, S; Strompl, C; Timmis, KN and Abraham, WR (2005) Stable isotope probing reveals the dominant role of *Burkholderia* species in aerobic degradation of PCBs. FEMS Microbiol Ecol 52(2): 207-217

[90] Jeon, CO; Park, W; Padmanabhan, P; DeRito, C; Snape, JR and Madsen, EL (2003) Discovery of a bacterium, with distinctive dioxygenase, that is responsible for in situ biodegradation in contaminated sediment. Proc Natl Acad Sci 100(23): 13591-13596

[91] Hanson, JR; Macalady, JL; Harris, D and Scow, KM (1999) Linking toluene degradation with specific microbial populations in soil. Appl Environ Microbiol 65(12): 5403-5408

[92] Pelz, O; Chatzinotas, A; Zarda-Hess, A; Abraham, W-R and Zeyer, J (2001) Tracing toluene-assimilating sulfate-reducing bacteria using 13C-incorporation in fatty acids and whole-cell hybridization. FEMS Microbiol Ecol 38: 123-131

[93] Smit, E; Leeflang, P; Glandorf, B; van Elsas, JD and Wernars, K (1999) Analysis of fungal diversity in the wheat rhizosphere by sequencing of cloned PCR-amplified genes encoding 18S rRNA and temperature gradient gel electrophoresis. Appl Environ Microbiol 65(6): 2614-2621

[94] Griffiths, RI; Manefield, M; Ostle, N; McNamara, N; O'Donnell, AG; Bailey, MJ and Whiteley, AS (2004) 13CO2 pulse labelling of plants in tandem with stable isotope probing: methodological considerations for examining microbial function in the rhizosphere. J Microbiol Methods 58(1): 119-129

[95] Rangel-Castro, JI; Killham, K; Ostle, N; Nicol, GW; Anderson, IC; Scrimgeour, CM; Ineson, P; Meharg, A and Prosser, JI (2005) Stable isotope probing analysis of the influence of liming on root exudate utilization by soil microorganisms. Environ Microbiol 7(6): 828-838

CONSTRUCTED WETLANDS FOR PHYTOREMEDIATION

Rhizofiltration, phytostabilisation and phytoextraction

MARINUS L. OTTE AND DONNA L. JACOB
Wetland Ecology Research Group, School of Biological and Environmental Science, College of Life Sciences, University College Dublin, Belfield, Dublin 4, Ireland Tel:+353-17162019, Fax:+353-17161153, E-mail: marinus.otte@ucd.ie

1. Introduction

The utilisation of wetlands for remediation of polluted soils and waters via rhizofiltration, phytostabilisation and phytoextraction has been increasing steadily over the past decades. The use of wetlands for quality improvement of wastewater, referred to as rhizofiltration, is the best known and most researched application of constructed wetlands. Wetlands as a cover over polluted soils and sediments, such as mine tailings, are referred to as phytostabilisation. It is a less common utilisation of wetlands, no doubt because of its more limited range of applications compared to rhizofiltration. The least applied and studied utilization of wetlands is that of phytoextraction, in which removal of substances is carried out through uptake by plants.

This paper aims to give an overview of the utilization of wetlands for the three approaches in phytoremediation described above. It is not intended to be an exhaustive review, but to give an impression of the versatility of wetlands and the wide range of potential applications.

2. The importance of biogeochemistry to wetlands for phytoremediation

2.1. GENERAL BIOGEOCHEMISTRY OF WETLAND SOILS AND SEDIMENTS

Of paramount importance to using wetlands for phytoremediation purposes, particularly rhizofiltration and phytostabilisation are the biogeochemical characteristics of wetlands. Wetland substrates are typically waterlogged, at least during part of the annual cycle [1]. Because oxygen diffuses through water 10,000 times slower than through air, the supply of oxygen to wetland substrates is much slower than to dryland substrates. Upon flooding, the oxygen levels in well aerated dryland soils rapidly decrease, and generally

M. Mackova et al. (eds.), Phytoremediation Rhizoremediation, 57–67.

wetland soils are low in or devoid of oxygen. These anaerobic conditions do not just require special adaptations in plants and sediment infauna, but also in the microbial community. Depending on the chemical composition of the substrate, a wide range of terminal electron acceptors other than oxygen are used for respiration by these organisms [2, 3]. Which terminal electron acceptor is used depends on their availability and that of other substances. This is best explained by considering what happens upon flooding of a well-aerated, dry soil, because under those conditions the redox reactions that occur due to the respiration activity of micro-organisms follow a thermodynamic sequence, from more oxidised to more reduced [2, 3]. After the disappearance of oxygen, nitrate is reduced to N_2 or N_2O. Because this process leads to the formation of gases and the subsequent removal of nitrogen from the substrate, this process is referred to as denitrification. After all nitrate has been removed, ferric iron, Fe(III), may be reduced to ferrous iron, Fe(II). Upon removal of most ferric iron, sulphate (SO_4^{2-}) may be reduced to sulphide (S^{2-}), and when most sulphate is removed carbon dioxide (CO_2) may be reduced to methane (CH_4). This is a highly simplified explanation of what is, in fact, a highly complex process, but these are the reactions that are most important for the purpose of this paper. The reactions are more or less exclusive, in that denitrification will not take place in the presence of oxygen, iron reduction will not happen in the presence of nitrate, and so forth. This is because the reduction of, for example, ferric to ferrous iron, is thermodynamically unfavourable in the presence of nitrate. Like all soils and sediments, wetland substrates contain a wide range of inorganic and organic redox-sensitive compounds, and which reactions take place not only depends on the thermodynamic characteristics, but also on the different micro-organisms competing for oxidisable substrates [4]. Therefore, while the redox potential of a wetland sediment gives some indication of the chemical status, it cannot be used to assess the actual chemical composition. As a general rule however, the following observations apply:

- flooding of any soil or sediment leads to rapid denitrification, and wetland sediments typically contain low levels of nitrate,
- most wetlands contain relatively high concentrations of ferrous iron, and
- because of the continuous influx of sulphate from flooding with seawater in coastal and estuarine wetlands (salt marshes, coastal lagoons, mud flats), their sediments are typically rich in sulphide, but rarely produce methane.

The pH of soils and sediments is also affected by waterlogging [3]. Upon waterlogging of acid soils, the pH increases towards neutral, mainly due to the reduction of Fe(III) to Fe(II), while in alkaline soils the pH decreases towards neutral due to the build-up of CO_2 and subsequent formation of carbonic acid. As a result, wetland substrates tend to be circum-neutral. Exceptions are wetlands that have organic-matter rich, ferric-iron poor soils, such as peat bogs, which tend to be more acidic, and wetlands with substrates that are very poor in organic matter, such as wetlands formed on metal mine tailings, which may be alkaline.

2.2. PLANTS AND HETEROGENEITY OF WETLAND SOILS AND SEDIMENTS

Wetland substrates are normally highly heterogeneous. While the bulk soil may be chemically reduced, the surface layer of the substrate is typically oxidised, particularly if the substrate is regularly exposed to the air, for example due to fluctuating water levels. This layer can be very thin (a few millimetres), as is for example typically found in coastal mudflats, but its importance in element cycling must not be underestimated. More important to substrate heterogeneity, however, is the presence of plants. Roots penetrating the soil often form a very dense network and take up a large volume. They affect the soil immediately surrounding them (the rhizosphere) due to organic root exudates and organic matter derived from die-back [e.g. 5, 6, 7] and due to radial oxygen loss [e.g. 8, 9]. The above-ground biomass as well contributes significantly to input of organic matter into wetland substrates [10]. The result is a three-dimensional patchwork of areas in the substrate greatly varying in organic matter content, redox status and elemental content. While organic compounds are supplied by the plants in one area, they are consumed by oxidation/reduction reactions in other areas, perhaps only millimetres away. Similarly, while ferrous iron is being oxidised very near the root surface, ferric iron is being reduced by anaerobic bacterial activity at the interface between oxidised rhizosphere and chemically reduced bulk substrate. The substrate heterogeneity makes it difficult to predict the mobility and bioavailability of nutrients and pollutants in wetlands, at least at the scale of the rhizosphere [11]. Despite this however, wetlands have proven to be very effective in retaining pollutants from water and in stabilising contaminated sediments.

3. Rhizofiltration

3.1. RANGE OF POLLUTANTS AND EFFICIENCY

Rhizofiltration refers to the use of wetlands for retention of pollutants from water. While the prefix "rhizo" suggests that the activity of the plants in these systems is limited to the roots, it is in fact the interaction of the whole plants with the substrates that is the key to the success of wetlands for this application. Natural and constructed wetlands have been successfully utilised for the removal of a wide range of pollutants, including plant macronutrients, metals, organic compounds and biological contaminants, such as coliform bacteria and parasites, and they also improve pH, Biological Oxygen Demand (BOD) and Chemical Oxygen Demand, (COD), for examples see Table 1.

While an overwhelming amount of reports exists to show that wetlands can successfully remove a wide range of pollutants, not all pollutants are equally removed by all types of wetlands. In fact, a wetland may retain one set of pollutants while acting as a source of another. Birch and co-workers [18] for example, reported efficient removal of Cr, Cu, Pb, Ni and Zn, but also negative removal efficiencies of -84% for Fe and -294% for Mn. Therefore, wetlands generally act as sinks of pollutants, but can also act as sources. Whether or not a wetland acts as a sink will depend on its size and structure (e.g. availability of binding sites for the pollutant, presence of appropriate micro-flora and plant species). Samecka-Symerman *et al.* [17] and Best *et al.* [21], for example, showed that the removal efficiency of wetlands strongly depended on the type of plants present in the systems. The removal efficiency will also depend on the season, as has been shown for both natural and constructed wetlands [13, 24, 25, 26]. As a general rule though, wetlands show net removal of pollutants over longer periods of time, as long as they are large enough to provide enough treatment capacity relative to the pollutant loading rate.

Table 1. Examples of applications and efficiency of wetlands for quality improvement of water

Type of pollutant	**Efficiency (% removal)**	**References**
Macronutrients		
Ammonium-N	16-67	12, 14
Nitrate/nitrite	40	13
Total Kjeldahl nitrogen (TKN)	49-81	14, 20
Organic nitrogen	82	14
Soluble Reactive Phosphate (SRP)	56	13
Total-P (TP)	44-68	12, 13
Metals and metalloids		
Al	81-97	17
As	65	15
Ba	70-95	17
Cd	58-71	17
Co	39-98	16
Cu	49-65	17, 18
Fe	91-97	16, 17
Mn	91-99	16, 17
Ni	22-67	16, 17, 18
Se	69	22
Sr	24-51	17
Zn	52-95	15, 17, 18
V	100	17
Organic compounds, including explosives	63-100	19, 20, 21
Coliform bacteria	26-98	18
Eggs of human parasites	94-100	23
COD	81	20
BOD_5	72-89	12, 14, 20
Suspended solids (SS)	43-94	12, 14

3.2. PROCESSES INVOLVED IN REMOVAL OF POLLUTANTS

Why wetlands are so successful in removing a wide range of pollutants form wastewater becomes clear when the underlying processes are considered. Key to these processes is the heterogeneity of the substrate. The removal of nitrogen is a good example to illustrate this. Mineralisation of nitrogen from plant-derived organic compounds leads initially to the formation of ammonium. In the aerobic environment of the surface layer of the sediments and in the rhizosphere of plants, ammonium may be further oxidised to nitrate. The latter process (nitrification) does not occur under the anaerobic conditions of the bulk wetland substrate and so inverse gradients in the nitrate and ammonium concentrations are formed between the aerobic (high in nitrate, low in ammonium) and anaerobic areas (low in nitrate, high in ammonium) in the substrate. As a consequence, ammonium diffuses from anaerobic to aerobic areas, where it is converted to nitrate. Nitrate, in turn, diffuses from aerobic into anaerobic areas, where it is subsequently converted to nitrous oxide (N_2O) or gaseous nitrogen (N_2) by denitrifying bacteria. Because both N_2O and N_2 are gases, denitrification effectively leads to the removal of nitrogen from the system. In a similar fashion, and equally dependent on alternation of aerobic and anaerobic areas in the substrate, sulphur forms volatile compounds (e.g. H_2S and dimethyl sulphide), which leads to an efflux of sulphur from wetland substrates (for detailed diagrams of the cycles of N and S associated with wetlands, see [1]). The interactions between plants, bacteria [27] and the wetland substrate, particularly by the formation of an oxidised rhizosphere in an otherwise chemically reduced bulk substrate; also favour the volatilisation via the lacunal spaces of macrophytes of elements that can be methylated, such as Se [22] and Hg [28]. The efficacy of wetlands for the removal of a wide range of organic compounds is also due to the heterogeneity of wetlands. Both anaerobic and aerobic degradation of organic compounds occurs [e.g. 29, 30, 31], while removal of the lighter compounds can at least in part be explained by volatilisation [30, 32].

The plant-induced heterogeneity in redox status of wetland substrates is also the main driver for the immobilisation and retention of metals such as Fe, Cd, Cu and Zn. In the anaerobic, chemically reduced bulk soil, many metals form insoluble, and therefore immobile, sulphides, provided of course that sulphides are formed in sufficient supply [33]. In the rhizosphere, the roots affect the redox status by the supply of oxygen through radial oxygen loss [8,9] and by supplying habitat for iron-oxidizing bacteria [34], both of which in turn lead to the oxidation and precipitation of iron and manganese oxides in the rhizosphere and on the roots (iron plaque) [35, 36]. Many metals and metalloids have high binding affinities for iron and manganese oxides, and iron and manganese deposits on the roots and in the rhizospheres of wetland plants therefore contain relatively high concentrations (compared to the bulk substrate) of those elements [e.g. 37, 38, 39].

The slow flow rates through wetlands are the main reason why suspended solids are removed – they simply precipitate out of suspension. The micro-pores of the substrate act as filters and this too contributes to the retention of suspended solids and eggs of parasites from the water. Removal of coliform bacteria is attributed to some extent to filtration as well [40], but the incidence of sunlight and UV-radiation is thought to be a more important factor [41].

4. Phytostabilisation

4.1. WHY COVER POLLUTED SOILS AND SEDIMENTS UNDER WETLANDS?

Wetlands for phytostabilisation of polluted soils and sediments have so far been limited to mining activities, for example for coal [42, 43], lignite [44, 45] and uranium [46]. One reason to cover soils and sediments under water may be the prevention of erosion and dispersal of pollutants due to wind and dust blows. Even in humid climates occasional droughts may cause dust blows, as happened at the abandoned lead-zinc mines of Silvermines, Ireland [47]. Another reason is that, as described for rhizofiltration above, the biogeochemistry of wetlands favours the chemical immobilisation of pollutants, particularly metals. The anaerobic, reduced conditions of the bulk substrate may lead to the formation of insoluble metal sulphides, while the aerobic, oxidised conditions of the plant rhizospheres and the surface layer favour immobilisation due to adsorption on and co-precipitation with iron and manganese oxides and phosphates. The plants also contribute by supplying organic matter. Pollutants are taken up by the roots and returned to the substrate bound to organic matter after the plants die off, or they may bind to plant-derived organic matter (through die-back or exudation) in the substrates [48]. While all these processes favour immobilisation, there is of course the problem that plants, and other organisms, may also mobilise pollutants.

4.2. MOBILISATION OF POLLUTANTS BY WETLAND ORGANISMS

While most studies agree that wetlands lead to overall net immobilisation, organisms may mobilise pollutants on a smaller scale, both in time and in space. The burrowing activity of sediment infauna, for example, may lead to re-suspension of sediments into the water column and oxidation of reduced compounds and subsequent release of metals into the water column [49]. Similarly, the growth of plant roots into the anaerobic, reduced bulk soil will, at least temporarily, mobilise metals from sulphides in the rhizosphere due to the oxidising activity of the roots and associated micro-organisms [11, 50, 51, 52]. While this is a matter of concern, there are no reports of mass mobilisation of pollutants from wetland covered soils or sediments.

4.3. METAL TOLERANCE IN WETLAND PLANTS

Particularly in the case of pollutants that are potentially toxic to the plants, such as metals, the question arises whether or not the wetland plants selected for the purpose of phytostabilisation should be tolerant. This problem is particularly well described for remediation of dry mine tailings, where only metal-tolerant populations of certain species, mostly grasses, can be used for revegetation of mine wastes. Classic studies on this topic were for example carried out by Bradshaw and co-workers [53]. However, it has only relatively recently been established that wetland plants typically show innate tolerance to metals.

McNaughton and co-workers [54] in 1974 were the first to report on tolerance to zinc, cadmium and lead in *Typha latifolia* and observed that this trait existed in this species without the development of specific metal-tolerant populations. Tolerance to copper and nickel in the same species was confirmed ten years later by Taylor and Crowder [55]. More recently, Ye and co-workers [56] confirmed the findings for *T. latifolia* and the same research team made similar observations in *Phragmites australis* [57, 58]. Then Otte and co-workers reported that populations of floating sweetgrass, *Glyceria fluitans*, which had not previously been exposed to zinc, were capable of growing well in zinc-lead mine tailings containing highly elevated levels of metals [59]. Research by the same group has now established the existence of innate metal tolerance in many wetland plants [60, 61, 62, 63, 64]. This has been confirmed for several other species by Wong and co-workers [65]. What this means is that wetland plants for applications in both rhizofiltration or phytostabilisation of metal-contaminated water or soils need not be sourced from specific metal-tolerant populations, as is the case for dryland conditions [53], but that they can be obtained from any population, anywhere. This is an important consideration, as it reduces the costs of supply of suitable plants and it avoids depletion of specific plant populations.

5. Phytoextraction

5.1. REMOVAL OF POLLUTANTS VIA PLANTS

Phytoextraction refers to the removal of pollutants from soils or water through plant uptake [66]. In this approach, the pollutants are not simply recycled back into the wetland, as is the case in phytostabilisation, but they are removed from the system by harvesting the plants or by volatilisation through the plants. The latter only applies to substances that can form gaseous compounds, such as Se [22, 67] and Hg [28]. For phytoextraction through uptake and harvest to be effective, uptake by the plants must be high relative to the amount of pollutant present. This is the case when the amount of pollutant is clearly defined, for example when relatively small volumes of water are contaminated. While the concentrations of pollutants resulting from uptake in aquatic plants may not be very high, the sheer volumes of the plants may be enough to successfully remove the pollutants from the water within a foreseeable period of time [68, 69, 70]. This is different, however, when attempts are made to use emergent plants for removal of pollutants from soils or sediments. In that case, the amount of pollutants present in the soils is often not clearly defined and much larger than that taken up by the plants annually. Some plants are much better at translocating pollutants to the harvestable parts (typically the shoots) than others [e.g. 71, 72, 73], but the amounts removed are typically not enough to reduce the levels of pollution in the soils within reasonable time [74].

One particular approach to phytoextraction, that of phytomining, has not had much attention in general, let alone in relation to wetlands, yet may have great potential. Phytomining uses the uptake and accumulation in plants to extract valuable metals from diffuse sources for commercial exploitation. The plants are harvested after growing in moderately metal-rich soils for time and the metals extracted and concentrated from the plant biomass. While the effort is not worth the gain for readily available metals like zinc and lead, Brooks and co-workers [75] show that even at current prices, it may in fact be worthwhile for precious metals, like gold, under certain circumstances. According to the Science Citation Index, the search for "wetland and phytomining" returns no matching reports. However, recent work by Otte and Jacob [76] suggests that plants from mineral rich wetlands, such as those associated with hot springs, show exceptional accumulation of less studied, yet valuable metals, such as Th, U and W, suggesting that there is potential for the use of plants wetland plants for phytomining.

6. Conclusions

The efficacy of wetlands for applications in phytoremediation, particularly for rhizofiltration, is a proven fact. Not only can they be applied to a wide range of pollutants, but they also provide suitable habitat for wildlife [77]. At the same time, while other reviews agree with this conclusion [78, 79, 80], care must be taken in the design, and failures have occurred [81]. Design mistakes may include building the system too small (rhizofiltration), with too little organic matter (rhizofiltration, phytostabilisation), and/or creating a system that is too homogeneous.

There is much scope for further development of wetlands for phytoremediation, particularly in the areas of phytostabilisation and phytomining. Where the latter is concerned, bioengineering (wetland) plants to improve uptake of valuable metals is likely to be worthwhile undertaking [82].

References

[1] Mitsch, WJ and Gosselink JG (2000) Wetlands. 3rd edition, John Wiley & Sons, New York
[2] Gambrell RP and Patrick WH, Jr (1978) Chemical and microbiological properties of anaerobic soils and sediments, in DD Hook and RMM Crawford, Eds., Plant life in anaerobic environments. Ann Arbor Publishers, Michigan, USA, Chapter 13, pp. 375-424
[3] Ponnamperuma, FN (1984) Effects of flooding on soils, in TT Kozlowski, Ed., Flooding and plant growth. Academic Press, Orlando, Fl, USA, pp. 9-45
[4] Van Bodegom, PM and Scholten, JCM (2001) Microbial processes of CH_4 production in a rice paddy soil: Model and experimental validation. Geochim Cosmochim Ac 65: 2055-2066
[5] Bertin, C; Yang, XH and Weston, LA (2003) The role of root exudates and allelochemicals in the rhizosphere. Plant Soil 256: 67083
[6] Lu, YH; Watanabe, A and Kimura, M (2004) Contribution of plant photosynthates to dissolved organic carbon in a flooded rice soil. Biogeochemistry 71: 1-15
[7] Saarnio, S; Wittenmayer, L and Merbach, W (2004) Rhizosphere exudation of *Eriophorum vaginatum* L. – Potential link to methanogenesis. Plant Soil 267: 343-355
[8] Purnobasuki, H and Suzuki, M (2004) Aerenchyma formation and porosity of a mangrove plant, *Sonneratia alba* (Lytharaceae). J Plant Res 117: 465-472
[9] Bezbaruah, AN and Zhang, TC (2005) Quantification of oxygen release by bulrush (*Scirpus validus*) roots in a constructed treatment wetland. Biotechnol Bioeng 89: 308-318
[10] Anderson, CJ and Cowell, BC (2004) Mulching effects on the seasonally flooded zone of west-central Florida, USA, wetlands. Wetlands 24: 811-819
[11] Jacob, DL and Otte, ML (2003) Conflicting processes in the wetland plant rhizosphere: metal retention or mobilization? Water Air Soil Poll Focus 3: 91-104

[12] Klomjek, P and Nitisoravut, S (2005) Constructed treatment wetland: a study of eight plant species under saline conditions. Chemosphere 58: 585-593
[13] Fink DF and Mitsch WJ (2004) Seasonal and storm event nutrient removal by a created wetland in an agricultural watershed. Ecol Eng 23: 313-325
[14] Yirong, C and Puetpaiboon, U (2004) Performance of constructed wetland treating wastewater from seafood industry. Water Sci Technol 49: 289-294
[15] Beining, BA and Otte, ML (1997) Retention of metals and longevity of a wetland receiving mine leachate, in JE Brandt; JR Galevotic; L Kost and J Trouart, Eds., Proc. 14th Annual National Meeting - Vision 2000: An Environmental Commitment. American Society for Surface Mining and Reclamation, Austin, Texas, May 10-16, 1997, pp. 43-46
[16] Ye, ZH; Whiting, SN; Lin, ZQ; Qian, JH and Terry, N (2001) Removal and distribution of iron, manganese, cobalt, and nickel within a Pennsylvania constructed wetland treating coal combustion by-product leachate. J Environ Qual 30: 1464-1473
[17] Samecka-Symerman, A; Stepien, D and Kempers, AJ (2004) Efficiency in removing pollutants by constructed wetland purification systems in Poland. J Toxicol Env Heal A 67: 265-275
[18] Birch, GF; Matthai, C; Fazeli, MS and Suh, J (2004) Efficiency of a constructed wetland in removing contaminants from stormwater. Wetlands 24: 459-466
[19] Keefe, SH; Barber, LB; Runkel, RL and Ryan, JN (2004) Fate of volatile organic compounds in constructed treatment wetlands. Environ Sci Technol 38: 2209-2216
[20] Ji, GD; Sun, T; Zhou, QX; Sui, X; Chang, SJ and Li, PJ (2002) Constructed subsurface flow wetland for treating heavy oil-produced water of the Liaohe oilfield in China. Ecol Eng 18: 459-465
[21] Best, EPH; Sprecher, SL; Larson, SL; Fredrickson, HL and Bader FD (1999) Environmental behavior of explosives in groundwater in groundwater from the Milan army ammunition plant in aquatic and wetland plant treatments. Removal, mass balances and fate in groundwater of TNT and RDX. Chemosphere 38: 3383-3396
[22] Lin, ZQ and Terry, N (2003) Selenium removal by constructed wetlands: quantitative importance of biological volatilization in the treatment of selenium-laden agricultural drainage water. Environ Sci Technol 37: 606-615
[23] Stott, R; May, E and Mara, DD (2003) Parasite removal by natural wastewater treatment systems: performance of waste stabilisation ponds and constructed wetlands. Water Sci. Technol. 48: 97-104
[24] Gauci, V; Fowler, D; Chapman, SJ and Dise, NB (2004) Sulphate deposition and temperature controls on methane emission and sulfur forms in peat. Biogeochemistry 71: 141-162
[25] Qiu, S; McComb, AJ; Bell, RW and Davis, JA (2004) Phosphorus dynamics from vegetated catchment to lakebed during seasonal refilling. Wetlands 24: 828-836
[26] Wiessner, A; Kappelmeyer, U; Kuschk, P and Kastner, M (2005) Influence of the redox condition dynamics on the removal efficiency of a laboratory-scale constructed wetland. Water Res 39: 248-256
[27] Benoit, JM; Gilmour, CC and Mason, RP (2001) The influence of sulfide on solid-phase mercury bioavailability for methylation by pure cultures of *Desulfobulbus propionicus* (1pr3). Environ Sci Technol 35: 127-132
[28] Lindberg, SE; Dong, WJ; Chanton, J; Qualls, RG and Meyers, T (2005) A mechanism for bimodal emission of gaseous mercury from aquatic macrophytes. Atmos Environ 39: 1289-1301
[29] Guerin, TF (1999) The anaerobic degradation of endosulfan by indigenous microorganisms from low oxygen soils and sediments. Environ Pollut 106: 13-21
[30] Lahvis, MA; Baehr, AL and Baker, RJ (1999) Quantification of aerobic biodegradation and volatilization rates of gasoline hydrocarbons near the water table under natural attenuation conditions. Water Resour Res 35: 753-765
[31] Armenante, PM; Kafkewitz, D; Lewandowski, GA and Jou, CJ (1999) Anaerobic aerobic treatment of halogenated phenolic compounds. Water Res 33: 681-692
[32] Keffe, SH; Barber, LB; Runkel, RL and Ryan, JN (2004) Fate of volatile organic compounds in constructed wastewater treatment wetlands. Environ Sci Technol 38: 2209-2216
[33] Allen, HE; Fu, G and Deng, B (2003) Analysis of acid-volatile sulfide (AVS) and simultaneously extracted metals (SEM) for the estimation of potential toxicity in aquatic sediments. Environ Toxicol Chem 12:1441-53
[34] Emerson, D; Weiss, JV and Megonical, JP (1999) Iron-oxidizing bacteria are associated with ferric hydroxide precipitates (Fe-plaque) on the roots of wetland plants. Appl Environ Microb 65: 2758-2761

[35] Mendelssohn, IA; Kleiss, BA and Wakeley, JS (1995) Factors controlling the formation of oxidized root channels: a review. Wetlands 15: 37-46

[36] Wang, T and Peverly, JH (1996) Oxidation states and fractionation of plaque iron on roots of common reeds. Soil Sci Soc Am J 60: 323-329

[37] Otte, ML; Rozema, J; Koster, L; Haarsma, MS and Broekman, RA (1989) Iron plaque on roots of *Aster tripolium* L.: interaction with zinc uptake. New Phytol 111: 309-317

[38] St.-Cyr L and Campbell PGC (1996) Metals (Fe, Mn, Zn) in the root plaque of submerged aquatic plants collected in situ: relations with metal concentrations in the adjacent sediments and in the root tissue. Biogeochemistry 33: 45-76

[39] Doyle, MO and Otte, ML (1997) Organism-induced accumulation of iron, zinc and arsenic in wetland soils. Environ Pollut 96: 1-11

[40] Arias, CA; Cabello, A; Brix, H and Johansen, NH (2003) Removal of indicator bacteria from municipal wastewater in an experimental two-stage vertical flow constructed wetland system. Water Sci Technol 48: 35-41

[41] Mayo, AW (2004) Kinetics of bacterial mortality in granular bed wetlands. Phys Chem Earth 29: 1259-1264

[42] Cole, CA and Lefebvre, EA (1991) Soil and water characteristics of a young surface mine wetland. Environ Manage 15: 403-410

[43] Sistani, KR; Mays, DA and Taylor, RW (1995) Biogeochemical characteristics of wetlands developed after strip mining for coal. Commun Soil Sci Plan 26: 3221-3229

[44] King, SL (1996) The effects of flooding on bottomland hardwood seedlings planted on lignite mine spoil in east Texas. Tex J Sci 75-84

[45] Johns, D; Williams, H; Farrish, K and Wagner, S (2004) Denitrification and soil characteristics of wetlands created on two mine soils in east Texas, USA. Wetlands 24: 57-67

[46] Pappin Willianen, S; Beckett, P and Courtin, G (1998) Progress in establishing wetland plants in permanently-flooded uranium tailings. Min Pro Ext Met Rev 19: 47-60

[47] Farrell, L (2002) Rehabilitation of Silvermines. Extractive Industry of Ireland Annual Review 3: 41-43

[48] Weis, JS and Weis, P (2004) Metal uptake, transport and release by wetland plants: implications for phytoremediation and restoration. Environ Int 30: 685-700

[49] Banta, GT and Andersen, O (2003) Bioturbation and the fate of sediment pollutants - Experimental case studies of selected infauna species. Vie Milieu 53: 233-248

[50] Stolz, E and Greger, M (2002) Cottongrass effects on trace elements in submerged mine tailings. J Environ Qual 31: 1477-1483

[51] Jacob, DL and Otte, ML (2004) Influence of *Typha latifolia* and fertilization on metal mobility in two different Pb-Zn mine tailings types. Sci Total Environ 333: 9-24

[52] Jacob DL and Otte ML (2004) Long-term effects of submergence and wetland vegetation on metals in a 90-year old abandoned Pb-Zn mine tailings pond. Environ Pollut 130: 337-345

[53] Smith, RA and Bradshaw, AD (1979) The use of metal tolerant plant populations for the reclamation of metalliferous wastes. J Appl Ecol 16: 595-612

[54] McNaughton, SJ; Folsom, TC; Lee, T; Park, F; Price, C; Roeder, D; Schmitz, J and Stockwell, C (1974) Heavy metal tolerance in *Typha latifolia* without the evolution of tolerant races. Ecology 55: 1163-1165

[55] Taylor, GJ and Crowder, AA (1984) Copper and nickel tolerance in *Typha latifolia* clones from contaminated and uncontaminated environments. Can J Bot 62: 1304-1308

[56] Ye, ZH; Baker, AJM; Wong, MH and Willis, AJ (1997) Zinc, lead and cadmium tolerance, uptake and accumulation by *Typha latifolia*. New Phytol 136: 469-480

[57] Ye, ZH; Baker, AJM; Wong, MH and Willis, AJ (1997) Zinc, lead and cadmium tolerance, uptake and accumulation by common reed, *Phragmites australis* Cav Trin ec Steudel. Ann Bot 80: 363-370

[58] Ye, ZH; Wong, MH; Baker, AJM and Willis, AJ (1998) Comparison of biomass and metal uptake between two populations of *Phragmites australis* grown in flooded and dry conditions. Ann Bot 82: 83-87

[59] McCabe OM and Otte ML (2000) The wetland grass *Glyceria fluitans* for revegetation of mine tailings. Wetlands 20: 548-559

[60] Moran BM and Otte ML (2004) Innate zinc tolerance in the wetland grass *Glyceria fluitans*. Phyton 44: 95-108

[61] Matthews DJ; Moran BM and Otte ML (2004) Zinc tolerance, uptake and accumulation in the wetland plants *Eriophorum angustifolium*, *Juncus effusus* and *Juncus articulatus*. Wetlands 24: 859-869

[62] Matthews DJ; Moran BM; McCabe PF and Otte ML (2004) Zinc tolerance, uptake, accumulation and distribution in plants and protoplasts of five European populations of the wetland grass *Glyceria fluitans*. Aquat Bot 80: 39-52
[63] Matthews DJ; Moran BM and Otte ML (2005) Screening the wetland plant species *Alisma plantago-aquatica*, *Carex rostrata* and *Phalaris arundinacea* for innate tolerance to zinc and comparison with *Eriophorum angustifolium* and *Festuca rubra* Merlin. Environ Pollut 134: 343-351
[64] Matthews DJ; Gallagher TF and Otte ML (2005) An analysis of genetic variation in zinc-tolerant European populations of the wetland grass *Glyceria* fluitans using amplified fragment length polymorphism (AFLP) analysis. Aquat Bot *In press*
[65] Deng, H; Ye, ZH and Wong, MH (2004) Accumulation of lead, zinc and copper and cadmium by twelve wetland plant species thriving in metal contaminated sites in China. Environ Pollut 132: 29-40
[66] Salt, DE; Blaylock, M; Kumar, NPBA; Dushenkov, V; Ensley, BD; Chet, I and Raskin, I (1995) Phytoremediation: a novel strategy for the removal of toxic metals from the environment using plants. Biotechnology 13: 468-474
[67] Lin, ZQ; de Souza, M; Pickering, IJ and Terry, N (2002) Evaluation of the macroalga, muskgrass, for the phytoremediation of selenium-contaminated agricultural drainage water by microcosms. J Environ Qual 31: 2104-2110
[68] Kamal, M; Ghaly, AE; Mahmoud, N and Cote, R (2004) Phytoaccumulation of heavy metals by aquatic plants. Environ Int 29: 1029-1039
[69] Mkandawire, M; Tauert, B and Dudel, EG (2004) Capacity of *Lemna gibba* L. (Duckweed) for uranium and arsenic phytoremediation in mine tailing waters. Int J Phytoremediat 6: 347-362
[70] Al Rmalli, SW; Harrington, CF; Ayub, M and Haris, PI (2005) A biomaterial based approach for arsenic removal from water. J Environ Monitor 7: 279-282
[71] Stolz, E and Greger, M (2002) Accumulation properties of As, Cd, Cu, Pb and Zn by four wetland plant species growing in submerged mine tailings. Environ Exp Bot 47: 271-280
[72] Fritioff, A and Greger, M (2003) Aquatic and terrestrial plant species with potential to remove heavy metals from stormwater. Int J Phytoremediat 5: 211-224
[73] Gothberg, A; Greger, M; Holm, K and Bengtsson, BE (2004) Influence of nutrient levels on uptake and effects of mercury, cadmium, and lead in water spinach. J Environ Qual 33: 1247-1255
[74] Hinton, TG; Knox, AS; Kaplan, DI and Sharitz, R (2005) Phytoextraction of uranium by native trees in a contaminated wetland. J Radioanal Nucl Ch 264: 417-422
[75] Brooks, RR; Chambers, MF; Nicks, LJ and Robinson, BH (1998) Phytomining. Trends Plant Sci 3: 359-362
[76] Otte, ML and Jacob, DL (2005) Chemical fingerprinting of plants from contrasting wetlands – salt marsh, geothermal and mining-impacted. Phyton, *in press*
[77] Horstman, AJ; Nawrot, JR and Woolf, A (1998) Mine-associated wetlands as avian habitat. Wetlands 18: 298-304
[78] Williams, JB (2002) Phytoremediation in wetland ecosystems: progress, problems and potential. Crit Rev Plant Sci 21: 607-635
[79] Kennedy, G and Mayer, T (2002) Natural and constructed wetlands in Canada: an overview. Water Qual. Res J Can 37: 295-325
[80] Li, Q-F; Li, Z-A; Ren, H; Du, W-B; Tian, S-N and Peng S-L (2004) The role of wetland plants and soil in decontamination of heavy metals. J Trop Subtrop Bot 12: 273-279
[81] Barton, CD and Karathanasis, AD (1999) Renovation of a failed constructed wetland treating acid mine drainage. Environ Geol 39: 39-50
[82] Dushenkov, S; Skarzhinskaya, M; Glimelius, K; Gleba, D and Raskin, I (2002) Bioengineering of a phytoremediation plant by means of somatic hybridization. Int J Phytoremediat 4: 117 126

INFLUENCE OF HELOPHYTES ON REDOX REACTIONS IN THEIR RHIZOSPHERE

ARNDT WIESSNER, PETER KUSCHK, UWE KAPPELMEYER, OLIVER BEDERSKI, ROLAND-ARNO MÜLLER AND MATTHIAS KÄSTNER
Department of Remediation Research, UFZ Centre for Environmental Research Leipzig-Halle, Permoserstraße 15, 04318 Leipzig, Germany
Tel.: +49-341-235-2821; Fax: +49-341-235-2492.
E-mail:peter.kuschk@ufz.de

1. Introduction

In recent decades, wetland sites have become increasingly important for wastewater treatment [1, 2]. Treating wastewater in semi-natural plant systems is a technique which can in principle be applied in natural wetlands such as marshes, moors and wet fields, in artificial ponds and lagoons, and in constructed wetlands. Constructed wetlands exist in a number of different basic designs featuring different flow characteristics [3, 4]. Today the use of constructed wetlands, particularly to treat wastewater from decentralized sources making for cheap and simple treatments (including in combination with other technologies) has become the state of the art, and general design and operation guidelines have been published [5, 6].

The processes inside the rhizosphere in principle and the role of the plants within the complex system of plants, microorganisms, solids and pollutants are crucial for most natural treatment wetlands and for relatively simple constructed wetlands with horizontal subsurface slow flow characteristics [7]. Judged by experiments and practical experience, species of helophytes (marsh plants) work best of all in semi-natural wastewater treatment systems. This is because helophytes have certain characteristics of growth physiology that guarantee their survival even under waterlogged rhizosphere conditions. Although all the plant species listed in Table 1 are suitable, reed along with types of rushes and cattails are the ones most frequently used. Recently, the suitability of fast-growing trees such as willows has also been examined [8].

M. Mackova et al. (eds.), Phytoremediation Rhizoremediation, 69–82.

Table 1. Plant species frequently used in constructed wetlands (adapted from [7])

Scientific name	**English name**
Acorus calamus L.	sweet flag
Carex spp.	sedges
Glyceria maxima (Hartm.) Holmb	reed grass
Iris pseudacorus L.	yellow flag
Juncus spp.	rushes
Phragmites australis (Cav.) Trin. ex Steud.	common reed
Scirpus spp.	bulrushes
Typha angustifolia L.	narrow-leaved cattail
Typha latifolia L.	broad-leaved cattail

One of the decisive advantages of wetlands for wastewater treatment is the simultaneous effective use of very different turnover processes. The wetland rhizosphere allows fundamentally different oxidation–reduction conditions to exist in the same 'reactor' at the same time due to aerobic and anaerobic microbiological processes and varied physical and chemical processes. The reasons for these multireactive conditions are the characteristics of the waterlogged soil, the enormous biodiversity of the micro-organisms established under these conditions, and the plants' specific adaptational responses to the flooding of their rhizosphere.

The objectives of this paper are in particular to shed light on the influence of the helophytes on the redox reactions in their rhizosphere and to present the latest results of investigations into the redox dynamics in the rhizosphere of treatment wetlands.

2. Characterisation of the rhizosphere of helophytes

The conditions in the waterlogged soils of treatment wetlands can be defined as extreme for the survival of macrophytes. When soil is flooded, a chain of physical, chemical and biological processes initiate reducing soil conditions [9-13]. The exclusion of atmospheric gases results in oxygen depletion in the soil following chemical changes, including the accumulation of CO_2, N_2, H_2 and methane, and the microbial reduction of nitrate, nitrite, manganese, iron and sulphate, as well as the microbial accumulation of acetic and other small-chain fatty acids. The redox state becomes more negative [10, 11, 14], reaching an Eh lower than –300 mV [15]. The wastewater may also cause extreme conditions in the rhizosphere of the plants depending on its pH (acidic or alkaline), the presence of toxic compounds such as phenols, tensides, biocides and heavy metals, etc., and its salinity [7].

These extreme reducing conditions result in particular in various stress symptoms for the wetland plants owing to the restricted oxygen supply of plant tissues [16-20], changes in the availability and/or concentrations of essential nutrients [13], the formation of phytotoxic compounds [9, 11, 21] causing injury to the plants [10, 22], the inhibited growth of the roots [23, 24] or even the entire plants [18-20, 25-27,], and other adverse effects on plants [18-20, 28-30]. Furthermore, reduced conditions have been found to initiate internal water stress [31, 32], decreases in photosynthetic rates [16, 33, 34], and reduced biomass growth [35-37].

Although many findings have been published on plant responses under flooded soil conditions [17, 36, 38-44], our understanding of the relationship between the redox state of the soil and the physiology of wetland plants is far from complete [13].

3. Responses of helophytes to waterlogged soil conditions

3.1. INTERNAL GAS TRANSPORT MECHANISM

Despite the extreme conditions in waterlogged soils, most wetland plants are able to survive, grow and reproduce there. In contrast to typical land plants, wetland plants have anatomical and physiological attributes necessary for their long-term survival under waterlogged soil conditions [17-20, 44]. The degree of adaptation is specific to individual species and their survival varies from a few hours to several months [45]. This adaptation is based on an extensive gas (oxygen) transport mechanism that enables the plants to provide their roots with atmospheric or internal photosynthetic oxygen for respiration and to oxidize reduced compounds in the rhizosphere [17]. This process has been considered as a major mechanism for coping with soil anaerobiosis [17-20, 22, 36, 38-43, 46-48]. Gas transport of air (oxygen) from aboveground parts of the plants to the underground parts for respiration and release into the flooded rhizosphere occurs by special internal tissues forming open channels with low flow resistance, known as aerenchyma [17]. Depending on the degree of adaptation, aerenchyma can account for as much as 60% of the total tissue volume [49]. In addition, this tissue enhances the potential for oxygen to be transported to the remote underground parts of the plant by diminishing the internal volume of respiring tissues and oxygen consumption [17-20, 50]. The genesis of aerenchyma structures by cell lysis or cell formation as well as their anatomical peculiarities has been thoroughly studied physiologically and anatomically [17, 42, 51-53].

Diffusion is one mechanism for moving the oxygen within the plants [54]. In addition, intensive convective gas flow induced by pressure gradients and known as "ventilation" or "through flow" exists in many wetland plants [18-20, 55-60]. The pressure gradients are formed by low pressure in oxygen-consuming tissues of the plants caused by different solubilities of the oxygen consumed and the carbon dioxide formed in this process, and by high pressure in the plant's leaves resulting from the inflow of atmospheric gases. The higher pressure in the leaves causes air to flow throughout the entire body of the plant, including the whole root system, by entering the aerenchyma tissues and leaving the plant through older leaves with lower stomatal conductance [49, 61]. The types and combination of the transport mechanisms (diffusion and/or convection) involved are specific to each plant species. Intensive through flow has been observed in *Typha latifolia* (cattail) and *Phragmites australis* (reed) for example [57, 62]. The inflow of air into the leaves to produce high internal pressure is mainly induced by temperature and humidity gradients between the inside of the leaves and the ambient air [18, 20, 49, 52, 58, 61-64].

Different mechanisms of oxygen transport to and through the plants, including the processes of the temperature and humidity induced turnover of air and its effects, are described in detail in [50] and [51]. The processes involved in ventilation in plants have been studied since the mid-19th century [65]. This interest has been increasingly revived since the 1980s, at least in connection with the growing attention paid to the biotechnological usage of helophytes to clean wastewater in constructed wetlands [65].

3.2. OXYGEN RELEASE BY ROOTS INTO RHIZOSPHERE OF HELOPHYTES

Oxygen reaching the root aerenchyma is respired by the root tissues and diffuses towards to the root apex and radially to the rhizosphere, to be consumed in the surrounding soil [54]. The oxygen flux from the root aerenchyma to the rhizosphere, known as "radial oxygen loss" (ROL), is determined by the concentration gradient, the consumption of oxygen by the cells along the radial path, and the physical resistance to oxygen diffusion [54, 56].

As a result, 30–40% of the oxygen supplied via the aerenchyma to the root apex is lost to the rhizosphere [54]. The radial oxygen loss causes oxygenation and therefore significant chemical and biological changes within the rhizosphere relevant to microbial populations [66], nutrient availability [67-70], and concentrations of potentially toxic substances [71-74]. However, the oxygen release from the root system to the rhizosphere mainly occurs at the apex, as well as at lateral parts which tend to be shorter, thinner and of lower porosity than the main axes [75, 76]. The lateral parts were described as the major source of oxygen loss to the rhizosphere in some species [77]. The oxygen released into the rhizosphere protects the young, sensitive parts of the root system against reduced soil compounds and helps the plants penetrate the reduced soil [19, 54, 78-81].

The oxygen released may result in a protective "layer" on the root surface [82]. This "layer" may have a thickness of between 1 and 4 mm depending on the reduced state of the root environment and it is characterized by a redox gradient ranging from about –250 mV as frequently measured in reduced rhizospheres to about +500 mV directly at the root surface [82]. Oxygen is more or less continuously released from the root system, counterbalancing chemical and biological oxygen consumption. This ability of helophytes to oxygenate their rhizosphere is of particular interest for biotechnological application [83, 84].

Different methods to estimate oxygen flow rates have been used, mainly in plant-physiological investigations [50]. Rates of 126 μmol O_2/h g root dry mass for *Juncus ingens* (giant rush) and 120–200 μmol O_2/h g root dry mass for *Typha latifolia* (cattail) determined by the titanium-citrate method are of technological relevance [85, 86]. Furthermore, model calculations for *Phragmites australis* (reed) resulting in oxygen input rates of 5–12 g O_2/m^2 patch area per day [86] and investigations with individual plantlets of different species in hydroponic cultures resulting in the highest oxygen release rates of 1.4 mg O_2/h plantlet for *T. latifolia* [84] highlight from a more biotechnological view the considerable potential of helophytes to release oxygen. Some studies have revealed that the redox state of the rhizosphere has a significant effect on the intensity of oxygen release of various helophytes, with oxygen release rates increasing as the soil Eh becomes more reduced [85, 87, 88, 84]. In hydroponics model investigations, plantlets of various species showed the highest intensity of oxygen release at –250 mV < Eh < –150 mV and for extremely reduced conditions of Eh < –250 mV the release intensities were found to be lower [84].

Additional investigations emphasize the importance of the above and underground portions of the plants for the intensity of oxygen release [89]. The release rates by the roots of *T. latifolia* and *J. effusus* were found to be determined by the growth state of the aboveground part of the plants and relatively independent from the size of the root systems [89]. To clarify the influence of ambient conditions (air temperature and air humidity) determining the gas input at the leaves on the intensity of oxygen release into the rhizosphere, additional investigations should be performed.

The ventilation inside the helophytes causes air flow through the plants, i.e. the transport of mainly O_2 and also N_2. The behaviour of the atmospheric nitrogen (given radial losses from the roots) and its importance inside the rhizosphere has still not been investigated.

Gas exchange at the appropriate parts of the root system not only causes the oxygenation of the rhizosphere but also enables the flux of gases from the rhizosphere together with gases generated inside the plant tissues to the atmosphere by internal diffusion and/or ventilation. Although this exchange has been investigated for ethylene, carbon dioxide and methane, for example [90-92], information about nitrogen generated by denitrification inside the rhizosphere is still lacking. The processes involved depend on the species [83, 93-100]. In connection with the gas exchange, determining the phytohormone ethylene as a marker for plant-physiology studies is of particular interest [52, 92, 101-108]. Carbon dioxide may be generated by microbial mineralization in the soil or be derived from respiration in underground tissues of the plants [95, 109] and can also be fixed in photosynthesis by the plants [110]. Because of the role of methane as a "greenhouse gas" – after all, up to 90% of methane emissions from flooded soils may be transferred by the plants to the atmosphere [98] – interest in the process of methane release by plants has grown in recent years [111].

4. Release of organic compounds from plants

Knowledge about the input of carbon from plants into the rhizosphere (rhizodeposition) mainly comes from agricultural research. The quantity of C-compounds released by agricultural crops has been estimated at 10–40% of net photosynthetic production [112]. The composition of the exudates is highly diverse and species-specific. For example, sugars and vitamins such as thiamine, riboflavine and pyridoxine etc., organic acids such as malate, citrate, amino acids, benzoic acids, and phenolic compounds have all been identified [113]. Rhizodeposition may initiate the mobilization of nutrients [114], allelopathic effects [113], and the stimulation of microbial growth and activity inside the rhizosphere [112, 115-117].

Generally speaking, our understanding of the composition of root exudates of helophytes is very limited. Various substituted aromatic derivatives with hydroxyl, methoxyl, aldehyde and carboxyl groups were found in rhizome extracts of *Scirpus lacustris* (bulrush) [118]. These compounds can be used as a carbon source by micro-organisms in the rhizosphere, causing oxygen depletion and decreasing the redox state. If not enough oxygen is available, other electron acceptor such as nitrate, nitrite, Fe^{3+}, sulphate or carbonate (methanogenesis) are used. The resulting metabolic products, especially H_2S, influence the redox conditions of the rhizosphere.

5. Dynamics of redox conditions and the removal of contaminants inside the rhizosphere of helophytes – results of model investigations

Because of the usually low hydrodynamics, rapid microbial consumption by more or less adsorbed micro-organisms and fast chemical reactions (fast oxidation in the strongly reduced surroundings) changes in redox reactions inside the rhizosphere due to gas exchange and the release and uptake of compounds mainly affect the area close to the root surface. Evaluating such micro-gradient processes near the root surface in constructed wetlands is very difficult due to the diurnal dynamics of different interacting redox reactions, such as nitrification, denitrification, the mineralisation of carbon substances, methanogenesis, sulphate reduction, sulphur/sulphide oxidation etc. [2, 7, 13].

Despite the overall reduced redox conditions detectable in the rhizosphere of constructed wetlands, highly effective oxidation processes can also be determined by evaluating the inflow and outflow concentrations of oxidizable contaminants. In addition to oxidation processes enhanced by the input of atmospheric oxygen at the surface of the wetland [119], diurnal micro-gradient oxidation processes near the rhizoplane clearly also play an important role. Furthermore, temporal gradients overlapping the diurnal gradients corresponding to changes in the oxygen release at the rhizoplane have to be considered, too, due to daily and annual variations of plant-physiological and/or ambient gas exchange conditions (temperature, humidity, illumination etc).

To shed more light on the dynamics of the redox conditions enhanced by the plants and the changes in micro-gradient redox reactions, investigations in a specially designed laboratory-scale reactor [120] were carried out [121]. This enabled the treatment of an artificial wastewater in a planted (*J. effusus*) gravel bed to be evaluated, and daily variations in the redox state depending on the intensity of illumination by daylight were observed, as shown in Figure 1.

Particularly in summer, these variations ranging from Eh $\approx$–150 to –230 mV at night to Eh $\approx$ +50 to +350 mV at midday and in the early afternoon could only have been caused by changes in the light-enhanced ventilation inside the plants, resulting in changes in the oxygen release by the roots. This effect was interpreted as fundamental proof of the daily temporal variation of the redox state in the areas of the rhizosphere near the roots. Further investigations are needed to evaluate the correlations between these short-term redox changes and certain removal processes. The pH was also found to change daily to a small but still significant extent in response to the intensity of the daylight (Figure 2).

This pH variability has to be interpreted as a reaction to changes in the removal processes, including the consumption of nitrate, as well as changes to carbon dioxide transport out of the rhizosphere that are enhanced by the plants.

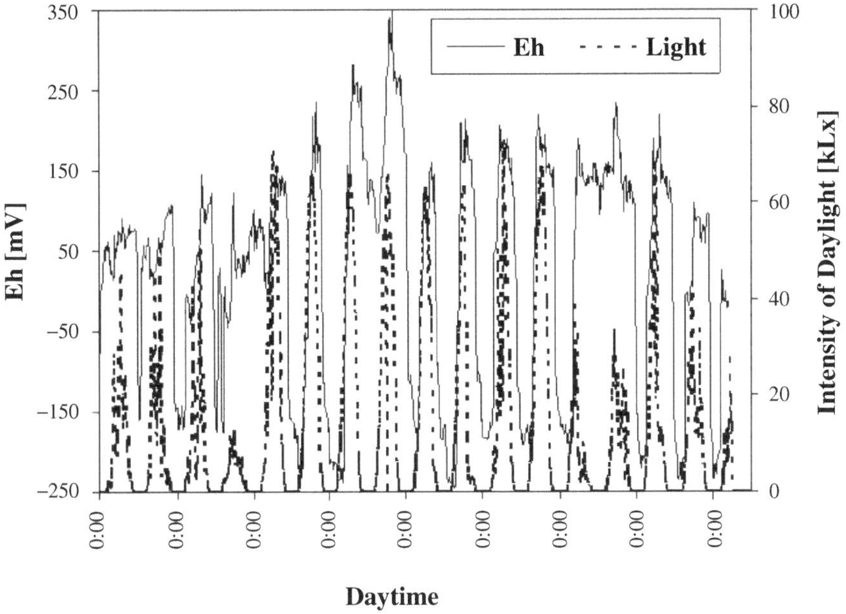

Figure 1. Changes of the redox state in the rhizosphere related to the intensity of daylight shown for several days in mid-summer (adapted from [121]).

The correlation between the gas exchange by the plants and carbon removal inside the rhizosphere also can be derived from a long-term seasonal evaluation, as shown in Figure 3. The increase in global radiation during the summer causes more intensive internal ventilation inside the plants. The resulting higher intensities of oxygen release into the rhizosphere on the other hand cause higher intensities of carbon removal and minimum outflow concentrations of COD, respectively. Contrasting behaviour is observed during darker seasons.

6. Outlook

Especially in natural and in simple horizontal subsurface flow constructed treatment wetlands; the interactions of the helophytes and their rhizosphere are of importance for the efficiency of the removal processes. The gas exchange by the plants is mainly responsible for influencing the redox conditions and the removal processes inside the rhizosphere. Above all, oxygenation initiates diurnal and temporal gradients of redox states and oxygen availability in the area around the roots and enables different chemical and biological oxidation and anoxic processes to take place simultaneously close together.

Our knowledge of these correlations in the rhizoplane and the roots' environment is by no means complete. However, they must be understood if treatment wetlands are to be ideally designed and operated.

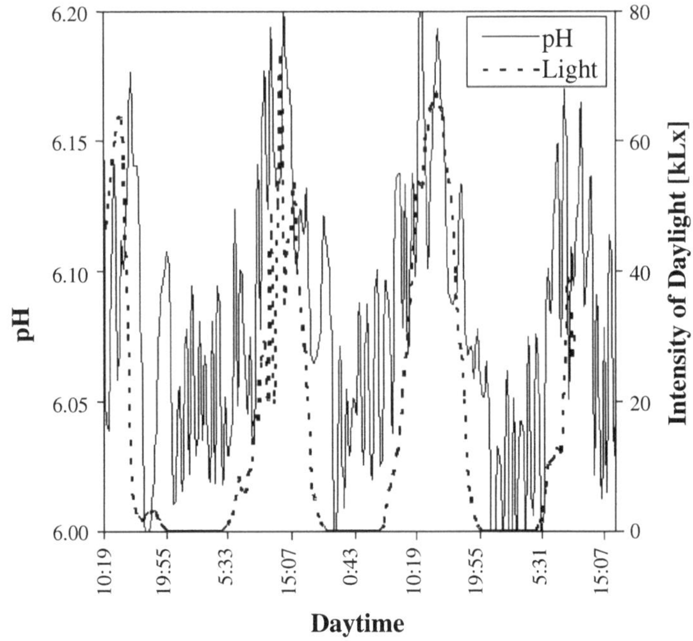

Figure 2. Changes of the pH in the rhizosphere correlated to the redox state shown for several days in mid-summer (adapted from [121]).

The problem is that because technologists mainly evaluate overall removal efficiencies by "black box" investigations of treatment wetlands, too little is known about the main processes and their phytobiological, microbiological and chemical interactions in the rhizosphere. Moreover, plant biologists extensively investigate the fundamental anatomical and physiological aspects of plants' gas exchange independently of technical application. Future biotechnological investigations should be based on current biological and technological expertise and should favour a combination of methods. Such investigations should be focused on: i) quantifying the amounts of gas transported into and out of the rhizosphere; ii) evaluating the variability of oxygen release by the plants into their rhizosphere; iii) investigating micro-gradient processes inside the rhizosphere by using suitable laboratory-scale techniques; iv) evaluating the microbial diversity inside the rhizosphere.

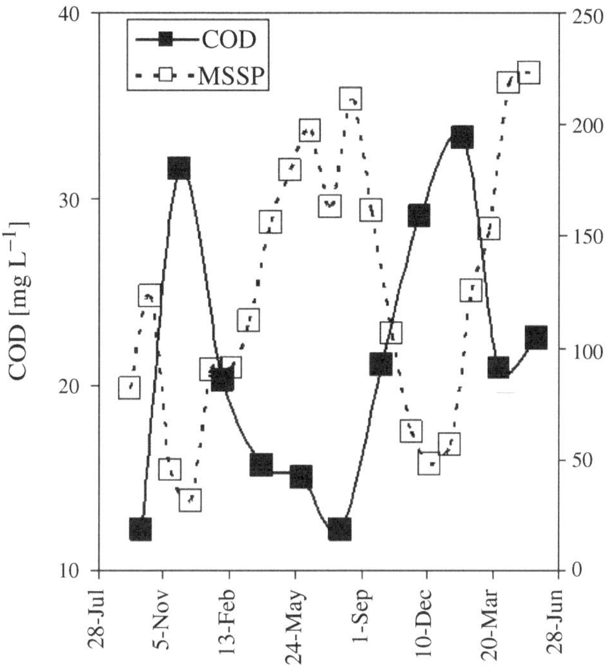

Figure 3. Mean COD outflow concentrations and the monthly sum of sunshine periods (MSSP) for a two years operation period (adapted from [121]).

References

[1] White, KD and Burken, JG (1999) Natural treatment and on-site processes. Water Environment Research 71(5): 676-685

[2] Williams, JB (2002) Phytoremediation in wetland ecosystems: Progress, problems, and potential. Critical Reviews in Plant Sciences 21(6): 607-635

[3] Kadlec, R (1987) Nothern natural wetland water treatment systems, in KR Reddy and WH Smith, Eds., Aquatic plants for water treatment and resource recovery. Magnolia Publ., Orlando, FL, USA, 83-98

[4] Wissing, F (1995) Wasserreinigung mit Pflanzen. Verlag Eugen Ulmer, Stuttgart, Germany

[5] Anonymous (1998) ATV instructions. Principles for the Dimensioning, Construction and Operation of Plant Beds for Communal Wastewater with Capacities up to 1000 Total Number of Inhabitants and Population Equivalents. Society for Promotion of Wastewater Treatment Technology e.V. (GFA), Hennef, Germany

[6] Kadlec, R; Knight, R; Vymazal, J; Brix, H; Cooper, P and Haberl, R (2000) Constructed Wetlands for Pollution Control. IWA Publishing, London, UK

[7] Stottmeister, U; Wiessner, A; Kuschk, P; Kappelmeyer, U; Kästner, M; Bederski, O; Müller, RA and Moormann, H (2004) Effects of plants and microorganisms in constructed wetlands for wastewater treatment. Biotechnology Advances 22(1-2): 93-117

[8] Greenway, M and Bolton, KGE (1996) From wastes to resources - turning over a new leaf: Melaleuca trees for wastewater treatment. Environmental Res. Forum 5-6: 363-366

[9] Ponnamperuma, FN (1972) The chemistry of submerged soil. Advances in Agronomy 24: 29-96

[10] Gambrell, RP and Patrick, WH (1978) Chemical and microbiological properties of anaerobic soils and sediments, in DD Hook and RMM Crawford, Eds., Plant life in anaerobic environments. Ann Arbor Science, Michigan, USA, 375-423
[11] Ponnamperuma, FN (1984) Effects of flooding on soils, in TT Kozlowski, Ed., Flooding and Plant Growth. Academic Press, Orlando, FL, USA, 1-44
[12] Gambrell, RP; DeLaune, RD and Patrick, WH (1991) Redox processes in soil following oxygen depletion, in MB Jackson, DD Davies and H Lambers, Eds, Plant Life Under Oxygen Deprivation: Ecology, Physiology, and Biochemistry. SPB Academic Publisher, The Hague, Netherlands, 101-117
[13] Pezeshki, SR (2001) Wetland plant responses to soil flooding. Environmental and Experimental Botany 46(3): 299-312
[14] Patrick, WH and DeLaune, RD (1977) Chemical and biological redox systems affecting nutrient availability in the coastal wetlands. Geoscience and Man 17: 131-137
[15] Turner, FT and Patrick, WH (1968) Chemical changes in waterlogged soils as a result of oxygen depletion, in Transactions of the 9th congress of the International Society of Soil Science 4: 53-56
[16] DeLaune, RD; Pezeshki, SR and Pardue, JH (1990) An oxidation-reduction buffer for evaluating physiological response of plants to root oxygen stress. Environ Exp Bot 30: 243-247
[17] Armstrong, W; Braendle, R and Jackson, MB (1994) Mechanisms of flood tolerance in plants. Acta Botanica Neerlandica 43: 307-358
[18] Armstrong, J; Armstrong, W; Beckett, PM; Halder, JE; Lythe, S; Holt, R and Sinclair, A (1996) Pathways of aeration and the mechanisms and beneficial effects of humidity- and Venturi-induced convections in *Phragmites australis* (Cav) Trin ex Steud. Aquatic Botany 54(2-3): 177-197
[19] Armstrong, J; Armstrong, W and VanderPutten, WH (1996) *Phragmites* die-back: Bud and root death, blockages within the aeration and vascular systems and the possible role of phytotoxins. New Phytologist 133(3): 399-414
[20] Armstrong, W; Armstrong, J and Beckett, PM (1996) Pressurised aeration in wetland macrophytes: Some theoretical aspects of humidity-induced convection and thermal transpiration. Folia Geobotanica and Phytotaxonomica 31(1): 25-36
[21] DeLaune, RD; Patrick, WH and Buresh, RJ (1978) Sedimentation rates determined by Cs-137 dating in a rapidly accreting salt marsh. Nature 275: 532-533
[22] Drew, MC and Lynch, JM (1980) Soil Anaerobiosis, Microorganisms, and Root Function. Annual Review of Phytopathology 18(1): 37-66
[23] Tanaka, A; Mulleriyawa, RP and Yasu, T (1968) Possibility of hydrogen sulphide induced iron toxicity of the rice plant. Soil Science and Plant Nutrition 4: 1-6
[24] Allam, AI and Hollis, JP (1972) Sulphide inhibition of oxidases in rice roots. Phytopathology 62: 634-639
[25] King, GM; Klug, MJ; Wiegert, RG and Chalmers, AG (1982) Relation of soil water movement and sulfide concentration to *Spartina alterniflora* production in a Georgia saltmarsh. Science 218: 61-63
[26] Ingold, A and Havill, DC (1984) The influence of sulphide on the distribution of higher plants in salt marshes. Journal of Ecology 72: 1043-4054
[27] Havill, DC; Ingold, A and Pearson, J (1985) Sulphide tolerance in coastal halophytes. Vegetatio 62: 279-285
[28] Rao, DN and Mikkelsen, DS (1977) Effects of acidic, propionic, and butyric acids on rice seedling growth and nutrition. Plant and Soil 47: 323-334
[29] Armstrong, J; Afreen Zobayed, F and Armstrong, W (1996) *Phragmites* die-back: Sulphide- and acetic acid-induced bud and root death, lignifications, and blockages within aeration and vascular systems. New Phytologist 134(4): 601-614
[30] Armstrong, J and Armstrong, W (1999) *Phragmites* die-back: toxic effects of propionic, butyric and caproic acids in relation to pH. New Phytologist 142(2): 201-217
[31] Kramer, PJ (1940) Causes of decreased absorption of water by plants in poorly aerated media. Am J Bot 27: 216-220
[32] Hiron, RWP and Wright, STC (1973) The role of endogenous abscisic acid in the response of plants to stress. Journal of Experimental Botany 24: 769-781
[33] Pezeshki, SR (1993) Differences in patterns of photosynthetic responses to hypoxia in flood-tolerant and flood-sensitive tree species. Photosynthetica 28: 423-430
[34] Pezeshki, SR and Anderson, PA (1997) Response of three bottomland woody species with different flood-tolerance capabilities to various flooding regimes. Wetland Ecol Manag 4: 245-256

[35] Kludze, HK and DeLaune, RD (1994) Methane emission and growth of *Spartina patens* in response to soil redox intensity. Soil Science Society of America Journal 58: 1838-1845

[36] Pezeshki, SR (1994) Plant responses to flooding, in RE Wilkinson, Ed, Plant-Environment Interactions. Marcel Dekker, New York, NY, USA, 289-321

[37] Kludze, HK and DeLaune, RD (1995) Straw application effects on Methane and oxygen exchange and growth in rice. Soil Science Society of America Journal 59: 824-830

[38] Hook, DD and Crawford, RMM (1978) Plant life in anaerobic environments. Ann Arbor Science Publishers, Michigan, MA, USA

[39] Kozlowski, TT (1984) Flooding and Plant Growth. Academic Press, New York, NY, USA

[40] Kozlowski, TT (1984) Plant responses to flooding of soil. Bioscience 34: 1626-167

[41] Drew, MC (1990) Sensing soil oxygen. Plant Cell and Environment 13: 681-693

[42] Drew, MC (1997) Oxygen deficiency and root metabolism: Injury and acclimation under hypoxia and anoxia. Annual Review of Plant Physiology and Plant Molecular Biology 48: 223-250

[43] Kozlowski, TT (1997) Response of woody plants to flooding and salinity. Tree Physiology Monograph 1: 1-29

[44] Vartapetian, BB and Jackson, MB (1997) Plant adaptations to anaerobic stress. Annals of Botany 79: 3-20

[45] Crawford, RMM and Braendle, R (1996) Oxygen deprivation stress in a changing environment. Journal of Experimental Botany 47(295): 145-159

[46] Teal, JM and Kanwisher, JW (1966) Gas transport in the marsh grass *Spartina alterniflora*. Journal of Experimental Botany 17: 355-361

[47] Armstrong, W; Justin, SHFW; Beckett, PM and Lythe, S (1991) Root adaptation to soil waterlogging. Aquatic Botany 39(1-2): 57-73

[48] Perata, P and Alpi, A (1993) Plant responses to anaerobiosis. Plant Science 93(1-2): 1-17

[49] Grosse, W and Schröder, P (1986) Pflanzenleben unter anaeroben Umweltbedingungen, die physikalischen Grundlagen und anatomischen Voraussetzungen. Ber Dtsch Bot Ges 99: 367-81

[50] Colmer, TD (2003) Long-distance transport of gases in plants: a perspective on internal aeration and radial oxygen loss from roots. Plant Cell and Environment 26(1): 17-36

[51] Allen, LH (1997) Mechanisms and rates of O_2 transfer to and through submerged rhizomes and roots via aerenchyma. Soil and Crop Science Society of Florida Proceedings 56: 41-54

[52] Jackson, MB and Armstrong, W (1999) Formation of aerenchyma and the processes of plant ventilation in relation to soil flooding and submergence. Plant Biology 1(3): 274-287

[53] Soukup, A; Votrubova, O and Cizkova, H (2000) Internal segmentation of rhizomes of *Phragmites australis*: protection of the internal aeration system against being flooded. New Phytologist 145(1): 71-75

[54] Armstrong, W (1979) Aeration in higher Plants. Advances in Botanical Research 7: 225-332

[55] Dacey, JWH (1981) Pressurized Ventilation in the Yellow Waterlily. Ecology 62(5): 1137-1147

[56] Armstrong, W and Beckett PM (1987) Internal aeration and the development of stelar anoxia in submerged roots: A multishelled mathematical model combining axial diffusion of oxygen in a cortex with radial losses to the stele, the wall layers and the rhizosphere. New Phytologist 105: 221-245

[57] Armstrong, J and Armstrong, W (1991) A convective through-flow of gases in *Phragmites australis* (Cav.) Trin. ex Steud. Aquatic Botany 39(1-2): 75-88

[58] Armstrong, J; Armstrong, W and Beckett, PM (1992) *Phragmites australis*: venturi- and humidity-induced pressure flows enhance rhizome aeration and rhizosphere oxidation. New Phytologist 120: 197-207

[59] Sorrell, BK; Brix, H and Orr, PT (1997) *Eleocharis sphacelata*: Internal gas transport pathways and modelling of aeration by pressurized flow and diffusion. New Phytologist 136(3): 433-442

[60] Vretare, V and Weisner, SEB (2000) Influence of pressurized ventilation on performance of an emergent macrophyte (*Phragmites australis*). Journal of Ecology 88(6): 978-987

[61] Grosse, W (1989) Thermoosmotic air transport in aquatic plants affecting growth activities and oxygen diffusion to wetland soils, in DA Hammer, Ed, Constructed Wetlands for Wastewater Treatment: Municipal, Industrial and Agricultural. Lewis Publisher, Chelsea, UK, 416-469

[62] Bendix, M; Tornbjerg, T and Brix, H (1994) Internal gas transport in *Typha latifolia* L. and *Typha angustifolia* L. 1. Humidity-induced pressurization and convective throughflow. Aquatic Botany 49(2-3): 75-89

[63] Grosse, W (1996) The mechanism of thermal transpiration (equals thermal osmosis). Aquatic Botany 54(2-3): 101-110

[64] Woermann, D; Metzner, H and Grosse, W (2002) Humidity-induced convection of air across porous membranes. Journal of Membrane Science 206(1-2): 69-85

[65] Grosse, W; Armstrong, J and Armstrong, W (1996) A history of pressurised gas-flow studies in plants. Aquatic Botany 54(2-3): 87-100

[66] Gilbert, B and Frenzel, P (1998) Rice roots and CH_4 oxidation: The activity of bacteria, their distribution and the microenvironment. Soil Biology and Biochemistry 30(14): 1903-1916

[67] Reddy, KR; Patrick, WH and Lindau, CW (1989) Nitrification-denitrification at the plant root-sediment interface in wetlands. Limnology and Oceanography 34: 1004-1013

[68] Kirk, GJD and Bajita, JB (1995) Root-induced iron oxidation, pH changes and zinc solubilisation in the rhizosphere of lowland rice. New Phytologist 131: 129-137

[69] Saleque, MA and Kirk, GJD (1995) Root-induced solubilisation of phosphate in the rhizosphere of lowland rice. New Phytologist 129: 325-336

[70] Christensen, KK and Wigand, C (1998) Formation of root plaques and their influence on tissue phosphorus content in *Lobelia dortmanna*. Aquatic Botany 61(2): 111-122

[71] Laanbroek, HJ (1990) Bacterial cycling of minerals that affect plant growth in waterlogged soils: a review. Aquatic Botany 38(1): 109-125

[72] Begg, CBM; Kirk, GJD; MacKenzie, AF and Neue, HU (1994) Root-induced iron oxidation and pH change in the lowland rice rhizosphere. New Phytologist 128: 469-477

[73] Mendelssohn, IA; Keiss, BA and Wakeley, JS (1995) Factors controlling the formation of oxidised root channels: a review. Wetlands 15: 37-46

[74] St-Cyr, L and Campbell, PGC (1996) Metals (Fe, Mn, Zn) in the root plaque of submerged aquatic plants collected *in situ*: relations with metal concentrations in the adjacent sediments and in the root tissue. Biogeochemistry 33: 45-76

[75] Sorrell, BK; Mendelssohn, IA; McKee, KL and Woods, RA (2000) Ecophysiology of wetland plant roots: A modelling comparison of aeration in relation to species distribution. Annals of Botany 86(3): 675-685

[76] Bouma, TJ; Koutstaal, BP; van Dongen, M and Nielsen, KL (2001) Coping with low nutrient availability and inundation: root growth responses of three halophytic grass species from different elevations along a flooding gradient. Oecologia 126(4): 472-481

[77] Armstrong, W; Armstrong, J and Beckett, PM (1990) Measurment and modelling of oxygen release from roots of *Phragmites australis*, in P Cooper and BC Findler, Eds, The Use of Constructed Wetlands in Water Pollution Control. Pergamon Press, Oxford, UK, 41-54

[78] Green, MS and Etherington, JR (1977) Oxidation of ferrous iron by rice (*Oriza sativa*) roots: a mechanism for waterlogging tolerance? Journal of Experimental Botany 28: 678-690

[79] Jaynes, ML and Carpenter, SR (1986) Effects of Vascular and Nonvascular Macrophytes on Sediment Redox and Solute Dynamics. Ecology 67(4): 875-882

[80] Koncalova, H (1990) Anatomical adaptations to waterlogging in roots of wetland graminoids: limitations and drawbacks. Aquatic Botany 38(1): 127-134

[81] Chabbi, A; McKee, KL and Mendelssohn, IA (2000) Fate of oxygen losses from *Typha domingensis* (Typhaceae) and *Cladium jamaicense* (Cyperaceae) and consequences for root metabolism. American Journal of Botany 87(8): 1081-1090

[82] Flessa, H (1991) Redoxprozesse in Böden in der Nähe von wachsenden und absterbenden Pflanzenwurzeln. Verlag Marie L. Leidorf, Buch am Erlbach, Germany

[83] Brix, H; Sorrell, BK and Schierup, HH (1996) Gas fluxes achieved by in situ convective flow in *Phragmites australis*. Aquatic Botany 54(2-3): 151-163

[84] Wiessner, A; Kuschk, P; Kästner, M and Stottmeister, U (2002) Abilities of helophyte species to release oxygen into rhizospheres with varying redox conditions in laboratory-scale hydroponic systems. International Journal of Phytoremediation 4(1): 1-15

[85] Sorrell, BK and Armstrong, W (1994) On the difficulties of measuring oxygen release by root systems of wetland plants. Journal of Ecology 82: 177-183

[86] Jespersen, DN; Sorrell, BK and Brix, H (1998) Growth and root oxygen release by *Typha latifolia* and its effects on sediment methanogenesis. Aquatic Botany 61(3): 165-180

[87] Kludze, HK and DeLaune, RD (1996) Soil redox intensity effects on oxygen exchange and growth of Cattail and Sawgrass. Soil Science Society of America Journal 60(2): 616-621

[88] Sorrell, BK (1999) Effect of external oxygen demand on radial oxygen loss by *Juncus* roots in titanium citrate solutions. Plant Cell and Environment 22(12): 1587-1593

[89] Wiessner, A; Kuschk, P and Stottmeister, U (2002) Oxygen release by roots of *Typha latifolia* and *Juncus effusus* in laboratory hydroponic systems. Acta Biotechnologica 22(1-2): 209-216

[90] Smith, KA and Russell, RS (1969) Occurrence of ethylene, and its significance in anaerobic soil. Nature 222: 769-771

[91] Smith, KA and Restall, SWF (1971) The occurrence of ethylene in anaerobic soil. Journal of Soil Science 22: 430-443

[92] Visser, EJW; Bogemann, GM; Blom, C and Voesenek, L (1996) Ethylene accumulation in waterlogged *Rumex* plants promotes formation of adventitious roots. Journal of Experimental Botany 47(296): 403-410

[93] Dacey, JWH (1979) Methane efflux from lake sediments through water lilies. Science 203: 1253-1255

[94] Dacey, JWH and Klug, MJ (1982) Tracer transport in *Nuphar*: $^{18}O_2$ and $^{14}CO_2$ transport. Physiologia Plantarum 56: 361-366

[95] Higuchi, T; Yoda, K and Tensho, K (1984) Further evidence for gaseous CO_2 transport in relation to root uptake of CO_2 in rice plant. Soil Science and Plant Nutrition 30: 125-136

[96] Sebacher, DI; Harriss, RC and Bartlett, KB (1985) Methane emissions to the atmosphere through aquatic plants. Journal of Environmental Quality 14: 40-46

[97] Sorrell, BK and Boon, PI (1994) Convective gas flow in *Eleocharis sphacelata* R. Br.: methane transport and release from wetlands. Aquatic Botany 47(3-4): 197-212

[98] Shannon, RD; White, JR; Lawson, JE and Gilmour, BS (1996) Methane efflux from emergent vegetation in peatlands. Journal of Ecology 84(2): 239-246

[99] Butterbach-Bahl, K; Papen, H and Rennenberg, H (1997) Impact of gas transport through rice cultivars on methane emission from rice paddy fields. Plant Cell and Environment 20(9): 1175-1183

[100] Yavitt, JB and Knapp, AK (1998) Aspects of methane flow from sediment through emergent cattail (*Typha latifolia*) plants. New Phytologist 139(3): 495-503

[101] Smith, KA and Robertson, PD (1971) Effect of ethylene on root extension of cereals. Nature 234(5325): 148-9

[102] Jackson, MB and Campbell, DJ (1975) Movement of ethylene from roots to shoots, a factor in the response of tomato plants to waterlogged soil conditions. New Phytologist 74: 397-406

[103] Drew, MC; Jackson, MB and Giffard, S (1979) Ethylene-promoted adventitious rooting and development of cortical air spaces (aerenchyma) in roots may be adaptive responses to flooding *Zea mays* L. Planta 147: 83-88

[104] Konings, H and Jackson, MB (1979) A relationship between rates of ethylene production by roots and the promoting or inhibiting effects of exogenous ethylene and water on root elongation. Zeitschrift für Pflanzenphysiologie 92: 385-397

[105] Jackson, MB (1989) Regulation of aerenchyma formation in roots and shoots by oxygen and ethylene, in DJ Osborne and MB Jackson, Eds, Cell Separation in Plants: Physiology, Biochemistry and Molecular Biology. H35. Springer-Verlag, Berlin, 263-274

[106] Liu, J; Mukherjee, I and Reid, DM (1990) Adventitious rooting in hypocotyls of sunflower (*Helianthus annuus*) seedlings. III. The role of ethylene. Physiol Plant 78(2): 268-276

[107] Jackson, MB (1991) Ethylene in root growth and development, in AK Mattoo and JC Shuttle, Eds., The Plant Hormone Ethylene. CRC Press, Boca Raton, FL, USA, 259-278

[108] Visser, EJW; Nabben, RHM; Blom, C and Voesenek, L (1997) Elongation by primary lateral roots and adventitious roots during conditions of hypoxia and high ethylene concentrations. Plant Cell and Environment 20(5): 647-653

[109] Brix, H (1990) Uptake and photosynthetic utilization of sediment-derived carbon by *Phragmites australis* (Cav.) Trin. ex Steudel. Aquatic Botany 38(4): 377-389

[110] Constable, J and Longstreth, DJ (1994) Aerenchyma Carbon Dioxide can be assimilated in *Typha latifolia* L. Leaves. Plant Physiol. 106(3): 1065-1072

[111] Lelieveld, J; Crutzen, PJ and Bruhl, C (1993) Climate effects of atmospheric methane. Chemosphere 26(1-4): 739-768

[112] Helal, HM and Sauerbeck, D (1989) Carbon turnover in the rhizosphere. Zeitschrift für Pflanzenernährung und Bodenkunde 152: 211-216

[113] Miersch, J; Krauss, G-J and Sclee, D (1989) Allelochemische Wechselbeziehungen zwischen Pflanzen - eine kritische Wertung. Wiss Ztg Univ Halle 38: 59-74

[114] Hoffland, E; van de Boogaard, R; Nelemans, J and Findenegg, G (1992) Biosynthesis and root exudation of citric and malic acids in phosphate-starved rape plants. New Phytologist 122: 675-680
[115] Donnelly, PK; Hegde, RS and Fletcher, JS (1994) Growth of PCB-degrading bacteria on compounds from photosynthetic plants. Chemosphere 28(5): 981-988
[116] Horswell, J; Hodge, A and Killham, K (1997) Influence of plant carbon on the mineralisation of atrazine residues in soils. Chemosphere 34(8): 1739-1751
[117] Moormann, H; Kuschk, P and Stottmeister, U (2002) The effect of rhizodeposition from helophytes on bacterial degradation of phenolic compounds. Acta Biotechnologica 22(1-2): 107-112
[118] Kaitzis, G (1970) Mikrobiozide Verbindungen aus *Scirpus lacustris* L. (Ein Beitrag zur Ökochemie des Wurzelraumes). Universität Göttingen, Göttingen, Germany.
[119] Wu, MY; Franz, EH and Chen, SL (2001) Oxygen fluxes and ammonia removal efficiencies in constructed treatment wetlands. Water Environment Research 73(6): 661-666
[120] Kappelmeyer, U; Wiessner, A; Kuschk, P and Kästner, M (2002) Operation of a Universal Test Unit for Planted Soil Filters - Planted Fixed Bed Reactor. Engineering in Life Sciences 2(10): 311-315
[121] Wiessner, A; Kappelmeyer U; Kuschk, P and Kästner M (2005) Influence of the redox condition dynamics on the removal efficiency of a laboratory-scale constructed wetland. Water Research 39: 248-256

EXPLOITATION OF FAST GROWING TREES IN METAL REMEDIATION

PAVEL TLUSTOŠ, DANIELA PAVLÍKOVÁ, JIŘINA SZÁKOVÁ, ZUZANA FISCHEROVÁ AND JIŘÍ BALÍK
Czech University of Agriculture, Department of Agrochemistry and Plant Nutrition, 165 21 Prague 6, Czech Republic, E-mail: tlustos@af.czu.cz

1. Advantages of using fast growing trees for phytoextraction

The cultivation of fast-growing tree species for remediation purpose and the production of renewable energy from contaminated biomass is an approach to the use of post-mining polluted areas which offers an alternative to more traditional types of land use [1-4]. Compared to herbaceous species, fast-growing trees have several advantageous characteristics, such as a deeper root system, a high productivity and transpiration activity.

Bungart and Hüttl [1, 2] investigated growth dynamics and biomass production of 8-yeard-old hybrid poplar clones (*Populus* spp., section *Tacamahaca*). Aboveground biomass production ranged from 24 to 49 t dry matter per hectare at age 8. The high-yield clones were Rap (46.7 t per ha) and Beaupré (48.6 t per ha). Differences in mean aboveground biomass production between the hybrid poplar clones were largely the result of genetic and ecophysiological causes. Investigation of plant nutrition and the water budget indicated that these two factors might have an influence on the productivity of the clones, measured in terms of the accumulated aboveground biomass. Armstrong *et al.* [5] found the yield of poplar dry biomass from 3.6 to 13.6 t per ha per year. The highest yields were achieved at the site with the highest average annual rainfall. Moffat *et al.* [6] confirmed significant effect of irrigation for poplar biomass production. Application of sewage sludge was not effective in increasing biomass yield in this experiment. According to Sebastiani *et al.* [3] poplar clones responded to waste treatment through increasing plant growth, although the investigated clones differed in their response.

Field experiments with four willow species (*Salix amygdalina, S. viminalis, S. americana, S. purpurea*) irrigated by municipal wastewater were performed by Kowalik and Randerson [7]. Annual yields for 2000 mm annual irrigation in two-year plant rotation were 5 – 14.4 t dry matter per ha. According to Cannell *et al.* [8] the potential production of stem dry biomass for a 1- or 2-year-old uncoppiced stand of *Salix viminalis* could be more than 12 $t.ha^{-1}$ per year. Yield of willow dry matter in one-year cutting cycle (14.1 $t.ha^{-1}$ per year) was significantly increased to 16.1 and 21.6 $t.ha^{-1}$ per

M. Mackova et al. (eds.), Phytoremediation Rhizoremediation, 83–102.

year when harvest was performed in two and three years cycle, respectively. The highest yield was found for *Salix viminalis* 082 clone cut in three years cycle, amounted up to 26.4 $t.ha^{-1}$ per year [9]. Hytönen and Issakainen [10] investigated biomass production of birches (*Betula pubescens*) and according to their results birch is not suitable for biomass production when using annual harvest.

The concentration of elements for optimum nutrition presented in Table 1 describes the range values refer to tree populations less than 9 years old. Clonal differences in biomass production seemed to be related to differences in the use-efficiency of nutrients for the purposes of internal N and P translocation. Bungart and Hüttl [2] recommended fertilization of poplar clones in the form of the application of an extended NPK fertilization regime after the culmination of height and diameter stem growth in the seventh year in order to achieve a higher and more stable biomass production. Given the local conditions, Bungart and Hüttl [2] suggest focusing on cultivation of a polyclonal mixture of high-yielding hybrid poplar clones for biomass production.

Table 1. Range of values (mg g-1 DM) for optimum nutrition of Populus spp. [2]

N	P	K	Ca	Mg
		$mg.g^{-1}DM$		
17 – 30	1.0 – 4.4	7 – 20	3.0 – 17	1.4 – 4.0

Planting density according to Adegbidi *et al.* [11], and Bergkvist and Ledin [12] did not significantly affect annual biomass production, nutrient removal and nutrient use efficiency. The most important factors taken into consideration for yield optimisation of stem wood are the number of plants per area and the length of the cutting cycle. The results of Bergkvist and Ledin [12] show during the first cutting cycle of willow, that the yield was positively correlated with the number of plants per ha up to 20000, but not with higher numbers. Tharakan *et al.* [4] investigated variation among 3-year-old 30 willow clones and recommended to grow willow shrubs in densities of 10000 – 20000 plants per ha.

Fast-growing species used in short-rotation forestry are dependent on a substantial and continuous availability of nutrients and water to give high yield. Increased biomass yields in willow and poplar stands have been observed as a result of fertilization with N, P and K, soil amendment with lime and irrigation. Rytter [13] published that the annual willow uptake of N during the last 2 years of the study was nearly 200 $kg.ha^{-1}$ per year and the N concentration in fully developed leaves near the shoot tip varied between 2.9 – 4.4% N. The results of Adegbidi *et al.* [11] showed that annual biomass production of willow clones of 15 – 22 $t.ha^{-1}$ removed 75 – 86 kg N, 10 – 11 kg P, 27 – 32 kg K, 52 – 79 kg Ca and 4 – 5 kg $Mg.ha^{-1}$ per year.

Growth and associated nutrient uptake during the first rotation of a willow plantation is summarized in Table 2. The estimation of the nitrogen amounts required for leaf growth in well established willow plantations is in the rate of 150 to 200 $kg.ha^{-1}$ per year [14]. Between one and two thirds of the leaf N and P contents can be transported back from senescing leaves to woody tissues and stored for use in coming season. The genus *Alnus* differs from *Populus* and *Salix* and translocation of N from senescing leaves is small. In poplar plantations more than 50% of the aboveground N requirement and 30% of P can come from retranslocation the previous year [14]. Stems constitute the major storage organ for resorbed nutrients in willows when the process of nutrient withdrawal is allowed to be completed, but hybrid poplars has been associated the major storage pool of retranslocated leaf-N to large structural roots. Leaf litter from species belonging to the genera *Alnus, Populus* and *Salix* may be regarded as easily decomposed. A release of N in the order of 30% of the initial N-content from 1-year leaf litter is reported for alders and willows after one year of decomposition [15]. A nitrogen budget during the first rotation of a willow plantation on good agricultural soil is presented in Table 3.

Table 2. Calculated growth and nutrient accumulation in roots, stems and leaves during the first rotation of a Salix viminalis plantation [14]

	Planting year		
	Year 1	Year 2	Year 3
		Biomass ($t.ha^{-1}$)	
Leaves	3.5	5.0	5.2
Stems	8.0	11.0	13.0
Roots	1.8	2.4	2.6
		Nitrogen uptake ($kg.ha^{-1}$)	
Leaves	123	175	182
Stems	40	44	46
Roots	22	29	31
Total uptake	185	248	259
		Potassium and phosphorus uptake ($kg.ha^{-1}$)	
K - total uptake	133	186	200
P – total uptake	28	39	42

Sander and Ericsson [16] studied the vertical distribution of elements in the woody biomass of two willows. In the 5-year-old shoots, the concentration of all elements increased significantly from shoot base to shoot top (Table 4). The general increase in plant nutrient and heavy metal concentrations from the shoot base towards the top can be explained by distribution patterns of wood and bark. The concentration of plant nutrients is generally higher in bark than wood.

Nielsen [17] published optimal proportions by weight of the essential mineral nutrient components in a fertilizer for achieving maximum willow production (Table 5). Biomass production, nutrient removal and nutrient use efficiency in fast-growing tree plantations are strongly affected by clonal selection. Fertilization, irrigation and a long harvest cycle significantly increased annual biomass production. Nutrient use efficiency is increased by extension of harvest cycle and fertilizer application.

Table 3. Estimated nitrogen uptake, nitrogen recycling and nitrogen fertilizer requirements of a high yielding willow plantation on good agricultural soil during the first rotation

	Planting year		
	Year 1	Year 2	Year 3
		Nitrogen ($kg.ha^{-1}$)	
Total N uptake*	185	248	259
Soil**	50	50	50
Retranslocation	-	41	58
Leaf litter	-	27	66
Root litter	11	14	15
Atmosphere	10	10	10
N-leakage	?	?	?
Fertilizer needs	114	106	60
Recommended	80 - 110	60 - 80	60 - 80
Outtake at harvest			128

* from Table 2; ** mineral soil, N-content 5000 kg $.ha^{-1}$;

Table 4. Plant nutrient and heavy metal concentrations at different shoot levels in 5-year-old shoots of Salix viminalis on a clay soil. Shoots were sampled in late March to early April 1995 [16]

element	Level above ground (m)					
	0.5	1.5	2.5	3.5	4.5	5.5
			$mg.g^{-1}$			
P	0.53	0.48	0.53	0.70	1.11	2.38
K	1.5	1.7	2.1	2.8	3.7	5.1
Ca	3.4	3.3	3.7	4.3	5.5	7.8
Mg	0.47	0.44	0.47	0.56	0.79	1.35
S	0.31	0.31	0.36	0.46	0.65	1.13
			$mg.kg^{-1}$			
Mn	18.2	17.2	18.4	21.8	29.6	46.3
Zn	45.3	48.6	51.0	54.5	67.3	87.3
Cu	2.3	2.2	2.4	2.9	3.8	10.0
Ni	0.98	0.86	1.01	1.25	1.86	6.04
Cd	0.78	0.78	0.81	0.85	0.91	0.92

2. Metal uptake by fast growing trees and metal distribution in plants

The application of fast growing trees, especially willows and poplars *(Salicaceae)*, for phytoremediation use was discussed at the soils contaminated to deeper horizons [18]. The uptake and distribution of potentially risk elements were investigated predominantly in genes *Salix* spp. and *Populus* spp., and also *Betula* spp., *Alnus* spp. (*Betulaceae*), and *Acer* spp. *(Aceraceae)*, but possible remediation use was checked only in the case of *Salix* spp. The remediation potential of trees including their metal tolerance, metal uptake by trees grown on contaminated substrates, metal distribution within trees, and phytoremediation using trees was deeply reviewed by Pulford and Watson [19]. Analyses of leaves and twigs of *Salix* sp. and *Populus* sp. showed relatively high contents of elements in the first part of vegetation period because of higher nutrient uptake compared to the growth rate. During the vigorous growth dilution effect resulting in lower element contents was observed until the flowering stage where the minimum values for almost all elements were observed. Senescence of trees usually results in increase of metal levels due to concentration caused by loss of fluids [20]. The variations in element contents in leaves of trees are demonstrated in Figure 1. Ali *et al.* [21] demonstrated that *Salix acmophylla* was able to accumulate considerable amounts of Cu, Ni and Pb in different plant parts and exhibited high tolerance to these metals. The plants showed ability for their detoxification by antioxidant enzymes and cellular antioxidants like cysteine and thiols.

Table 5. Optimal proportions by weight of the essential mineral nutrient components in a fertilizer for achieving maximum willow production (the amount of nitrogen is set at 100 % and the other are relative to that) [17]

Macro nutrients		Micro nutrients	
element	proportion	element	proportion
Nitrogen	100	Iron	0.6
Potassium	72	Manganese	0.4
Phosphorus	14	Boron	0.1
Magnesium	8.5	Copper	0.03
Calcium	7	Zinc	0.06
Sulphur	9	Chlorine	0.003
		Molybdenum	0.007
		Sodium	0.003

Individual tree species differ significantly in ability to transport metals from roots to shoots and to accumulate them in individual plant parts. Pulford and Watson [19] concluded that within the tree Pb, Cr, and Cu are usually immobilized in roots while Cd, Ni, and Zn are more easily transported to shoots. Figure 2 indicates opposite pattern for Cr and Ni but different soil properties of different sites and/or different sampling time and climate conditions must be taken into account in this case.

Nissen and Lepp [22] found that the concentrations of Cu and Zn varied between plant parts and between species. There was no consistent pattern of metal concentration. Copper concentrations decreased in the order twigs > leaves > wood > bark, and those of Zn in the order leaves > bark > twigs > wood indicating the low retention of Zn in the xylem tissues. *Salix* x *sericans* and *Salix cinerea* had high concentrations of both metals; *S. purpurea* had high concentrations of Zn but low concentrations of Cu, whilst *Salix fragilis* and *Salix viminalis* had low concentrations of both metals. Concentration factors for both elements in shoot tissues were consistent, despite the variations in associated metal concentrations. Concerning cadmium the willow leaves contain approximately 1.6-fold higher level compared to wood representing 49% of total shoot cadmium content. Roots contained 90% of cadmium contents in shoots [23]. Distribution of arsenic within tree was leaves > branch bark > stem bark > wood, where stem position on the tree was significant for leaf arsenic content [24].

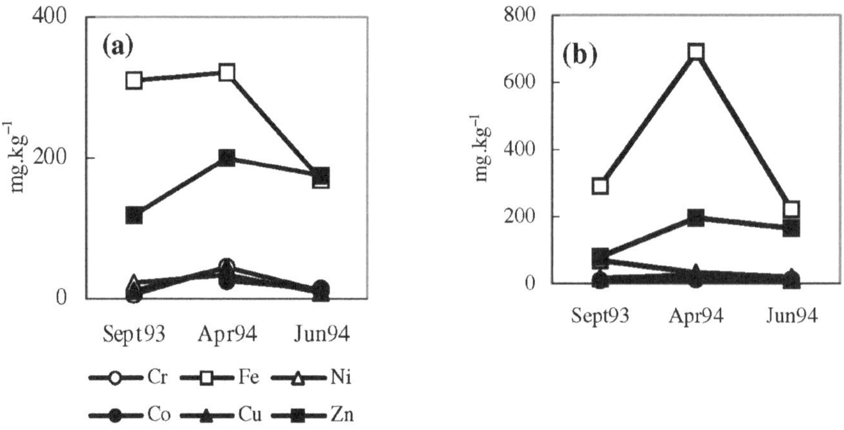

Figure 1. Seasonal element variations in (a) Salix spp. and (b) Populus nigra leaves according to Dinelli and Lombini [20].

Willows are generally characterized by high production of aboveground biomass and intensive uptake of cadmium [25] where close relationships of these parameters are evident. This dependence can be significantly affected by water uptake and transpiration of plants differing among individual clones [26]. High amount of risk elements in cultivation medium reduces yield of willow shoots (*S. viminalis*) compared to plants grown in uncontaminated environment [27]. The total cadmium uptake by plants is affected by both yield of aboveground biomass and element content in leaves and twigs, and these parameters must be taken into account for an evaluation of phytoremediation potential [25]. Willows characterized by higher production of twigs usually contain lower cadmium levels and simultaneously grow at less acidic soils with lower cadmium mobility compared to willows with lower production of twigs. The portion of leaves represent in the case of willows 17 - 37% of total shoot yield whereas cadmium uptake by willow leaves represents 21 - 48% of total cadmium uptake by aboveground biomass [25]. Luňáčková *et al.* [28] investigated the cadmium impact on roots of *Salix alba* and *Populus* × *euroamericana* cv. Robusta cuttings grown in 10 $\mu mol.l^{-1}$ solution of $Cd(NO_3)_2$ (direct treatment) or in Knop solution and afterwards in 10 $\mu mol.l^{-1}$ solution of $Cd(NO_3)_2$ (indirect treatment). The cumulative length, number and biomass of willow roots, pigment and starch contents, leaf net photosynthetic rate and dry mass/leaf area ratio of willow leaves were positively influenced by indirect treatment. However, indirectly treated poplars were more sensitive to Cd than directly treated ones. Indirect treatment lowered root Cd uptake in willow, Cd accumulation in cuttings of both species and Cd accumulation in poplar shoots. Structural changes caused by Cd were similar in both species and in both treatments. Root apices, rhizodermis and cortex were the most seriously damaged root parts. Pulford and Watson [19] summarized that survival of fast growing trees at contaminated site seems to be due to facultative tolerance (avoidance of highly contaminated substrate by roots and/or immobilization of elements in the root) and genetically transmitted tolerance systems were not evolved. However, tolerance of plants can be increased by acclimation of trees to elevated concentrations of risk elements.

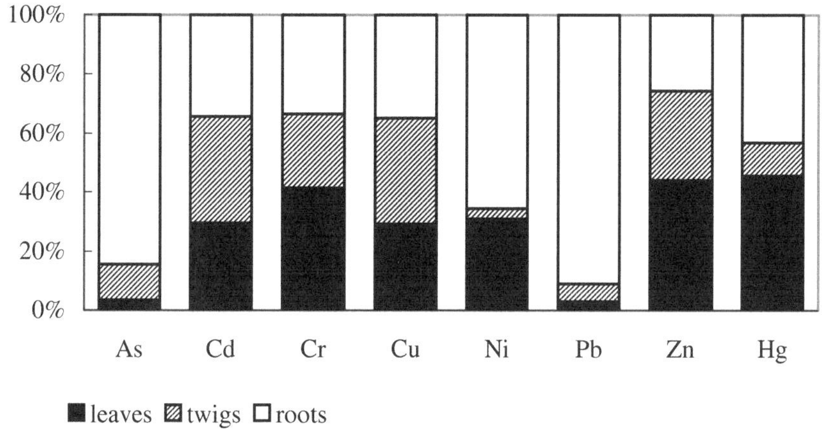

Figure 2. Distribution of selected elements into individual parts of Salix fragilis.

Perttu *et al.* [29] focused on some important processes that are likely to be driving variables for Cd fluxes in the soil – willow system. These included plant growth and processes that are linked with biomass production. In the prevailing northern European climate, where precipitation normally exceeds evapotranspiration, soil fluxes of elements are seldom directed towards the soil surface. Such fluxes therefore depend on living plants and their root uptake. The principal flows of Cd are: i) root uptake of subsoil Cd, transported to the foliage which eventually is shed as litter fall, ii) Cd leakage from deep penetrating roots passing the topsoil, iii) decomposition of roots that pass the topsoil with organically bound Cd, and iv) fertilisation with wood ashes, containing Cd. The first one is the most likely process for transporting Cd from the subsoil to the topsoil. The results indicated that the efficiency of Cd uptake is about the same, regardless of where the roots are situated. Most of the Cd taken up by the deep penetrating roots is attached and tightly bound to these; only about 5 - 11% is transported upwards to the shoots and 3 - 7% to the foliage. The latter will end up in the topsoil after litter fall. A very small part, probably less than 1%, of the Cd in the shoot ends up in the topsoil as the stool remains after final harvest. About 40% of the Cd taken up by the deep penetrating roots passes into the shoot part. Consequently, the root turn over rate is an important factor to consider when discussing Cd translocation. The Cd that is adhered to organic material (stems and roots) will remain there and eventually be released into the soil when the material is broken down and humified. No Cd is leaking out from the living roots penetrating into deeper soil layers. The result was expected as mineral and element uptake only occurs close to the root tips and older parts of the suberised root functions as a transport link to the above ground parts of the plant. Relocation of Cd within the root system, i.e. from deep penetrating roots into the topsoil roots is negligible as the element-uptake processes is from the root tip towards the foliage.

3. Metal accumulation capacity of fast growing trees

A large number of species and hybrids of *Salix* spp. suggests wide genetic variability within the genus and some species are known to colonise contaminated soil. For example *S. alba*, *S. dasyclados*, *S. viminalis* (containing up to 4.1 mg $Cd.kg^{-1}$ in stems and 7.3 mg $Cd.kg^{-1}$ in leaves), *S. cinerea*, and *S. caprea* naturally invade polluted dredged sediment disposal sites [19, 25, 30]. Some willow or poplar varieties do not retain metals in roots but transfer them to aboveground plant tissues [19, 23, 31]. The advantage of these species is their greater harvestable biomass in contrast to hyperaccumulators mostly with only a small aboveground part [32]. Poplar trees grown in mine spill affected soil were able to take up the Cd and Zn contents exceeding normal values in plants (up to 15 mg $Cd.kg^{-1}$ and 1312 mg $Zn.kg^{-1}$) [33]. These levels can be usually considered as phytotoxic for plants. Greger and Landberg [34] determined cadmium contents in shoots of 104 clones of *S. viminalis* grown in medium enriched by 1, 3, and 10 $\mu mol.l^{-1}$ solution of Cd in range between 0.2 and 8.5 $mg.kg^{-1}$. In the case of zinc the element contents in shoots varied between 14 and 1813 $mg.kg^{-1}$ confirming also the ability of willows and poplars to take up higher Cd and Zn contents compared to other plant species [35].

Plant accumulation capacity and remediation capability were tested in two years pot experiment [36]. Anthropogenically contaminated (by industrial activity) Cambisol containing 28 mg $As.kg^{-1}$, 5.46 mg $Cd.kg^{-1}$, 956 mg $Pb.kg^{-1}$, and 279 mg $Zn.kg^{-1}$ was applied in this case. Detailed characteristics of the site were described in Šichorová *et al.* [37]. Plants were selected from two groups: hyperaccumulators – *Arabidopsis halleri*, and *Thlaspi caerulescens*; and trees with high element accumulation capacity and a great biomass production – *Salix smithiana*, *Salix dasyclados*, *Salix caprea, Populus trichocarpa*, and *Populus nigra*. Figure 3 confirms good willingness of all tested species to transport Cd and Zn from roots to aboveground biomass. All species deposit As and Pb in roots preferentially. The highest shoot/root Cd ratio was recognised for *A. halleri*, *S. dasyclados*, *P. trichocarpa* and *P. nigra*. Significantly higher transport of Zn was progressed in both poplars compare to other species. This is in concordance with results of many authors they had described the highest concentration of Cd and Zn in poplars or willows leaves compare to other plant parts [19]. We are able to resume that poplar trees are really disposed to transport and deposit Cd and Zn in shoots, but their remediation factor (calculated as percentage of element removed from total amount of element in pot soil) is lower in contrast to both hyperaccumulators and *S. dasyclados*. The hyperaccumulation possibility of *A. halleri* and *T. caerulescens* for Cd and great accumulation ability for other studied elements was confirmed but very good accumulation was determined in tested trees, too. Especially willow *S. dasyclados* compensated lower metal content in shoots by higher biomass production compare to hyperaccumulators which have higher element content but lower aboveground biomass resulting in similar remediation capability. Although poplar trees showed the best willingness to transport Cd and Zn from roots to shoots, their remediation potential do not achieve *S. dasyclados* or hyperaccumulators level.

Madejon *et al.* [33] surveyed the content of eight trace elements (As, Cd, Cu, Fe, Mn, Ni, Pb and Zn) in leaves and stems of 25 white poplar (*Populus alba*) trees in the riparian forest of the Guadiamar River (S. Spain), one year after this area was contaminated by a mine spill, and 10 trees in non-affected sites. The spill-affected soils had significantly higher levels of available cadmium (mean of 1.25 $mg.kg^{-1}$), zinc (117 $mg.kg^{-1}$), lead (63.3 $mg.kg^{-1}$), copper (58.0 $mg.kg^{-1}$) and arsenic (1.70 $mg.kg^{-1}$), than non-affected sites. The concentration of trace element in poplar leaves was positively and significantly correlated with the soil availability for cadmium and zinc, and to a lesser extent for arsenic. Thus, these results confirmed that poplar leaves could be used as biomonitors for soil pollution of Cd and Zn, and moderately for As.

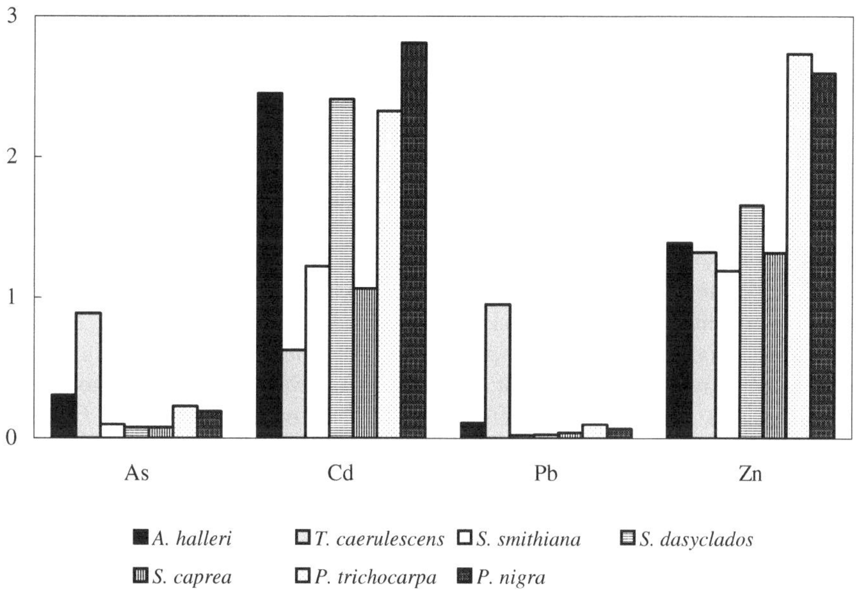

Figure 3. Average metal accumulation shoot/root ratio of different plant species [36].

Similarly Roselli *et al.* [38] compared phytoextraction capacity of five tree species grown at metal contaminated site (total element content in the soil was 557 ± 51 $mg.kg^{-1}$ of Cu, 620 ± 41$mg.kg^{-1}$ of Zn, and 1.8 ± 0.2 $mg.kg^{-1}$ of Cd), and determined fairly good bioconcentration factors (calculated as ratio of total element content in plant tissue and in soil) of *Salix viminalis* and *Betula pendula* for cadmium and zinc in contrast to other species as summarized in Table 6. In the case of copper, however, the bioconcentration factors were negligible regardless of tree species.

Table 6. Bioconcentration factors to the aboveground plant tissues for copper, zinc and cadmium for the trees growing on the contaminated site [33]

Plant species	Bioconcentration factor for Cu		Bioconcentration factor for Zn		Bioconcentration factor for Cd	
	leaves	twigs	leaves	twigs	leaves	Twigs
Alnus incana	0.05	0.03	0.07	0.06	a	A
Betula pendula	0.02	0.03	0.37	0.32	0.06	0.11
Fraxinus excelsior	0.01	0.04	0.02	0.05	a	a
Salix viminalis	0.01	0.03	0.37	0.28	0.83	0.72
Sorbus mougeotti	0.01	0.02	0.03	0.05	a	a

a...data under detection limit

For assessment of clonal variability of willow plants seven clones of high biomass production willows (*S. x smithiana* S-218, *S. x smithiana* S-150, *S. viminalis* S-519, *S. alba* S-464, *S. alba* "Pyramidalis" S-141, *S. dasyclados* S-406, *S. x rubens* S-391) were planted in a pot experiment for three vegetation periods in two soils differing in total content of risk elements [39] as characterized in Table 7. Because of serious symptoms of Zn phytotoxicity at extremely contaminated Fluvisol the experiment was terminated after second vegetation period at this soil. Comparing the remediation factors, reasonable phytoextraction potential of willows was obtained for cadmium and zinc at moderately contaminated Cambisol where aboveground biomass removed about 30% Cd and 5% Zn of total element content, respectively. Clones showed different ability to remove Cd and Zn, depending on soil type and contamination level: *S. x smithiana* (S-150) and *S.x.rubens* (S-391) demonstrated the highest phytoextraction effect for Cd and Zn. The differences in accumulation between the clones are due to the property of clone and not to the soil element concentration or properties [24, 35, 40]. The mechanisms behind the accumulation, transport, and tolerance are specific for each of the different metals and the uptake property of the clone is stable [34]. In extremely contaminated Fluvisol production of willow biomass was limited by phytotoxicity of zinc resulting in phytoextraction efficiency not exceeding 1% for both Cd and Zn (Figure 4). Concerning arsenic and lead, poor ability of willows to translocate these elements from roots to aboveground biomass led to low removal of these elements from soil (less than 1%) and application of these plants for cleaning of As and Pb contaminated soil is not reasonable. Similarly Nissen and Lepp [22] found the evidence for the exclusion of Cu from shoot tissue indicating low potential for depletion of the plant available soil Cu pool as a consequence of repeated cropping and for concentration in combustion residues.

The cultivation of willows can decrease bioavailable portion of soil cadmium even in deeper horizons. The annual removal of 3 – 4% of plant-available portion of Cd was demonstrated in dependence on mobilization/immobilization processes of cadmium bound in less mobile soil fractions [19, 25, 41-42]. Moreover, potential of willow trees to reduce element leaching to groundwater as well as to stabilize soil characteristics was already described [41-42]. Difference between cadmium uptakes by roots within soil profile was not observed [25]. Perttu *et al.* [29] also described the uptake of Cd in *Salix* from deeper soil layers. The actual decrease in the topsoil after harvesting is therefore less than what may be theoretically calculated. However, the risk of increasing Cd concentrations, because of the relocation from deeper soil layers, is small as long as the percentage is less than about 60% of total uptake. Normally, more than 70% of the active roots are situated in the topsoil and there is no indication that the roots penetrating into the subsoil are more effective than roots in the topsoil.

Table 7. Basic characteristics and total element contents in experimental soils [39]

Soil	pH	$C_{org.}$ (%)	As $mg.kg^{-1}$	Cd $mg.kg^{-1}$	Pb $mg.kg^{-1}$	Zn $mg.kg^{-1}$
Cambisol	6.1	2.1	37.5	4.73	1158	180
Fluvisol	5.3	1.8	64.1	30.5	2297	3718

Shann [43] investigated the "target-neighbour" cocropping approach to determine the effect of planting density on the uptake of metal (in this case, selenium) by known accumulating (mustard) and nonaccumulating, sensitive species (tomato). If the resource for which plants are competing is a toxin or a contaminant, density should determine the amount that each individual acquires from the substrate. The concentration of metal in plant tissue may be highest for the individual at low density, but total metal found in a stand of vegetation should occur at the density where biomass is maximal. If accumulating plants are to be managed in a manner that optimises the amount of metal removed from the substrate, then this maximal biomass density needs to be determined. In addition, the cocropping of an accumulator with a sensitive species will provide information on the possibility of growing crop species with accumulators, thus minimizing the amount of metal that is taken up by an individual. This could decrease metal phytotoxicity or could limit the tissue concentration of metals to a level considered safe for consumption. Liu et al. [44] investigated a cocropping technique using a known zinc hyperaccumulator *Sedum alfredii* with a grain crop, *Zea mays*. After a 3-month growth trial, the results indicated that when *Z. mays* is cocropped with *S. alfredii*, heavy metals accumulated in the grains were significantly reduced when compared to monoculture cropping. Cocropping improved the growth of both plant species and seems to be an effective approach to reduce the risk of contaminant uptake in edible crops.

Hyperaccumulator *Thlaspi caerulescens* is able to change conditions in rhizosphere shared with other plant species and subsequently affect the bioavailability of selected elements for adjacent plants. Hyperaccumulator cocropped with other plant species can either mobilize soil elements such as cadmium or on the contrary immobilize other elements [45]. The effect of cocropping of willow and hyperaccumulator *Thlaspi caerulescens* was investigated in pot experiment where the plants were planted either separately or in cocropping version. Moderately contaminated Cambisol and extremely contaminated Fluvisol (Table 7) were used as cultivation medium.

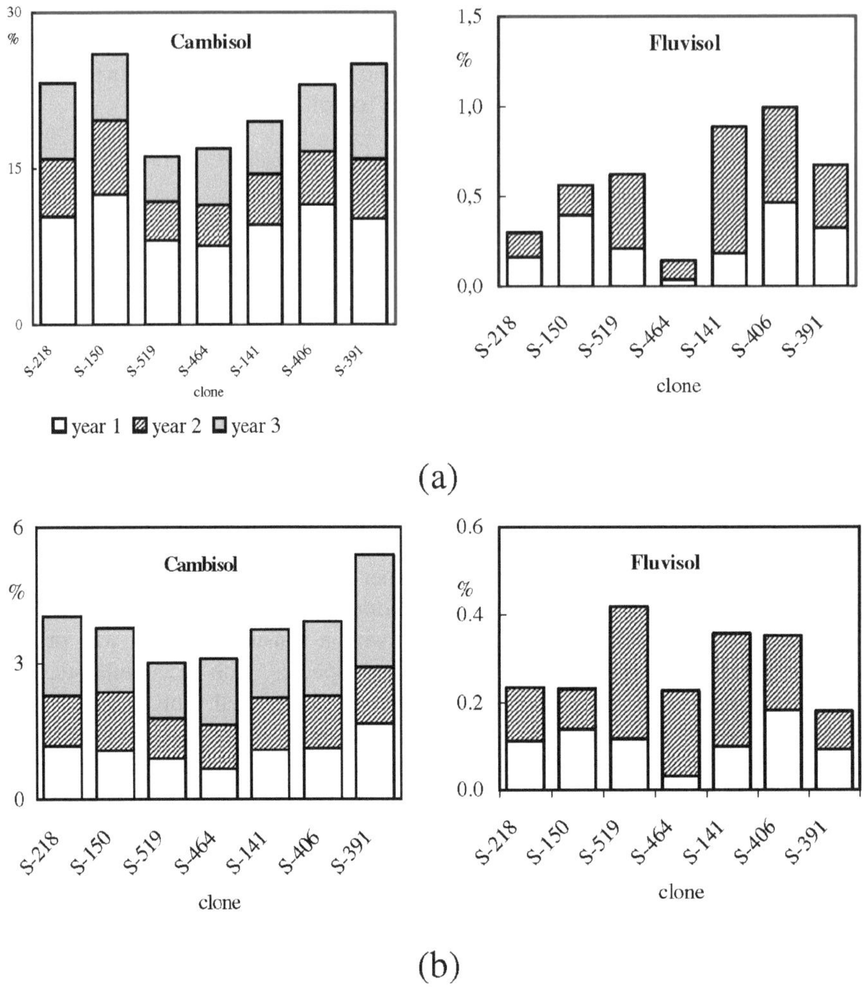

Figure 4. Total remediation factors (%) of cadmium (a) and zinc (b) obtained for aboveground biomass of willow plants during three years of cultivation in pot experiment at moderately contaminated Cambisol and extremely contaminated Fluvisol [39].

Concerning cadmium contents in plants the most significant differences were observed in the leaves of *T. caerulescens* grown at the soils with different level of contamination. Cocropping of *T. caerulescens* and willow did not result in change of cadmium content in aboveground biomass of *T. caerulescens* at moderately contaminated Cambisol whereas Cd content in penny-cress plants grown together with willow at extremely contaminated Fluvisol significantly decreased. The contents of lead and arsenic in penny-cress biomass dropped down significantly in cocropping variant at both investigated soils while zinc contents remained unchanged.

In the case of aboveground biomass of *Salix sp.* the cadmium contents in leaves significantly decreased in plants grown together with penny-cress plants regardless of experimental soil. For As, Pb, and Zn this effect was observed only at the extremely contaminated Fluvisol. Moreover, the growth of individually cultivated willows at the extremely contaminated soil was significantly limited due to phytotoxic effect of this soil whereas the willow trees cultivated together with *T. caerulescens* were able to survive. Phytoremediation factors of both plants (Figure 5) also demonstrate the beneficial effect of cocropping of willow and penny-cress plants at extremely contaminated site. Evidently, penny-cress plants accumulated sufficiently high amount of mobile cadmium and zinc to improve conditions of cultivation of willows especially at extremely contaminated soil. For arsenic and lead the removal of these elements from both soils remains very low in both individual and cocropping variants of planting.

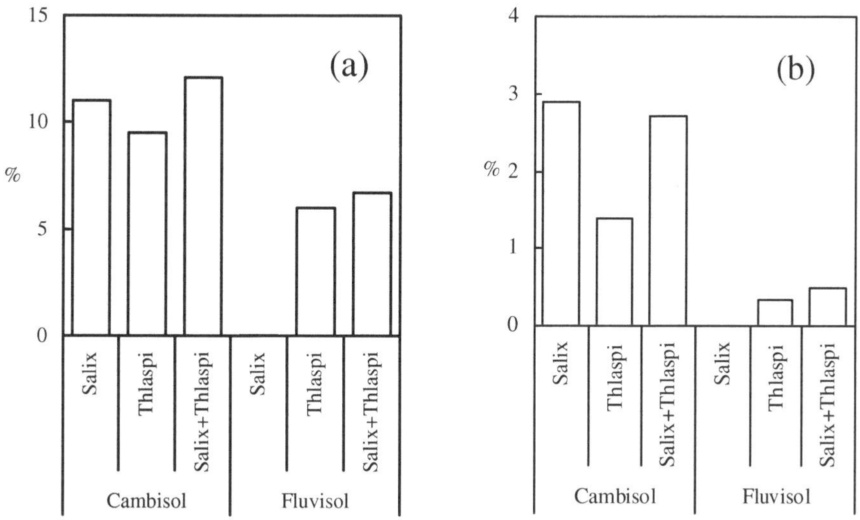

Figure 5. The effect of cocropping of Salix sp. and hyperaccumulator Thlaspi caerulescens on phytoremediation factors of cadmium (a) and zinc (b) at moderately contaminated Cambisol and extremely contaminated Fluvisol.

4. Modification of soil properties for improvement of tree phytoextraction potential

The present results suggest that fast-growing trees and especially willows have very promising potential for phytoremediation use because of their large biomass and good ability to accumulate risk elements [19]. However, some limitations reduce significantly these perspectives. In extremely contaminated soil the production of willow biomass is limited by phytotoxicity of elements resulting in phytoextraction efficiency not exceeding 1% for both Cd and Zn. For these soils, stabilization and immobilization of toxic elements is required before application of phytoremediation technologies [46]. The enhancement of soil organic carbon in soil belongs to the possible approaches tested for decreasing of soil element mobility [47]. Selected clone of high biomass production willow (*S. dasyclados* S-406) was planted in a pot experiment in two highly contaminated soils (Table 8) for one vegetation period [48]. The soils were supplied by nutrients and i) sewage sludge, ii) farmyard manure, and iii) wood chips in the rate representing 20 $t.ha^{-1}$. The addition of organic matter improved the soil properties of soil A resulting in significantly higher yield of aboveground biomass of willow plants whereas at soil B no changes and even decrease of biomass yield at wood chips treatment was observed. The total uptake of individual elements from experimental soils by aboveground biomass of willows taking into account the changes in plant yield in treated pots is summarized in Table 9. The highest effect of organic matter addition was reported for As at soil A where total uptake represented from 280 to 530% of control value. At the soil B the increasing uptake of arsenic was observed only at manure treated pots (180%). The enhancement of total uptake of Cd and Zn in treated pots was evident at both soils except the wood chips treatment at the soil B (due to lower plant yield at this treatment) and represented 120 – 180% of control value. In the case of Pb the soil treatment by organic matter did not lead to significant change in its uptake by willow plants at soil A and even decreased the uptake of this element at soil B. The mobility of elements in soil determined by single (0.01 $mol.l^{-1}$ $CaCl_2$) and sequential extraction procedure was not significantly affected by the addition of different forms of organic matter after one vegetation period and long-term investigation of these effects will be necessary in further research.

Table 8. Basic characteristics and total element contents in experimental soils [48]

Soil	Soil type	pH	$C_{org.}$ (%)	As ($mg.kg^{-1}$)	Cd ($mg.kg^{-1}$)	Pb ($mg.kg^{-1}$)	Zn ($mg.kg^{-1}$)
A	Fluvisol	5.3	1.8	64.1	30.5	2297	3718
B	Chernozem	6.8	1.8	118	40.4	29.3	2087

Hasselgren [49] also found that the stem biomass production was enhanced by sludge application rate and it also led to more uniform growth and a greater shoot number than in control plots. Rockwood *et al.* [24] also reported improved willow yields after application of sewage effluent where the wastewater application did not enrich the heavy metal (Cd, Cu, Pb, Zn) content in the wood. Tsakou *et al.* [50] observed no changes in both growth parameters and element content in poplar cultures grown on sewage sludge treated soil, as well.

Present results also indicated the specificity of phytoremediation potential of willows for cadmium and zinc whereas other elements in soil such as lead are not removed by these plants due to low mobility of lead in soil. Therefore, the possibilities to enhance lead mobility in soil and subsequently the uptake by plants are intensively investigated. Plant roots are able to release specific chelating agents to rhizosphere characterized by binding ability for metals [51, 52]. The effect of soil application of synthetic chelating agents on increasing uptake of Pb, Cd, and Zn by plants was confirmed by many authors

[53-61]. Following chemicals were tested for this purpose: EDTA (ethylendiamino-tetraacetic acid), CDTA (trans-1,2-diaminocyclohexane-N,N,N´,N´-tetraacetic acid), DTPA (diethylentriamino-pentaacetic acid), EGTA (ethylenglycol-bis(β-amino-ethylether)), HEDTA (*N*-(2-hydroxyethyl)-ethylen-diamino-tetraacetic acid), NTA (nitrilotriacetic acid) [57, 61-63]. The effectivity of these agents depends on soil properties, experimental plant, and investigated element but EDTA is considered as the most effective one [61, 62]. The EDTA – metal complexes are taken up by plant roots and transported to aboveground biomass *via* xylem [61]. The EDTA application can increase the effectivity of phytoextraction even at the soils multicontaminated by more elements [64].

Table 9. Total uptake of elements by aboveground biomass of willows according to the individual treatments (mg) [48]

Treatment	As (mg)		Cd (mg)		Pb (mg)		Zn (mg)	
	soil A	soil B	soil A	soil B	soil A	soil B	soil A	soil B
Control	32.8	76.0	1382	1183	120	28.4	39981	17444
Sludge	117	70.2	2480	1407	174	12.6	71456	22803
Manure	174	134	2163	2141	90.3	19.4	49558	31650
Wood chips	94.0	65.3	2031	801	148	16.2	66005	11777

Robinson *et al.* [23] investigated the effect of chelating agents on cadmium uptake by two clones of poplar and one clone of willow. Although two weeks after application of chelating agents cadmium content significantly increased in leaves, in the end of experiment no significant differences were found between treated and untreated variants. These results indicate that if chelating agents are used to enhance plant Cd uptake in a phytoremediation operation, then the plants should be harvested shortly after their addition to take immediate advantage of the surge in uptake. Moreover, two of the chelating agents (2 g.kg^{-1} EDTA and 0.5 g.kg^{-1} NTA of soil) also resulted in a significant reduction in growth, as well as abscission of leaves. Plants treated with 0.5 g.kg^{-1} EDTA showed reduced growth and leaf discoloration. Schmidt [61] also observed reduced plant growth after application of chelates during vigorous growth of plants. Thus, the application of chelates is recommended after achievement of required size of plants. The results of Wenzel *et al.* [65] indicate that metal uptake and translocation from roots to shoots is limited in the absence of EDTA. Split chelate agent applications are generally more effective to enhance Pb and Zn shoot concentrations in canola (*Brassica napus*) than the same amounts of EDTA added at once (Figure 6). The plateau or maximum is reached at about 0.4 g EDTA kg^{-1} soil for Zn. Lead accumulation in shoots further increases at larger EDTA application rates. It is less effective to apply the same amount of EDTA at once 12 days before harvest, and a plateau is obtained even at about 0.25 g EDTA kg^{-1} soil [65].

On a soil containing 5 mg Cd.kg^{-1}, trees harvested shortly after a pulse of EDTA would have a Cd concentration in the dry aboveground biomass of 53 mg.kg^{-1}. The biomass production of poplars and willows under optimal conditions is 30 tones per hectare per annum. Thus, a single crop would remove 1.06 kg of Cd per hectare. This equates to 106 years of regular P fertiliser addition. A soil containing 5 mg.kg^{-1} Cd in the top 100 mm of soil contains 6 kg of Cd per hectare, assuming a bulk density of 1.2. Four crop rotations of willows or treated poplars would reduce the soil Cd burden to 2.35 $mg.kg^{-1}$ [23].

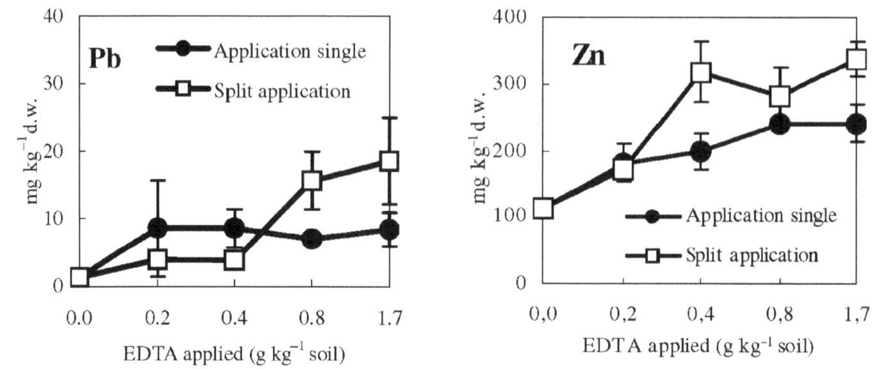

Figure 6. Lead and zinc contents in shoots of Brassica napus in the pot experiment as a function of EDTA applied and mode of application (single versus split) [65].

5. Future prospective for technological application of phytoextraction at contaminated sites

Summarizing the representative set of papers dealing with evaluation of phytoextraction capacity of different plant species there is very well described a potential of tested plant groups and their capacity to remediate contaminated areas. Unfortunately majority was done in laboratory or in a small-scale experiments and there is very difficult to estimate their responses in real conditions up to date. Rockwood *et al.* [24] summarized the possible beneficial phytoremediation use of fast-growing trees for following purposes: i) where former industrial land has little potential for development, fast-growing trees have an opportunity to introduce the land into economic use; ii) fast-growing trees can be grown at agricultural land contaminated by atmospheric deposition or waste disposal where food crops are not permitted, and iii) fast-growing trees can be grown at the agricultural land contaminated from rock phosphate fertilizers where the soil quality can be improved within 10 - 15 years.

To extend the environmental friendly and socio-economically highly accepted technology into regular use there is necessary to meet several criteria for successful phytoextraction of metals from contaminated sites as follows: i) The recognition, extend and distribution of contaminants: Fast growing trees show promising results only for several elements mainly for mobile ones like cadmium and zinc presented in medium level of contamination so far. *Salix* spp. as well as *Populus* spp. showed high phytotoxicity at soils showing high content of total Zn (more than 2000 $mg.kg^{-1}$) with relation to soil properties. There is necessary to provide immobilization treatment at first or exclude phytoextraction as suitable method. Sensitivity to other elements is not seen in such extent [46]. There is also necessary to recognize distribution of contaminants within profile, because phytoextraction is limited by the length of roots and their penetration through the profile. Contamination of deep subsoil less than 80 cm as well as topsoil from 0 to 10 cm is not suitable for phytoextraction. ii) Suitable conditions for plant growth: The yield of aboveground biomass belongs among crucial parameters affecting the remediation potential. Agricultural soils usually have relatively suitable conditions for plant growth, but there are many other contaminated areas without suitable soil fertility caused by soil properties or lack of nutrients. In such conditions it is required to improve soil properties first. The application of fertilizers in relation to soil is able to maximize yield and substantially increase removal of metals. iii) Planting of trees and their harvest: Plant development and maximization of yield can be discussed from the point of view of wood harvest. The use of poplar or willow for phytoremediation would require annual harvesting to avoid the recycling of leaf-bound Cd in autumn [23]. The annual wood biomass yield showed the highest if cutting is done once in two or three year interval unfortunately leaves containing high amount of metals are lost, therefore annual harvest of wood with leaves show higher total removal of metals without any risk of contaminant penetration into surrounding environment. Annual harvest increases the need of nutrients applied for maximum growth. iv) Post harvest manipulation: Burning of harvested biomass can reduce the volume of biomass and only 1 - 3% of total dry matter mass is left as the ash. The risk of possible metal losses during burning is not satisfactory solved. The great advantage of burning is the production of renewable energy. Gasification (i.e. pyrolysis), which occurs under reducing conditions, seems to be a better method than incineration under oxidizing conditions to increase volatilization and, hence subsequently recovery, of Cd and Zn from plants. It would also allow the recycling of the bottom ash as fertilizer [66].

Acknowledgements

Financial support for these investigations was provided by MSM project No. 6046070901.

References

[1] Bungart, R and Hüttl, RF (2001) Production of biomass for energy in post-mining landscapes in Lusatia and nutritient dynamics. Biomass Bioenergy 20: 181-187

[2] Bungart, R and Hüttl, RF (2004) Growth dynamics and biomass accumulation of 8-yeard-old hybrid poplar clones in a short-rotation on a clayey-sandy mining substrate with respect to plant nutrition and water budget. Eur J Forest Res 123: 105-115

[3] Sebastiani, L; Scebba, F and Tognetti, R (2004) Heavy metal accumulation and growth responses in poplar clones Eridano (*Populus deltoids x maximowiczii*) and I-214 (*P. x euramericana*) exposed to industrial waste. Environ Exp Bot 52: 79-88

[4] Tharakan, PJ; Volk, TA; Nowak, CA and Abrahamson, LP (2005) Morphological traits of 30 willow clones and their relationship to biomass production. Can J For Res 35: 421-431

[5] Armstrong, A; Johns, C and Tubby, I (1999) Effects of spacing and cutting cycle on the yield of poplar grown as an energy crop. Biomass Bioenergy 17: 305-314

[6] Moffat, AJ; Armstrong, AT and Ockleston, J (2001) The optimization of sewage sludge and effluent disposal on energy crops of short rotation hybrid poplar. Biomass Bioenergy 20: 161-169

[7] Kowalik, PJ and Randerson, PF (1994) Nitrogen and phosphorus removal by willow stands irrigated with municipal waste water – a review of the polish experience. Biomass Bioenergy 6: 133-139

[8] Cannell, MGR; Milne, R; Sheppard, LJ and Unsworth, MH (1987) Radiation interception and productivity of willow. J Appl Ecol 24: 261-278

[9] Szczukowski, S; Tworkowski, J; Klasa, A and Stolarski, M (2002) Productivity and chemical composition of wood tissues of short rotation willow coppice cultivated on arable land. Rostl Výr 48: 413-417

[10] Hytönen, J and Issakainen, J (2001) Effect of repeated harvesting on biomass production and sprouting of *Betula pubescens*. Biomass Bioenergy 20: 237-245

[11] Adegbidi, HG; Volk, TA; White, EH; Abrahamson, LP; Briggs, RD and Bickelhaupt, DH (2001) Biomass and nutrient removal by willow clones in experimental bioenergy plantations in New York State. Biomass Bioenergy 20: 399-411

[12] Bergkvist, P and Ledin, S (1998) Stem biomass yields at different planting designs and spacings in willow coppice systems. Biomass Bioenergy 14: 149-156

[13] Rytter, R-M (2001) Biomass production and allocarion, including fine-root turnover, and annual N uptake in lysimeter-grown basket willows. For Ecol Manage 140: 177-192

[14] Ericsson, T (1994) Nutrient cycling in energy forest plantations. Biomass Bioenergy 6: 115-121

[15] Slapokas, T. and Granhall, U. (1991). Decomposition of litter in fertilized short-rotation forests on a low-humified peat bog. For Ecol Manage 41: 143-165

[16] Sander, M-L and Ericsson, T (1996) Vertical distribution of plant nutrients and heavy metals in *Salix viminalis* stems and their implications for sampling. Biomass Bioenergy 14: 57-66

[17] Nielsen, KH (1994) Environmental aspects of using waste waters and sludges in energy forest cultivation. Biomass Bioenergy 6: 123-132

[18] Saxena, PK; KrishnaRaj, S; Dan, T; Perras, MR and Vettakkorumakankav, NN (1999) Phytoremediation of heavy metal contaminated and polluted soils, in Prasad, MNV and Hagemeyer, J (Eds.), Heavy Metal Stress in Plants - From Molecules to Ecosystems, Springer-Verlag Berlin Heidelberg, pp. 305-329

[19] Pulford, ID and Watson, C (2002) Phytoremediation of heavy metal-contaminated land by trees - A review. Environ Intern 29: 529-540

[20] Dinelli, E and Lombini, A (1996) Metal distributions in plants growing on copper mine spoils in Northern Apennines, Italy: the evaluation of seasonal variations. Appl Geochem 11: 375-385

[21] Ali, MB; Vajpayee, P; Tripathi, RD; Rai, UN; Singh, SN and Singh, SP (2003) Phytoremediation of lead, nickel, and copper by *Salix acmophylla* Boiss.: Role of antioxidant enzymes and antioxidant substances. Bull Environ Contam Toxicol 70: 462-469

[22] Nissen, LP and Lepp, NW (1997) Baseline concentrations of copper and zinc in shoot tissues of a range of *Salix* species. Biomass Bioenergy 12: 115-120

[23] Robinson, BH; Mills, TM; Petit, D; Fung, LE; Green, SR and Clothier, BE (2000) Natural and induced cadmium-accumulation in poplar and willow: Implications for phytoremediation. Plant Soil 227: 301-306

[24] Rockwood, DL; Naidu, CV; Carter, DR; Rahmani, M; Spriggs, TA; Lin, C; Alker, GR; Isebrands, JG and Segrest, SA (2004) Short-rotation woody crops and phytoremediation: Opportunities for agroforestry? Agroforest Syst 61: 51-63

[25] Klang-Westin, E and Eriksson, J (2003) Potential of *Salix* as phytoextractor for Cd on moderately contaminated soils. Plant Soil 249: 127-137

[26] Greger, M and Landberg, T (2001) Investigations on the relation between biomass production and uptake of Cd, Cu, and Zn, in Greger, M; Landberg, T and Berg, B (Eds.): *Salix* clones with different properties to accumulate heavy metals for production of biomass. Akademitryck AB, Edsbruk, Sweden, ISBN 91-631-1493-3, 19-27

[27] Greger, M. and Landberg, T. (2001). Tolerance to and uptake of metals in different clones of *Salix viminalis* grown in wastewater, in Greger, M; Landberg, T and Berg, B (Eds.): *Salix* clones with different properties to accumulate heavy metals for production of biomass. Akademitryck AB, Edsbruk, Sweden, ISBN 91-631-1493-3, 28-37

[28] Luňáčková, L; Šottníková, A; Masarovičová, E; Lux, A and Streško, V (2003) Comparison of cadmium effect on willow and poplar in response to different cultivation conditions. Biol. Plant. 47, 403-411

[29] Perttu, K; Eriksson, J; Greger, M; Göransson, A; Blombäck, K; Klang-Westin, E and Landberg, T (2003) Förråd och flöden av kadmium i systemet mark-Salix. (Content and fluxes of cadmium in the soil – willow – system). Rep. ER 19:2003, Energimyndigheten, Box 310, SE-63104 Eskilstuna, Sweden
[30] Vandecasteele, B; De Vos, B and Tack, FMG (2002) Cadmium and zinc uptake by volunteer willow species and elder rooting in polluted dredged sediment disposal sites. Sci Tot Environ 299; 191-205
[31] Vysloužilová, M; Tlustoš, P and Száková, J (2003) Cadmium and zinc phytoextraction potential of seven clones of *Salix spp.* planted on heavy metal contaminated soils. Plant Soil Environ 49: 542-547
[32] Cobbett, CS and Meagher, RB (2002) *Arabidopsis* and the genetic potential for the phytoremediation of toxic elements and organic pollutants, in: Somerville, C and Meyerowitz, E (Eds.): The *Arabidopsis* Book, American Society of Plant Biologists, USA, 1543-8120
[33] Madejón, P; Maratón, T; Murilloa, JM and Robinson, B (2004) White poplar (*Populus alba*) as a biomonitor of trace elements in contaminated riparian forests. Environ Pollut 132: 145-155
[34] Greger, M and Landberg, T (1999) Use of willow in phytoextraction. Int J Phytoremed 1: 115-123
[35] Riddell-Black D (1994) Willow vegetation filters for municipal wastewaters and sludges. A biological purification system. Proc Study Tour, Conference and Workshop in Sweden, Department of Ecology and Environmental Research, Swedish University of Agricultural Sciences, Uppsala, 145-151
[36] Fischerová, Z; Tlustoš, P; Száková, J and Šichorová, K (2005) A comparison of phytoremediation capability of selected plant species for cadmium and lead. Proc 8th ICOBTE, Adelaide, Australia, 274-275
[37] Šichorová, K., Tlustoš, P., Száková, J., Kořínek, K., and Balík, J. (2004). Horizontal and vertical variability of heavy metals in the soil of a polluted area. Plant Soil Environ 50: 525-534
[38] Roselli, W; Keller, C and Boschi, K (2003) Phytoextraction capacity of trees growing on a metal contaminated soil. Plant Soil 256: 265-272
[39] Tlustoš, P; Száková, J; Vysloužilová, M and Kolář, J (2005) The phytoremediation potential of fast growing trees on anthropogenically contaminated soil. Proc 8th ICOBTE, Adelaide, Australia, 816-817
[40] Greger, M and Landberg, T (1995) Use of willow clones with high Cd accumulating properties in phytoremediation of agricultural soils with elevated Cd levels. Proc 3rd ICOBTE, Paris, France, 505-511
[41] Eriksson, J and Ledin, S (1998) Changes in phytoavailability and concentration of cadmium in soil following long term *Salix* cropping. Water Air Soil Pollut 114: 171-184
[42] King, R; Royle, A; Putwain, P and Dickinson, N (2005) The potential of phytoremediation: Monitoring changes in contaminant bioavailability of a sediment in a 3-year field trial. Proc 8th ICOBTE, Adelaide, Australia, 260-261
[43] Shann, JR (1995) The role of plants and plant/microbial systems in the reduction of exposure. Environ Health Perspect 103: 13-15
[44] Liu, XM; Wu, QT and Banks, MK (2005) Effect of simultaneous establishment of *Sedum alfredii* and *Zea mays* on heavy metal accumulation in plants. Int J Phytoremed 7: 43-53
[45] Gove, B; Hutchinson, JJ; Young, SD; Craigon, J and McGrath, SP (2002) Uptake of metals by plants sharing a rhizosphere with the hyperaccumulator *Thlaspi caerulescens*. Int J Phytoremed 4: 267-281
[46] Vysloužilová, M; Tlustoš, P; Száková, J and Pavlíková, D (2003) As, Cd, Pb and Zn uptake by different *Salix* spp. grown at soils enriched by high loads of these elements. Plant Soil Environ 49: 191-196
[47] Halim, M; Conte, P and Piccolo, A (2003) Potential availability of heavy metals to phytoextraction from contaminated soils induced by exogenous humic substances. Chemosphere 52: 265-275
[48] Száková, J; Tlustoš, P; Pavlíková, D; Najmanová, J and Balík, J (2005) The response of extractable fractions of potentially toxic elements on alteration of organic matter content in contaminated soil. Proc 8th ICOBTE, Adelaide, Australia, 106-107
[49] Hasselgren, K (1999) Utilization of sewage sludge in short-rotation energy forestry: A pilot study. Waste Manage Res 17: 251-262
[50] Tsakou, A; Roulia, M and Christodoulakis, NS (2003) Growth parameters and heavy metal accumulation in poplar tree cultures (*Populus euroamericana*) utilizing water and sludge from a sewage treatment plant. Bull Environ Contam Toxicol 71: 330-337
[51] Kinnersely AM (1993). The role of phytochelates in plant growth and productivity. Plant Growth Regul 12: 207-217
[52] Marschner, H (1995) Mineral Nutrition of Higher Plants. 2nd Edition, Academic Press, London, ISBN 0-12-473543-6, 889
[53] Huang, JW; Chen, J; Berti, WR and Cunningham, SD (1997) Phytoremediation of lead-contaminated soils: Role of synthetic chelates in lead phytoextraction. Environ Sci Technol 31: 800-805

[54] Huang, JW and Cunningham, SD (1996) Lead phytoextraction: species variation in lead uptake and translocation. New Phytol 134: 75-84
[55] Vassil, AD; Kapulnik, Y; Raskin, I and Salt, DE (1998) The role of EDTA in lead transport and accumulation by Indian mustard. Plant Physiol 117: 447-453
[56] Wenzel, WW; Adriano, DC; Salt, D and Smith, R (1999) Phytoremediation: A plant-microbe-based remediation system, in: Adriano, DC; Bollag, J-M; Frankenberger, Jr, WT and Sims RC (Eds.): Bioremediation of Contaminated Soils, Madison, Wisconsin USA, 457-508
[57] Bricker, TJ; Pichtel, J; Brown, HJ and Simmons, M (2001) Phytoextraction of Pb and Cd from a superfund soil: Effects of amendments and croppings. J Environ Sci Health 36; 1597-1610
[58] McGrath, SP; Zhao, F-J and Lombi, E (2001) Plant and rhizosphere processes involved in phytoremediation of metal-contaminated soils. Plant Soil 232: 207-214
[59] Pereira, BFF; Abreu, CA and Berton, RS (2005) EDTA effect on soil solution ionic speciation and lead availability to jack bean and sunflower in an oxisol. Proc. 8th. ICOBTE, Adelaide, Australia, 98-99
[60] Piechalak, A; Tomaszewska, B and Barałkiewicz, D (2003) Enhancing phytoremediative ability of *Pisum sativum* by EDTA application. Phytochemistry 64: 1239-1251
[61] Schmidt, U (2003) Enhancing phytoextraction: The effect of chemical soil manipulation on mobility, plant accumulation, and leaching of heavy metals. J Environ Qual 32: 1939-1954
[62] Chen, H and Cutright, T (2001) EDTA and HEDTA effects on Cd, Cr, and Ni uptake by *Helianthus annuus*. Chemosphere 45: 21-28
[63] Kos, B and Leštan, D (2003) Induced phytoextraction/in situ soil washing of lead using biodegradable chelate and permeable barriers. Environ Sci Technol 37: 624-629
[64] Lai, H-Y and Chen, Z-S (2005) The interaction of Cd, Zn, and Pb on the phytoextraction of Rainbow Pink growing in combined metals-contaminated soils. Proc. 8th. ICOBTE, Adelaide, Australia, 286-287
[65] Wenzel, WW; Unterbrunner, R; Sommer, P and Sacco, P (2003) Chelate-assisted phytoextraction using canola (*Brassica napus* L.) in outdoors pot and lysimeter experiments. Plant Soil 249: 83-96
[66] Keller, C; Ludwig, C; Davoli, F and Wochele J (2005) Thermal treatment of metal-enriched biomass produced from heavy metal phytoextraction. Environ Sci Technol 39: 3359-3367

USING HYPERACCUMULATOR PLANTS TO PHYTOEXTRACT SOIL Cd

AUTUMN S. WANG[1], RUFUS L. CHANEY[2], J. SCOTT ANGLE[1] AND MARLA S. MCINTOSH[1]
[1]University of Maryland at College Park, Maryland, USA, 20742, E-mail: wspring@wam.umd.edu, [2]USDA-ARS, Animal Manure and By-Products Lab, Beltsville, Maryland, 20705 USA

1. Why is Cd phytoextraction being developed?

Adverse effects of soil Cd on humans were observed in Toyoma, Japan in 1969, and subsequently found in many locations in Japan and China. In these cases, mining or smelting of non-ferrous metals (Zn, Pb, Cu, Ag) caused dispersal of mine tailings which contaminated rice paddies where farmers grew their own food for many years. Such contamination localized to the rice irrigation network is one example of how rice uptake to grain of soil Cd harmed humans [1]. Similar contamination has now been found in other countries where rice is produced by subsistence farm families and where it is likely that human disease is occurring or will occur in time.

Much has been learned about transfer of Cd from paddy soils to rice grain which helps explain why high incidence of Cd disease (renal proximal tubular dysfunction) has been seen among subsistence rice farm families, but not in Europeans or North Americans with even higher soil Cd concentrations. Chaney *et al.* summarized data from rice Cd research, and bioavailability of food-borne Cd to humans [2]. Further analysis of the Japanese epidemiologic data showed that urinary Cd had to exceed a threshold before any adverse effects on tubular function occurred, and these preceded any effect on bone [3]. These re-interpretations of the susceptibility of humans to dietary Cd illustrated the importance of malnutrition of Fe and Zn in subsistence rice farm families which causes them to absorb a much higher fraction of diet Cd than persons with more nutritious diets. Rice is also remarkable because grain Cd is greatly increased with no increase in grain Zn even though the soils contain 100-times more Zn than Cd [4, 5]. No other crop grown on geogenic Cd rich soils causes such high transfer of soil Cd to human diets in bioavailable forms. Tobacco is the other crop that can comprise risk to humans when the crop is grown on geogenic Cd contaminated soils. Because part of the Cd enters smoke and is inhaled and absorbed very efficiently in the lung, factors which would reduce bioavailability of ingested Cd cannot protect against absorption of tobacco smoke Cd. Tobacco with minimal yield reduction can reach as high as 25 mg Cd kg^{-1} dry weight in leaves, high enough to cause adverse effects in smokers after many years.

M. Mackova et al. (eds.), Phytoremediation Rhizoremediation, 103–114.

The bioavailability of Cd in specific foods has been studied for many years, and it is now evident that foods rich in supply of Fe, Zn, Ca, or phytate can inhibit absorption of Cd. Cadmium rich shellfish also carry high levels of Fe and Zn which prevent high absorption of shellfish Cd ingested by humans. High phytate and fibre in whole grains reduce Cd bioavailability. Because polished rice has such low levels of bioavailable Fe and Zn, intestinal processes to absorb Cd are increased substantially compared to persons with higher bioavailable Fe and Zn in their diets [6]. Thus, although many locations in Europe and North America have become contaminated by mining and smelting of Zn ores, none are now recognized to have caused adverse effects on humans through food-chain transfer of Cd from contaminated soils. This change in understanding of soil Cd risk thus focuses attention on rice soils, and soils which are contaminated by Cd sources with high Cd:Zn ratio such as Cd pigments, Cd plating, high Cd biosolids, phosphate mine wastes, and Cd-Ni battery wastes.

2. Where are those possible commercial applications for Cd phytoextraction?

Thus paddy rice land in Japan, China, Korea, Vietnam and Thailand which has been shown to be contaminated by geogenic Zn + Cd sources is the largest area of soils requiring Cd remediation. No method for inactivation of soil Cd has been identified that can persistently reduce Cd phytoavailability. Recent evaluation of the contamination of rice fields in Japan suggests that 500,000 ha would need Cd remediation if the Cd limit for marketed rice were set at 0.2 mg kg^{-1}, with about 100,000 ha needing remediation if the rice Cd limit is 0.4 mg kg^{-1}. The improved understanding of Cd effects on kidney function reported by Ikeda *et al.* [3] and new population studies reported by Horiguchi *et al.* [7] indicate that with present nutritional status of Japanese farm families, setting the rice limit at 0.4 mg Cd kg^{-1} would provide the needed protection for the most sensitive and exposed individuals.

One can estimate the cost of soil Cd phytoextraction remediation by removal and replacement of contaminated soil depth needed for rice soils which produce brown rice with Cd greater than 0.4 mg kg^{-1} in Japan by multiplying the area times the cost per ha [100,000 ha -times- $2,500,000 ha^{-1}] = $250 billion if the traditional soil removal and replacement approach were used as shown in the remediation of about 646 ha in Japan in the 1980s [8]. The extreme cost of remediation of this area by soil removal and replacement has delayed government decisions on soil Cd remediation. Fortunately, phytoextraction of this Cd would cost < 1% of soil removal. Rice farmers could be paid to produce high Cd *T. caerulescens* biomass during a short clean up period, and be paid based on the amount of Cd in delivered biomass (mass times concentration) to encourage best phytoextraction practice for production of the phytoextraction crop. Soils contaminated by sources with high Cd:Zn ratio comprise the remaining soils where remediation will be required to protect humans from accumulation of Cd in soils. In the case of geogenic Cd+Zn contamination, a crop such as lettuce will attain only about 4-5 mg Cd kg^{-1} DW when it reaches 400-500 mg Zn kg^{-1} DW and suffers Zn phytotoxicity. When Zn phytotoxicity is observed, growers can add limestone and prevent economic loss due to Zn and further reduce Cd concentration in the plant tissues. But when the Cd source has high Cd:Zn ratio, leafy crops can reach over 100 mg Cd kg^{-1} DW with no yield reduction and no visible symptoms of the crop contamination. Thus soils with high Cd:Zn ratio also require remediation and Cd phytoextraction appears to be the only economic method which is available.

3. Phytoextraction concept

The concept of using hyperaccumulator plants to accumulate high quantities of metals in plant biomass to remove heavy metals from contaminated soils was first suggested by Chaney [9]. In addition to the low cost, phytoextraction has several other important advantages over the traditional soil removal/replacement remediation methods. For example, it is *in situ*, preserves top soil, reduces the secondary waste stream, is environmental sustainable and the plant ash may also have economic value [10, 11]. It is important to note that only hyperaccumulator plants can make it practical. Non-hyperaccumulator plants have no practical value in phytoextraction [12]. Hyperaccumulator plants typically contain >100 mg Cd kg^{-1}, >1000 mg Ni kg^{-1}, or >10,000 mg Zn kg^{-1} in their leaf tissue (dry weight). Most plants suffer toxicity and experience yield reduction when leaves contain about 400-500 mg Zn kg^{-1} or 50-100 mg Ni kg^{-1}.

4. *Thlaspi caerulescens*

To date, about 400 taxa have been identified as hyperaccumulators [13]. The majority of them are Ni hyperaccumulators. *Thlaspi caerulescens* is primarily a Zn and Cd hyperaccumulator and is the most extensively studied. It actually requires abnormal amounts of Zn to be able to grow normally [14]. Concentrations can exceed 3% and 0.1% of Zn and Cd, respectively, in shoot dry matter without yield reduction. *Thlaspi caerulescens* is not uniform in metal accumulation capability [15]. The accumulation rates vary among populations [16, 17], and are influenced by the physical and chemical characteristics of soils. Some populations of *T. caerulescens* from the south of France exhibited extraordinary Cd hyperaccumulating ability where foliar Cd concentrations could reach 3000 mg kg^{-1} [18-20]. These Southern France populations appear to have the great potential to make Cd phytoextraction a reality.

5. Metal phytoavailability and *T. caerulescens* hyperaccumulation

Chemical fractionation procedures have been proposed as a means to identify plant available forms of heavy metals in soil. Different sequential extraction procedures (SEP) have been used to partition metals into fractions as soluble, exchangeable, absorbed, organically bound, oxide-bound, precipitated, occluded and residual [21, 22]. Researchers have for many years tried to correlate metals in these fractions with plant concentrations. Although SEP is useful as an indicator of metal bioavailability, correlation studies are of less value. Metal bioavailability only correlated with plant tissue concentration when it is a limiting factor for plant uptake due to low soil buffering capacity or low plant solubilization. In many cases, especially for *T. caerulescens*, metals released from formerly non-available forms comprised more than 50% of the metals accumulated in plants [23-25]. Thus, the dynamic cyclic process: depletion due to plant uptake and replenishment due to solubilization and desorption are generally not at equilibrium. Measured metal concentrations can only capture a "moment in time" while plant metal concentration is an accumulation of uptake over time. In mathematical terms, it is an integration of numerous "moments" of metal concentrations. This may explain similar studies gave varied results.

Progress in making phytoextraction a practical commercial technology is hindered by a lack of strategies to optimize plant uptake of metals. Although *T. caerulescens* has extraordinary ability to transfer high amount of Zn and Cd from soil into the shoot, its use for commercial remediation of contaminated soils is limited by the restricted phytoavailability of metals in a contaminated soil. Although the mechanisms of hyper-accumulation remain unclear, it is generally agreed that hyperaccumulation process involves three major steps: rapid uptake of metals by roots, high rates of translocation from roots to shoots, and huge storage capacity by vacuolar compartmentalization [15, 26, 27]. The first step is usually rate-limiting. Plant uptake of metals is confined by metal availability. Numerous observations have suggested that metal supply rate in soil is more limiting than plant uptake of the metals.

6. Mechanisms by which *T. caerulescens* scavenges metals

Instead of avoiding metal polluted spots, *Thlaspi caerulescens* roots preferentially colonize Zn and Cd-polluted areas [28]. The allocation and morphology of roots are strongly influenced by Zn and Cd content, form, and location in soil. When all roots were in homogeneous soil polluted with a soluble Zn salt ($ZnSO_4$), root growth was severely inhibited. The positive response of roots to metals is specific, only to Zn and Cd. There was no response to Pb localization [29]. This specificity and precision of distribution of the root system is considered an important factor in determining the efficient removal of metals.

Once roots have proliferated in metal rich soil, there is still a problem that *T. caerulescens* has to overcome: how to make the metals available? Rhizosphere acidification and release of root exudates are two common mechanisms by which plants modify the rhizosphere to acquire nutrients. A study by Luo *et al.* [30] investigated soil solution Zn and pH dynamics during phytoextraction by *T. caerulescens*. Soil solution pH decreased initially and then increased slightly in both planted and unplanted soil zones. From 60 to 84 days after transplanting, the pH of the rhizosphere soil solution was higher than that of non-rhizosphere soil solution. This indicated that rhizosphere acidification was not the primary mechanism for mobilization of Zn in soil for *T. caerulescens*. Similar result was found in a pot study [31]. Root exudates do not appear to play a role in metal mobilization by *T. caerulescens* hyperaccumulation, either [32].

On the contrary, it was repeatedly found that *T. caerulescens* was able to access less soluble Zn fractions in soil. In McGrath's study, decreases in the mobile fraction of Zn accounted for less than 10% of the total uptake of *T. caerulescens*; that is, more than 90% of the Zn must have come from the non-mobile fractions [31]. These authors also found that rhizosphere soils tended to have higher concentrations of mobile Zn than the non-rhizosphere soils. Similarly, in a study, the decrease of Zn in soil solution after growth accounted for only 1% of the total Zn uptake by *T. caerulescens*. The authors suggested that either *T. caerulescens* was highly efficient at mobilizing Zn which was not initially soluble, or the soil could replenish solution Zn rapidly due to high buffering capacity [23].

To test which one of the above two possible mechanisms was more important, Whiting *et al.* used co-cultivated plants to see if mobilization of Zn by *T. caerulescens* affected Zn concentrations of a co-cultivated indicator plants (*Thlaspi arvense* or *Festuca rubra*) provided that they shared the same rhizosphere. *Thlaspi caerulescens* did not increase Zn concentrations in either of the indicator plants, suggesting that *T. caerulescens* does not "strongly" mobilize Zn in its rhizosphere [25]. In another experiment, Whiting *et al.* used five Zn compounds of different solubility (ZnS, $Zn_3(PO_4)_2$, ZnO, $ZnCO_3$, and $ZnSO_4 \cdot 7H_2O$) to test how Zn hyperaccumulation was influenced by Zn bioavailability. In a Clough Wood soil, the use of Zn-sulphate resulted in the greatest total Zn in plant biomass, while in a Prayon soil, highest uptake was from the Zn-oxide fraction. In the unenriched and ZnS enriched treatments, about 70% and 50% of *T. caerulescens* biomass Zn came from previously non-soluble forms. But Zn hyperaccumulation in these two treatments was less than that from the other four treatments. This may indicate that the solubilization effect of Zn by *T. caerulescens* was not strong [24]. But comparing the nitrate-extractable Zn in day 0 and day 90, there was a significant increase in all of the five Zn-enriched treatments. In the Zn-sulphide treatment, there was an almost 10-fold increase in both soils. If this was caused by *T. caerulescens*, it was obviously a very strong solubilization effect. Unfortunately they did not study an un-planted treatment to exclude the possible effect of soil microbial oxidation of ZnS during incubation. The paper by Gérard *et al.* showed that *Thlaspi* absorbed Cd from the same labile pool as other plant species, only much faster [18]. It seems likely to us that the dense, fine root system of *Thlaspi caerulescens* gives the plant access to more soil volume within the rhizocylinder (the soil within the root hair distance from root surfaces). High root surface area coupled with rapid uptake of Cd and Zn lets this species achieve hyperaccumulation needed for phytoextraction.

7. Localization of Cd in leaves of *T. caerulescens*

In a recent approach, Cosio et al. reported that in both Ganges and Prayon ecotypes, Cd was found primarily at leaf edges. Scanning electron microscopy coupled with energy dispersive X-ray microanalysis (cryo-SEM-EDXMA) and tissue fractionation revealed similar distribution patterns of Cd in leaf cells of both ecotypes. Cadmium was in both inside the cells and in the cell walls of large epidermal cells as well as small epidermal cells [33].

8. Development of Cd phytoextraction technology

Chaney *et al.* identified the systematic approaches to develop commercial phytoextraction technique:

- collection of plant genetic diversity so that improved phytoremediation cultivars can be selected and/or bred;
- valid comparison of genotypic differences in yield and hyperaccumulation;
- breeding improved plant cultivars which are effective in metal-rich fields;
- identification of soil and plant management practices needed to attain high yields and high metal concentrations in the biomass including tillage, fertilization, soil conditioners, pH adjustment, herbicides, etc.;
- identification of methods to plant, grow, harvest and market the biomass;
- selection of methods to economically recover the metals from the biomass (e.g. a method to burn the biomass which retains the metals in a form that they can be sold as a high grade metal ore), and the biomass energy may be used for power generation;
- identification of methods to recover the metals from the ash;
- identification of farming systems which allow use of this technology to produce jobs and profits for growers as well as smelters [11]

Chaney *et al.* summarised the philosophy and progress of our team in developing Cd phytoextraction technology: "The approach...is based on our understanding of the science and the potential market for Cd and Zn phytoextraction. Because the Zn and Cd accumulated in biomass have much lower value than Ni, the value of metal in the biomass will not drive development and use of this technology. Biomass energy can reduce the cost of soil Cd cleanup, but is unlikely to make Zn+Cd phytoextraction profitable by phytomining. Rather, the value of the cleanup will drive the market. Risk from Zn can be controlled by soil pH management, but for Cd, risk from rice soils, tobacco soils, and soils with high Cd:Zn ratio must be remediated to be able to produce crops safe for consumption. Most of the Cd phytoextraction market will be in paddy rice land contaminated by emissions of Zn-Pb-Cu-Ag mining and smelting. It is now clear that paddy rice lands in Japan, China, Korea, Thailand and Vietnam have been contaminated enough to require remediation or change in land use." [12].

"We adopted the agricultural paradigm to develop a Cd phytoextraction technology. Comparing different models for Zn and Cd phytoextraction by crop and hyperaccumulator species, the value of southern France genotypes of *Thlaspi* is evident. We selected *Thlaspi caerulescens* for development because it was the largest Zn hyperaccumulator which also accumulated Cd. The "Prayon" population was initially studied by researchers because Alan Baker shared collected seed. Some other species which can accumulate Cd exist, but were even smaller than *Thlaspi* (e.g., *Arabidopsis halleri, Arabis gemmifera*), or did not accumulate favourable levels of Cd (*Dichapetalum gelonioides*)." [12].

"After our initial studies to understand the potential of *Thlaspi* to accumulate soil and solution Cd and Zn [34, 35], we evaluated 20 different *Thlaspi* genotypes in nutrient solutions and a smaller subset in contaminated soils. The genetic screening in nutrient solutions demonstrated high accumulation of Cd by southern France populations compared to other populations [36]. A similar outcome was observed when we grew *Thlaspi* genotypes in smelter contaminated field plots at Palmerton, PA [37] (Fig. 1). We concluded that natural variation in Cd accumulation by *Thlaspi* populations would support breeding of improved cultivars useful for practical Cd phytoextraction. Strong promise was also reported by Keller and Hammer [38] who grew *T. caerulescens* in several contaminated fields" [12].

"Phytoextraction Associates LLC, a new company created to commercialize Cd phytoextraction, is working with USDA-ARS to develop and commercialize improved *T. caerulescens* cultivars with higher yields and the high Cd:Zn accumulation of southern France genotypes. Based on the findings of Reeves *et al.* who analyzed *T. caerulescens* collected across Europe, the southern France genotypes are of special importance [20]. In the study of Perner *et al.* [16], eight plants were grown from seeds of each of 25 mother plants collected at these southern France Zn-Cd-contaminated sites. The wide variation among siblings (due to mixed inbreeding and out-crossing) suggests that Cd hyperaccumulation is a quantitative trait [39], while the southern France genotypes appear to have a single gene which causes 10-fold higher Cd accumulation than "Prayon" genotypes. Although more research is being conducted to understand these genetic differences, it is clear that normal plant breeding can be used to develop cultivars with higher yield and high annual Cd removal needed to achieve remediation of soils which require Cd remediation to protect human health" [12].

As indicated previously in this paper, restricted bioavailability of metals in contaminated soils is a limitation for phytoextraction efficiency. Researchers have tried different methods to increase metal solubility. Apparently, EDTA or other chelating

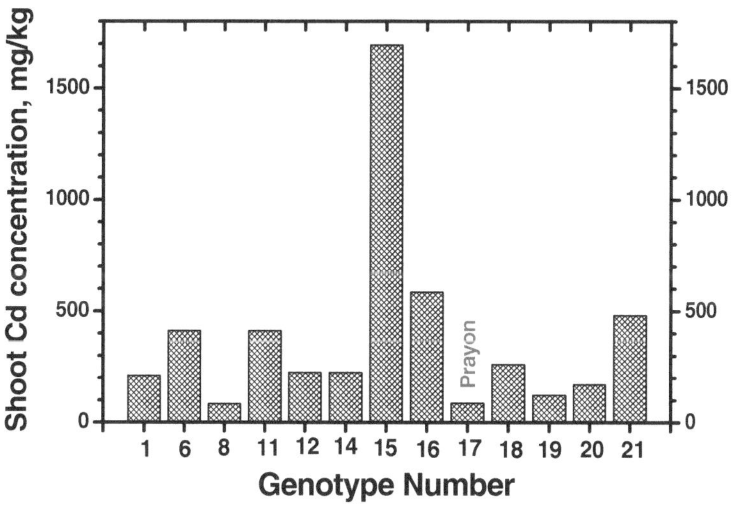

Figure 1. Cd accumulation into shoots of Thlaspi caerulescens ecotypes grown in the field at Palmerton, PA, where soil contains 15,000 mg Zn and 150 mg Cd kg^{-1} dry weight [37]. The "Prayon" ecotype is marked. All of the much higher Cd ecotypes came from the southern French populations. (Figure source: Chaney et al., 2004)

induced treatments is not a solution due to high cost and the concern of ground-water contamination. Instead, using S addition to reduce soil pH to enhance metal availability could be a good strategy. Elemental S will be oxidized into sulphate salts ($CaSO_4$, etc), excessive salts will be drained out of the soil by natural rainfall. We conducted a greenhouse study by using two contaminated soils differing in levels of Cd and Zn (5 mg Cd kg^{-1} and 450 mg Zn kg^{-1} in the lower metal soil and 25 mg Cd kg^{-1} and 1500 mg Zn kg^{-1} in the higher metal soil) and which were adjusted to 5 or 6 different pH levels (from 7.27 to 4.74). Reducing pH significantly increased 0.1 M $Sr(NO_3)_2$ extractable Cd and Zn concentrations in both soils. Further, pH adjustment had significant effect on both plant yield and metal concentrations in plant biomass. However, pH adjustment caused different responses in these two soils. For the higher metal soil, yield continued to increase with the decrease of pH, and was the highest at the lowest pH treatment. Cadmium and Zn concentration in plant tissue as well as the total metal in plant shoots were increased with the reduction of pH. The highest shoot metal concentration was observed at pH 5.28 (Fig. 2). For the lower metal soil, plant yield, metal concentration in plant tissue and total metal extracted by plant shoot all increased with decreasing pH, reached a peak at pH treatment of 6.07 and rapidly decreased at further reduced pH treatments. Soluble Al and Mn concentrations reached toxic levels in lower pH treatments in the lower metal soil. For the higher metal soil, from the highest pH (6.88) to the lowest pH (4.74), Al concentration only increased by about 30%. However, there was 8 to 11 fold increase in soluble Al concentration for the lower metal soil. Al toxicity in the lower metal soil restricted root development and hindered the root ability to accumulate metals [40].

The extraordinary Cd phytoextraction ability of *T. caerulescens* was further demonstrated in this experiment. In the optimum pH treatments, *Thlaspi caerulescens* extracted 45% and 37% of total Cd in the lower and higher metal soils, respectively, with just one 6-month planting. One of the major criticisms of phytoextraction is the long time requirement. In order to remove all contaminants or reduce their level to legislatively allowed limits, phytoremediation often requires decades to be effective. If constant uptake by *T. caerulescens* in subsequent croppings can be achieved-where

Keller and Hammer's study suggested that it is possible [38]- reducing pH would shorten the required remediation time from 5 yr (the original soil pH) to 3 yr (the optimum soil pH 5.28 in the higher metal soil and 6.07 in the lower metal soil) to remove all the Cd from both soils. For Zn, reducing pH would shorten the remediation time from 60 yr to 46 yr for the higher metal soil and 38 yr to 14 yr for the lower metal soil. We recommend focus on Cd phytoextraction because risks from soil Zn can be alleviated by liming, but soil Cd risks are not so easily remediated.

We concluded that reducing pH is an effective method to enhance Cd phytoavailability and *T. caerulescens* uptake for both Cd and Zn. It also can greatly shorten the time span for phytoremediation to be complete and has the potential to overcome the problems associated with the long time requirement of phytoremediation. Using S to reduce soil pH is advantageous over chelating-induced methods because it is cost-effective and with few toxic effects on plant growth. The proper and effective pH range for maximum metal uptake may differ for individual soils and therefore must be identified to ensure successful phytoremediation. For Cd phytoextraction, the higher

Cd:Zn accumulation in plants allows Cd removed at high rates without Zn phytotoxicity reducing yields (which limits annual Cd removed by "Prayon" types).

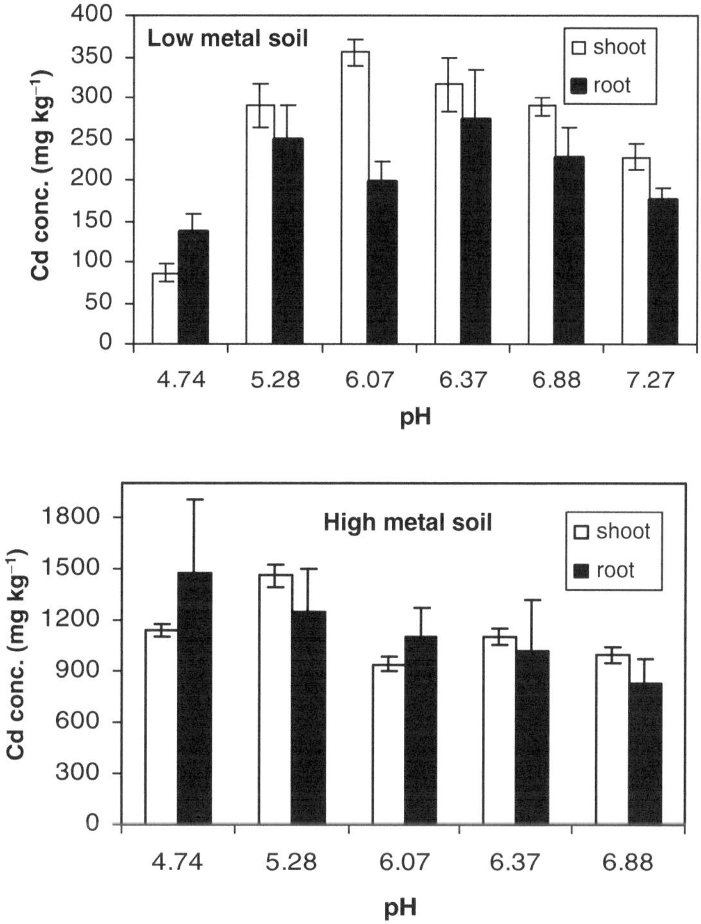

Figure 2. Means and standard errors of T. caerulescens tissue Cd concentration in higher metal and lower metal soils with different pH treatments.

9. Ecological considerations of Cd phytoextraction

Recently, there is increasing awareness of how to define a truly successful phytoextraction. The ultimate goal of soil remediation is to achieve a healthy soil ecosystem and crops safe for lifetime consumption. Metal removal is one of the methods to achieve this, not the remediation purpose itself. Therefore a successful phytoextraction include not only metal removal from soil, but also return of a healthy soil ecosystem. One of the major misunderstandings during phytoextraction is the sole focus on a simple calculation of how much pollutants were removed. The subject of whether the soil health is eventually recovered has been largely ignored. Therefore, it is possible that the process of remediation itself brings a more severe environmental problem (leaching of metal-EDTA to groundwater; reduction of soil microbial function, etc). It is important to realize that monitoring soil health throughout the remediation process is vital in remediation.

When investigating the effect of reducing soil pH to enhance Cd phytoextraction, we selected numerous soil quality indicators to monitor soil ecosystem health conditions responding to reducing pH treatment both before and after phytoextraction. Soil enzyme activities, nitrification, respiration, number of bacteria, actinomycetes, fungi, and rhizobia were tested both under low pH treatments and after pH neutralization. This research is the first attempt to examine whether and to what degree the soil biological activities and soil microbial populations return to the background levels after soil pH is re-adjusted to normal conditions after phytoextraction. Except for acid phosphatase activity and the fungal population, reducing pH significantly reduced all tested activities and microbial populations. However, soil retained the capacity to recover toward the condition before acidification treatment even at the lowest pH treatment. But full recovery may take a much longer time and may be very difficult to achieve original levels seen in the lowest pH treatment. It is recommendable to keep the pH values above 6.1 and 5.3 for lower and higher metal soils tested in this experiment, respectively. Above this value, most of the soil biological activities and all tested microbial populations can return to background levels within a short period. Moreover, metal leaching was negligible if soil pH were maintained at above these values. Based on our results, soil biological activities and microbial populations can be severely damaged when soil pH is reduced below a certain threshold. Precaution must be exercised when we try to enhance phytoextraction by reducing soil pH. Fortunately, our research revealed that it is not necessary to reduce pH too much. Optimum plant growth and phytoextraction can be achieved before pH being lowered to a value where possibly permanent soil ecologic damage would occur. The pH threshold can vary according to the soil type, mineral composition and other properties. There is no unique solution to all different kinds of soils [41].

More ecological considerations should be integrated into Cd phytoextraction process. Representative and sensitive soil health indicators may be used to monitor the soil ecosystem health and function changes accompanying phytoextraction processes after returning the soil to a pH required by most crops, and cropping to feed the soil microbes and aid population growth.

References

[1] Kobayashi, J (1978) Pollution by cadmium and the itai-itai disease in Japan. pp. 199-260, in FW Oehme, Ed., Toxicity of Heavy Metals in the Environment. Marcel Dekker, Inc., New York

[2] Chaney, RL and Ryan, JA (1994) Risk Based Standards for Arsenic, Lead and Cadmium in Urban Soils, Dechema, Frankfurt. ISBN 3-926959-63-0

[3] Ikeda, M; Ezaki, T; Tsukahara, T; Moriguchi, J; Furuki, K; Fukui, Y; Ukai, H; Okamoto, S and Sakura, H (2003) Threshold levels of urinary cadmium in relation to increases in urinary β2-microglobulin among general Japanese populations. Toxicol Lett 137: 135-141

[4] Chaney, RL; Reeves, PG; Ryan, JA; Simmons, RW; Welch, RM and Angle, JS (2004) An improved understanding of soil Cd risk to humans and low cost methods to remediate soil Cd risks. BioMetals 17: 549-553

[5] Simmons, RW; Pongsakul, P; Chaney, RL; Saiyasitpanich, D; Klinphoklap, S and Nobuntou, W (2003) The relative exclusion of zinc and iron from rice grain in relation to rice grain cadmium as compared to soybean: Implications for human health. Plant Soil 257: 163-170

[6] Reeves, PG and Chaney, RL (2002) Nutritional status affects the absorption and whole-body and organ retention of cadmium in rats fed rice-based diets. Environ Sci Technol 36, 2684-2692

[7] Horiguchi, H; Oguma, E; Sasaki, S; Miyamoto, K; Ikeda, Y; Machida, M and Kayama, F (2004). Dietary exposure to cadmium at close to the current provisional tolerable weekly intake does not affect renal function among female Japanese farmers. Environ Res 95: 20-31

[8] Iwamoto, A (1999) Restoration of Cd-polluted paddy fields in the Jinzu River basin - Progress and prospects of the restoration project. pp. 179-183, in K Nogawa; M Kurachi and M Kasuya, Eds., Advances in the Prevention of Environmental Cadmium and Countermeasures. Eiko Laboratory, Kanazawa, Japan

[9] Chaney, RL (1983) Plant uptake of inorganic waste constituents. pp. 50-77, in JF Parr; PB Marsh and JM Kla, Eds, Land Treatment of Hazardous Wastes. Noyes Data Corp., Park Ridge, NJ

[10] Baker, AJM; McGrath, SP; Reeves, RD and Smith, JAC (2000) Metal hyperaccumulator plants: a review of the ecology and physiology of a biological source for phytoremediation of metal-polluted soil. pp. 85-107, in N Terry and GS Bañuelos, Eds, Phytoremediation of Contaminated Soil and Water. Lewis Publishers, Boca Raton

[11] Chaney, RL; Li, YM; Angle, JS; Baker, AJM; Reeves, RD; Brown, SL; Homer, FA; Malik, M and Chin, M (2000) Improving metal hyperaccumulator wild plants to develop commercial phytoextraction systems: approaches and progress. pp. 131-160, in N Terry and GS Bañuelos, Eds, Phytoremediation of Contaminated Soil and Water. CRC Press, Boca Raton, FL

[12] Chaney, RL; Angle, JS; McIntosh, MS; Reeves, RD; Li ,YM; Brever, EP; Chen, KY; Roseberg, RJ; Perner, H; Synkowski, EC; Broadhurst, CL; Wang, AS and Baker, AJM (2004) Using hyperaccumulator plants to phytoextract soil Ni and Cd. Z Naturforsch C 60: 190-190

[13] Baker, AJM; Reeves, RD and Hajar, ASM (1994) Heavy metal accumulation and tolerance in British populations of the metallophyte *Thlaspi caerulescens* J and C Presl. (Brassicaceae). New Phytol 127: 61-68

[14] Shen, ZG; Zhao, FJ and McGrath, SP (1997) Uptake and transport of zinc in the hyperaccumulator *Thlaspi caerulescens* and the non-hyperaccumulator *Thlaspi ochroleucum*. Plant Cell Environ 20: 898-906

[15] Pollard, AJ; Harper, KD and Smith, JAC (2002) The genetic bases of metal hyperaccumulation in plants. Crit Rev Plant Sci 21, 539-566

[16] Perner H; Chaney, RL; Reeves, RD, Römheld, V, McIntosh, MS; Angle, JS and Baker, AJM (2004) Variation in Cd- and Zn-accumulation by French genotypes of *Thlaspi caerulescens* J and C Presl. grown on a Cd and Zn contaminated soil. Submitted to: Plant Soil

[17] Molitor, M; Dechamps, C; Gruber, W and Meerts, P (2005) *Thlaspi caerulescens* on nonmetalliferous soil in Luxembourg: ecological niche and genetic variation in mineral element composition. New Phytol 165: 503-512

[18] Gérard, E; Echevarria, G; Sterckeman, T and Morel, JL (2000) Cadmium availability to three plant species varying in Cd accumulation pattern. J Environ Qual 29: 1117-1123

[19] Lombi, E; Zhao, FJ; Dunham, SJ and McGrath, SP (2000) Cd uptake in populations of *Thlaspi caerulescens* and *Thlaspi goesingense*. New Phytol 145: 11-20

[20] Reeves, RD; Schwartz, C; Morel, JL and Edmondson, J (2001) Distribution and metal-accumulating behaviour of *Thlaspi caerulescens* and associated metallophytes in France. Int J Phytorem 3: 145-172

[21] Davidson, CM; Thomas, RP; McVey, SE; Perala, R; Littlejohn, D and Ure, AM (1994) Evaluation of a sequential extraction procedure for the speciation of heavy metals in sediments. Analytica Chimica Acta 291: 277-286

[22] Welter, E; Calmano, W; Mangold, S and Tröger, L (1999) Chemical speciation of heavy metals in soils by use of XAFS spectroscopy and electron microscopical techniques. Fresenius J Anal Chem 364: 238-244
[23] Knight, BP; Zhao, FJ; McGrath, SP and Shen, ZG (1997) Zinc and cadmium uptake by the hyperaccumulator *Thlaspi caerulescens* in contaminated soils and its effects on the concentration and chemical speciation of metals in soil solution. Plant Soil 197: 71-78
[24] Whiting, SN; Leake, JR; McGrath, SP and Baker, AJM (2001) Zinc accumulation by *Thlaspi caerulescens* from soils with different Zn availability: a pot study. Plant Soil 236: 11-18
[25] Whiting, SN; Leake, JR; McGrath, SP and Baker, AJM (2001) Assessment of Zn mobilization in the rhizosphere of *Thlaspi caerulescens* by bioassay with non-accumulator plants and soil extraction. Plant Soil 237:147-156
[26] Chaney, RL; Malik, M; Li, YM; Brown, SL; Brewer, EP; Angle, JS and Baker, AJM (1997) Phytoremediation of soil metals. Curr Opin Biotech 8: 279-284
[27] McGrath, SP; Zhao, FJ and Lombi, E (2002). Phytoremediation of metals, metalloids, and radionuclides. Adv Agron 75, 1
[28] Whiting, SN; Leake, JR; McGrath, SP and Baker, AJM (2000) Positive responses to Zn and Cd by roots of the Zn and Cd hyperaccumulator *Thlaspi caerulescens*. New Phytol 145: 199-210
[29] Schwartz, C; Morel, JL; Saumier, S; Whiting, SN and Baker, AJM (1999) Root development of the zinc-hyperaccumulator plant *Thlaspi caerulescens* as affected by metal origin, content and localization in soil. Plant Soil 208: 103-115
[30] Luo, YM; Christie, P and Baker, AJM (2000) Soil solution Zn and pH dynamics in non-rhizosphere soil and in the rhizosphere of *Thlaspi caerulescens* grown in a Zn/Cd contaminated soil. Chemosphere 41: 161-164
[31] McGrath, SP; Shen, ZG and Zhao, FJ (1997) Heavy metal uptake and chemical changes in the rhizosphere of *Thlaspi caerulescens* and *Thlaspi ochroleucum* grown in contaminated soils. Plant Soil 188: 153-159
[32] Zhao, FJ; Hamon, RE and McLaughlin, MJ (2001) Root exudates of the hyperaccumulator *Thlaspi caerulescens* do not enhance metal mobilization. New Phytol 151: 613-620
[33] Cosio, C; DeSantis, L; Frey, B; Diallo, S and Keller, C (2005) Distribution of cadmium in leaves of *Thlaspi caerulescens*. J Exp Bot 56: 765-775
[34] Brown, SL; Chaney, RL; Angle, JS and Baker, AJM (1994) Phytoextraction potential of *Thlaspi caerulescens* and Bladder Campion for zinc and cadmium-contaminated soil. J Environ Qual 23: 1151-1157
[35] Brown, SL; Chaney, RL; Angle, JS and Baker, AJM (1995) Zinc and cadmium uptake by hyperaccumulator *Thlaspi caerulescens* and metal tolerant *Silene vulgaris* grown on sludge-amended soils. Environ Sci Technol 29: 1581-1585
[36] Li, YM; Chaney, RL; Angle, JS; Chen, KY; Kerschner, BA and Baker, AJM (1996) Genotypical differences in zinc and cadmium hyperaccumulation in *Thlaspi caerulescens*. Agron Abstr 1996, 27
[37] Li, YM; Chaney, RL; Chen, KY; Kerschner, BA; Angle, JS and Baker, AJM (1997) Zinc and cadmium uptake of hyperaccumulator *Thlaspi caerulescens* and four turf grass species. Agron Abstr 1997, 38
[38] Keller, C and Hammer, D (2004) Metal availability and soil toxicity after repeated croppings of *Thlaspi caerulescens* in metal contaminated soils. Environ Pollut 131: 243-254
[39] Synkowski, EC; McIntosh, MS; Chaney, RL; Angle, JS and Reeves, RD (2004) RAPD variation in two *Thlaspi caerulescens* populations: Implications for breeding. Proc. Fourth International Serpentine Ecology Conference, Havana, Cuba
[40] Wang, AS; Angle, JS; Chaney, RL; Delorme, TA and Reeves, RD (2005) Soil pH effects on uptake of Cd and Zn by *Thlaspi caerulescens*. Plant Soil (In press)
[41] Wang, AS; Angle, JS; Chaney, RL; Delorme, TA and McIntosh, MS (2005) Ecological Risks of Reducing Soil pH during Zn and Cd Phytoextraction. Submitted to: J Environ Qual

ENHANCED HEAVY METAL PHYTOEXTRACTION

DOMEN LEŠTAN
Agronomy Department, Biotechnical Faculty, University of Ljubljana, Jamnikarjeva 101, 1000 Ljubljana, Slovenia – Fax 386 1 423 1161
Email: domen.lestan@bf.uni.lj.si

1. Introduction

Phytoextraction refers to the use of metal-accumulating plants, which are able to transport and concentrate inorganic contaminants, most importantly heavy metals but also metalloids and radionuclides, from the soil into their harvestable, above-ground parts. Metal-enriched plant biomass can be safely disposed of as a hazardous material or, if economically feasible, used for metal recovery. For phytoextraction to be possible, the metals must be in soil horizons within a plant's root zone, be bioavailable for plants and plants must have a genetic predisposition for compartmentalisation of extracted metals. Some metals are readily bioavailable for plants: Cd, Ni, Zn, As, Se, Cu, and some have a low bioavailability: Pb, Cr, U, Hg [1].

There are two general approaches to phytoextraction: continuous and chemically enhanced phytoextraction [2]. The first approach uses naturally hyperaccumulating plants with the ability to accumulate an exceptionally high metal content in the shoots. Hyperaccumulating plants usually hyperaccumulate only a specific metal and metals that are primarily accumulated (Ni, Zn and Cu) are not among the most important environmental pollutants. No plant species has yet been found that demonstrates a wide spectrum of hyperaccumulation [3]. Hyperaccumulators are also mostly slow growing, low biomass-producing species, lacking in good agronomic characteristics [4]. There is no evidence that natural hyperaccumulator plants can access a less soluble and bio-available pool of metals in soil.

No effective hyperaccumulating plant, with high Pb uptake and high biomass essential for efficient phytoextraction, has been reported so far for Pb, one of the most widespread and important metal contaminants. Vegetation growing in heavily contaminated areas often has less than 50 mg g^{-1} Pb in shoots [4] and Pb is mostly confined to the roots, with minimal transport to the green parts of plants. In fact, there is only one reliable report of Pb hyperaccumulation. Puschenreiter *et al.* [5] reported that the small alpine plant *Thlaspi goesingense* accumulated 2840 mg Pb kg^{-1} in its shoots. However, when we tested the same reported hyperaccumulator for phytoextraction of soil contaminated with 1170 Pb and 750 mg kg^{-1} Zn from Mežica Valley in Slovenia, contaminated by mining and the smelting industry, the concentration of Pb in leaves of

M. Mackova et al. (eds.), Phytoremediation Rhizoremediation, 115–132.

T. goesingense did not exceed 20 mg kg^{-1} and was statistically comparable to the Pb concentration in leaves of lettuce (*Lactuca sativa*) [6]. These results might indicate that the Pb hyperaccumulating capacity of *T. goesingense* is limited only to certain soil and contamination types.

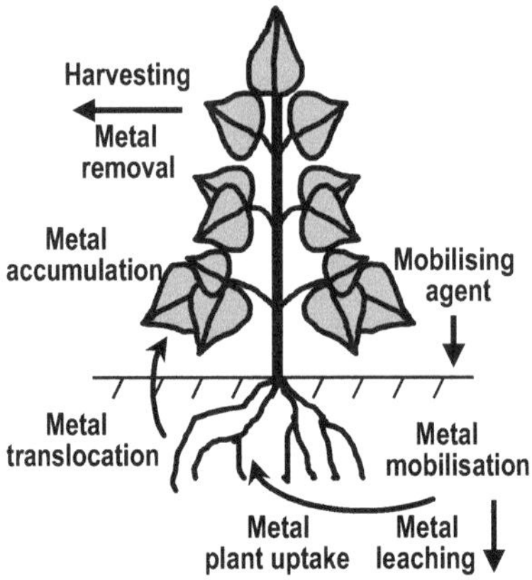

Figure 1. Conceptual representation of heavy metal phytoextraction enhanced by the addition of mobilizing (i.e. chelating) agent.

In non-hyperaccumulating plants, factors limiting their potential for phytoextraction include small root uptake and little root-to-shoot translocation of heavy metals. Chemically enhanced phytoextraction has been shown to overcome the above problems [7-9]. Common crop plants with high biomass can be triggered to accumulate high amounts of low bioavailable metals, when their mobility in the soil and translocation from the roots to the green part of plants was enhanced by the addition of mobilizing agents when the crop had reached its maximum biomass. The feasibility of chemically enhanced phytoextraction has been primarily studied for Pb and chelating agents as soil additives, less attention has been given to other metals and radionuclides or their mixtures.

2. Fractionation of metals in soil

The geochemical forms of metals in contaminated soils affect their solubility, which directly influences their availability to plants. Plant uptake of metals therefore shows marked dependence on the chemical speciation and soil fractionation of the metals. For example, Pichtel *et al.* [10] reported that a significant correlation exists between Pb uptake by dandelion (*Taraxacum officinale*) and bioavailable fractions of soil Pb soluble in the soil solution and exchangeable from soil colloids to the soil solution. The fraction of metals soluble in the soil solution consists of free hydrated ions, water-soluble organic and inorganic complexes and metals sorbed on dissolved organic matter. They are the most mobile forms of metals in soil and directly available to plants. A plausible explanation for the lack of Pb hyperaccumulating plants is therefore that in most soils capable of supporting plant growth, only a very small portion of the Pb in the soil is present in soil solution or exchangeable from soil colloids and thus does not allow substantial Pb mobility and plant uptake. Indeed, Pb from most contaminated sites tends to persist in the soil surface layers, although some Pb bound to the soluble part of the soil organic matter can be transported with percolates through the soil profile [11].

The fractionation of metals into soluble in soil solution and exchangeable from soil colloids (two bioavailable fractions), and metals strongly bound into the soil solid phase is probably controlled by many types of reactions, a) Adsorption/desorption reactions due to chemical bonding, complex formation, and ion-exchange. Partly covalent bonds of metals with mineral/humic colloids and hydrous oxides are formed. Coordinate bonds and complexes of metals with soluble soil organic matter and polycarboxylic organic acids of microbial origin are also formed. Reversible electrostatic exchange occurs between metals and other cations from the negatively charged surface of soil colloids. Adsorption/desorption reactions mostly depend on the pH of the soil solution. b) Precipitation of metals, generally with anions such as phosphate, carbonate, sulphate and as hydroxides. Precipitation is more unlikely in acidic soil. c) Penetration into the crystal structure of minerals and isomorphic exchange with cations and d) Biological mobilization and immobilization. Metals bioaccumulate and biomagnify in the food chain [12]. These reactions are presumably determined and constrained by soil properties: soil texture, content of organic matter, content and type of clay minerals and Al, Fe and Mn oxides, and prevailing physicochemical conditions in the soil: soil saturation, soil aeration, pH, and redox potential. A considerable number of studies have tried to infer these relationships. Sauve *et al.* [13], for example, reported that the concentration of dissolved Pb and activity of free Pb^{2+} ions were highly significantly correlated to soil pH and total Pb content. Janssen *et al.* [14] reported that pH was in fact the only soil parameter of significance for partitioning Pb, Zn and other metals between the soil solution and soil solid phases. This association between metal adsorption on exchangeable surfaces of soil colloids and pH is due partly to competition of H^+ for adsorption sites at low pH, resulting in decreased metal adsorption [11].

Sequential extraction analysis remains a widely used procedure for identifying the metal fractionation, despite some difficulties: sensitivity to procedural variables, limited selectivity of extractants, re-adsorption of metals at different phases during extraction, and overload of the chemical system if the content of metals is too high [15, 16]. The distribution pattern of Pb in different soil fractions from contaminated sites in Mežica Valley and the Celje region in Slovenia is shown in Table 1 [17-18]. We found Pb mostly bound to soil organic matter and carbonates or remaining in the residual form. A strong association of Pb with organic matter has also been described by Kabata-Pendias [19], based on the results of many investigators. Li and Thornton [20] reported a significant association of Pb with the soil carbonate phase. As also observed before by other authors [15, 16, 21], a very small fraction of the total Pb was determined in the soil solution and exchangeable from the soil colloids to the soil solution.

Table 1. A modified analytical procedure according to Tessier et al. [22] was used to determine fractionation of Pb in soils from Mežica Valley and the Celje region in Slovenia, contaminated by the metal smelting industry.

Fractions	Mežica Valley[1] (range of 12 samples, %)	Celje region[2] (range of 30 samples, %)
soil solution	0.05-0.57	0.0-0.08
exchangeable	0.01-6.4	0.0-1.6
carbonate	4.9-67.1	2.04-43.5
Fe and Mn oxides	0.11-8.3	0.0-16.1
organic matter	31.5-74.7	35.8-71.1
residual	5.9-23.8	10.4-53.4

[1]Soils pH range 3.7-7.3
[2]Soils pH range 4.6-7.5

3. Enhanced phytoextraction

In addition to the inherent metal characteristics and soil factors, which determine fractionation and, consequently, the bioavailability of metals, plants can themselves change metal mobility. Plants can regulate metal solubility by acidification of the rhizosphere due to the extrusion of H^+ from roots and by exuding their own chelating agents, phytosiderophores and organic acids; for example, malic and citric acids. A chelating agent is a substance whose molecules can form several coordinative bonds to a single metal ion (Figure 2). A chelating agent is therefore a multidentate metal ligand and forms complexes with metals. The localized excretion of plant chelating agents mobilizes nutrients such as Zn, Fe, Mn and other metals. Water-soluble chelating agents increase and maintain metal concentration in the soil aqueous phase.

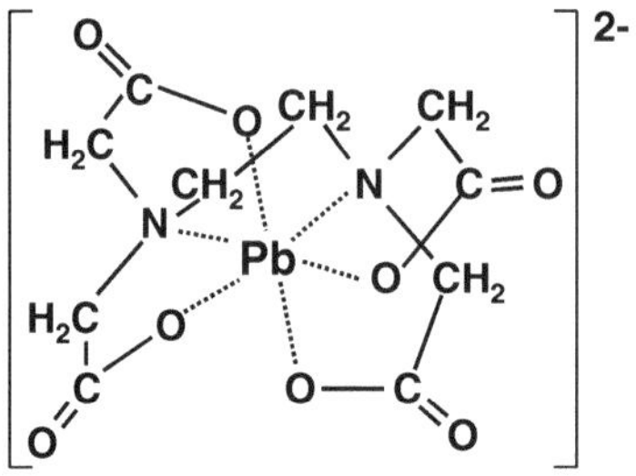

Figure 2. Pb complexes with EDTA. Dotted bonds to Pb are coordinate.

These plant mechanisms are sufficient to desorb readily bioavailable polluting metals, such as Cd, which predominantly form easily hydrolysable, low-energy bonds to the soil solid phase [23]. To make plants take up Pb and other low bioavailable metals, which are much more strongly bound to the soil solid phase than Cd, strong synthetic chelating agents have been used to bring metals into the soil solution. For example, the addition of Na salt of ethylenediamine-tetracetic acid (EDTA) to soil shifted Pb mostly from the carbonate fraction to the fraction soluble in the soil solution, in which the Pb concentration increased from 3 to 362 mg kg^{-1} (Figure 3).

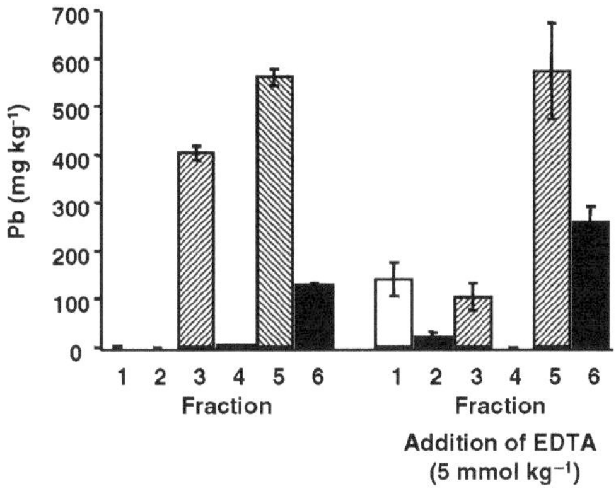

Figure 3. Fractionation of Pb in soil sample from contaminated site in Mežica Valley, Slovenia, before and after mobilization with 5 mmol kg^{-1} EDTA. Fractions: 1 soil solution, 2 exchangeable, 3 carbonate, 4 Fe and Mn oxides, 5 organic matter, 6 residual.

Theoretically, the metal-chelating efficiency of chelating agents depends on the stability constant (logK) of the metal-complex formation. Martell and Smith [24] compiled an extensive database of stability constants for different metals and chelating agents. Synthetic chelating agent such as EDTA and its structural analogues have a strong affinity for Pb and some other important metallic soil pollutants (Table 2). The resulting complexes are very stable, prevent precipitation and sorption of metals, and do not release their metal ions unless there is a significant drop in soil pH. However, the chelating agents are not specific to a particular metal in the soil and are subject to numerous interferences with other cations, most notably Fe and Ca, usually present in soil at much higher concentrations than the polluting metal [25, 26]. Furthermore, complex formation is controlled by the kinetic of all complexation (coordination) reactions, adsorption in the soil solid phases, mineral dissolution and possible degradation of the chelating agent or its metal complexes [27]. In practice, these interactions often limit the mobilization efficiency of chelating agents for a specific polluting metal in the soil.

Synthetic chelating agents, mostly EDTA, are widely used. In agriculture, for example, as ingredients of mineral fertilizers to increase the phytoavailability of Fe and other soil micronutrients, and to maintain the solubility of micronutrients in hydroponic solutions. Chelating agents are also used for *ex situ* chemical extraction of polluting metals from contaminated soils. Two technological approaches are in use; extraction of metals from soil slurry in reactors and leaching of metals from soil heaps or columns. The literature to date reports a number of chelating agents that have been tested for enhanced heavy metal phytoextraction. The ability of a chelating agent to facilitate phytoextraction does not necessarily always relate to its theoretical affinity for metals (Table 2). Blaylock *et al.* [8] compared five chelating agents for their ability to enhance Pb accumulation in the shoots of Indian mustard (*Brassica juncea*) and found EDTA to be the most effective. For Cd uptake by *B. juncea,* Blaylock *et al.* [8] tested

CDTA, DTPA, EDTA and EGTA and found the last named the most effective. Of the organic acids tested as chelating agents (acetic, citric and malic) Huang *et al.* [28] found citric acid to be the most effective in enhancing the U concentration in shoots of the same plant species. Zn binding by DTPA is so strong, that plants cannot use Zn from its complex and potentially suffer from Zn deficiency. We compared four chelating agents (citric acid, EDTA, DTPA and [S,S] isomer of EDDS) for enhanced phytoextraction of Cu into Chinese cabbage (*Brassica rapa*) and found [S,S]-EDDS to be the most efficient [29].

Table 2. Most important chelating agents tested for enhanced phytoextraction of metals from the soil and their stability constant of complex formation (logK at T 20-25 °C and ionic strength 0.1-1.0) for low bioavailable Pb, Cr and U in the soil, and readily available Fe and Ca (source: Martel and Smith [24]).

Name	Acronym	LogK				
		Pb^{2+}	Cr^{3+}	U^{4+}	Fe^{3+}	Ca^{2+}
ethylenediamine tetraacetic acid	EDTA	18.0	23.4	25.8	25.1	10.6
trans-1,2-diaminocyclohexane-*N,N,N',N'*-tetraacetic acid	CDTA	20.2	23.0	27.6	30.1	13.1
dietylenetriamine pentaacetic acid	DTPA	18.8	/	28.8	27.8	10.7
ethylenediamine dissucinic acid	EDDS	12.7	/	/	/	4.72
nitrilotriacetic acid	NTA	11.5	6.2	9,6[1]	15.9	6.3
N-hydroxyethylethylenedioamine triacetic acid	HEDTA	15.6	6.1	/	19.7	8.1
ethylenebis(oxyethylenetrinitrilo)-*N,N,N',N'* tetraacetic acid	EGTA	14.6	/	9.4[1]	20.5	10.8
citric acid	/	4.4	/	7.4[1]	11.2	3.5

[1] UO_2^{2+}

Several plants have been used in combination with chelating agents for enhanced phytoextraction. The ideal plant should be fast growing and produce a large biomass while accumulating high concentrations of polluting metals. It should also be tolerant enough to grow in contaminated soils and be resistant to the chelating agent. It should have known agronomics practice and produce usable fruit or biomass to generate some financial income after harvesting. *B. juncea* possesses several of these characteristics and is the most commonly used plant species in this remedial approach.

Chelating agents are almost exclusively used as mobilizing agents for enhanced phytoextraction. However, different mineral acids and salts have also been tested. Screening of potential mobilizing agents for radionuclide ^{137}Cs among chelating agents, reducing agents, mineral acids and salts, showed ammonium and potassium salts to be the most effective [30]. In a pot study, Lasat *et al.* [31] showed that the application of NH_4^+ (40 – 80 mmol kg^{-1}) increases the accumulation of ^{137}Cs in different plants by 2-12 times. Hammer *et al.* [32] tested sulphur as a soil amendment. Elemental sulphur is oxidized into sulphuric acid by lithoautotrophic soil microorganisms. This acidifies the soil and possibly mobilizes toxic metals. They used high biomass crops such as willow (*Salix viminalis*) to extract Cd and Zn from one calcareous and one acidic soil. However, the addition of elemental sulphur to the soil did not yield any additional benefit.

Once a metal is chelated by a chelating agent, the complex has to move from the bulk of the soil to the root's xylem. A threshold concentration of chelating agent is usually required to disrupt the physiological barriers that control metal root uptake under normal conditions [33]. Two pathways are then possible: the solution with complexed metals can enter the symplast by crossing the cell membranes, or move via the apoplast. Further translocation of metal complexes from the roots to the green parts of the plants is driven by plant transpiration. We reported that a single dose addition of 10 mmol EDTA kg^{-1} to soil contaminated with Pb, Zn and Cd increased their concentration in the aboveground biomass of *B. rapa* by 104.6, 3.2 and 2.3-times, respectively, while the concentration of the same elements in the roots was 1.7, 3.5 and 3-times lower compared to the corresponding plant tissues from control treatments [34]. These data indicate that metals are probably translocated as a complex with the chelating agent. Indeed, an ultra-structure study using a transmission electron microscope [35] located un-complexed Pb predominantly in the root tissue of *Chamaectisus palmensis* plants, while HEDTA and EDTA chelated Pb was mainly taken up by the shoots. In *B. juncea,* measurements of a Pb-EDTA complex in xylem confirmed that the majority of Pb was transported as metal complex in the transpiration stream [33, 36].

Soil contamination is seldom monometallic, and several polluting metals are usually simultaneously present in elevated concentrations. It would therefore be of great practical advantage if the use of a single chelating agent allowed phytoextraction of multi-contaminated soils. Many studies support this. For example, we reported that the addition of 10 mmol kg^{-1} [S,S]-EDDS to soil contaminated with 1100 mg Pb kg^{-1}, 800 mg Zn kg^{-1}, and 5.5 mg Cd kg^{-1}, enabled the uptake of 1053 ± 125, 211 ± 16 and 5.4 ± 0.8 mg kg^{-1} of Pb, Zn and Cd, respectively, in the biomass of hemp (*Cannabis sativa).* This was 105-, 2.3- and 31.7-times higher, respectively, than in the control treatment [37]. However, other authors have reported that the use of chelating agents such as EDTA and DTPA did not enhance, and in some cases even reduced, plant heavy metal uptake [1].

4. Efficiency of chelating agent enhanced phytoextraction

The efficiency of phytoextraction is determined by two key factors: biomass production and the metal bioconcentration factor. The bioconcentration factor is defined as the ratio of metal concentration in the plant shoot to metal concentration in the soil. For phytoextraction to be feasible, the bioconcentration factor of the plant must be greater than 1, regardless of how large the achievable biomass is. Another way to measure the efficiency of phytoextraction is by using the phytoextraction potential. This can be calculated from soil and plant metal concentrations and dry biomass plant yield, as the total amount of metal extracted per ha of soil, in a single phytoextraction cycle, and expressed as kg ha^{-1}. In order to be economically viable, plants for Pb phytoextraction should be able to accumulate at least 10,000 mg Pb kg^{-1} in their green parts (harvesting the roots is not practical) and achieve a dry biomass of 20 t ha^{-1} [7].

Shen et al. [38] used 3.0 mmol kg^{-1} EDTA to treat soil from a mining site in Hong Kong, heavily contaminated with over 10,000 mg kg^{-1} of Pb. Application of EDTA in three separate doses was the most effective and enabled the Pb concentration in the shoots of *B. rapa* to exceed 5000 mg kg^{-1} of dry plant biomass. As explained above, transpiration is believed to be a major force that drives heavy metal accumulation in plant shoots. It has been shown that EDTA soil treatment is toxic for plants and could decrease the plant transpiration rate [33]. Applying EDTA in three separate additions Shen *et al.* [38], therefore, presumably minimized its adverse effect on the transpiration rate. Barosci *et al.* [39] applied EDTA in multiple doses to provide time for plants to initiate their adaptation mechanisms and raise their damage threshold against EDTA phytotoxicity. In contrast to these results, we found application of EDTA in multiple doses to be less effective for phytoextraction of Pb, Zn and Cd by *B. rapa.* [34].

Other authors have reported even higher plant metal concentrations induced by chelating agent application. Blaylock *et al.* [8] used 3 week-old seedlings and measured more than 15,000 mg Pb kg^{-1} in the dry weight of shoots of *B. juncea* after a 10 mmol kg^{-1} EDTA addition. Huang *et al.* [36] determined 8960 mg kg^{-1} and 2410 mg Pb kg^{-1} in two-week old pea and corn shoots transplanted into a soil substrate pre-treated with 1.5 mmol EDTA kg^{-1}. However, experimental conditions were used in these studies that would be unrealistic in field conditions and full-scale remediation. These include the use of very young plants, a Pb soil fractionation that was favourable for phytoextraction and achieved by "artificial" soil contamination with water-soluble metal salts, and an experimental set up in which no losses of the Pb complex due to leaching occurred.

Since phytoextraction is a long-term technology, it is imperative to keep areas undergoing phytoremediation productive to achieve economically viable and socially acceptable decontamination. Industrial plants, i.e. energy crops or crops for bio-diesel production, are therefore the prime candidates for phytoextraction plants. The use of energy and/or bio-diesel crops for metal phytoextraction would give contaminated soil a productive value and decrease remediation costs. Furthermore, if the phytoextraction cost falls within the margin of interest, or even turns a profit, then the time needed for the operation becomes less important. We evaluated the Pb, Zn and Cd phytoextraction potential of a selection of potential energy plants (*C. sativa, Sorghum vulgare, Arundo donax*), bio-diesel plants (*Brassica napus, Raphanus sativus oleiformis, Sinapis alba*) and other plants (*Amaranthus spp., Linum usitatissimum, Trifolium pratense, Trifolium repens, Medicago sativa, Zea mays, B. rapa*) [37]. Chelating agents EDTA and [S,S]-EDDS were applied when some plants were in the juvenile vegetative and some in the adult vegetative growth phase. The most effective was treatment with 10 mmol kg^{-1} [S,S]-EDDS and *C. sativa,* in which the Pb phytoextraction potential reached 26.3 kg ha^{-1}. We used literature data for plant biomass yields in our calculation and potentials were thus probably overestimated. Older, full-grown plants are likely to concentrate less heavy metals than younger ones. With the obtained phytoextraction potential, the percentage of Pb phytoextracted in a single cycle was only approx. 0.6% of the total Pb present in the upper 30 cm of soil. The achieved Pb concentration in *C. sativa* was clearly far from concentrations required for efficient soil remediation within a reasonable time span. Pb concentrations 10-times higher than actually obtained (exceeding 10,000 mg kg^{-1} dry biomass) would be required to reduce soil Pb concentrations from an initial 1100 to 300 mg kg^{-1} Pb (the regulatory limit set by European Union Council, directive 86/278/EEC) in approx. 10-15 years.

There have been very few field demonstrations of chelating agent enhanced phytoextraction. Blaylock [40] showed a significant decrease in soil Pb concentration over two years at two sites in the United States using *B. juncea* and EDTA. However, the possible leaching of the Pb complex through the soil profile was not determined and it therefore remains uncertain how much Pb the plants really removed. It is well known that EDTA and other chelating agents act as chemical ploughs, redistributing surface contamination from the surface to lower soil horizons, which gives rise to concerns about the environmental safety of enhanced phytoextraction.

5. Concerns relating to the use of chelating agents for enhanced phytoextraction

The ideal chelating agent for enhanced phytoextraction should be specific for targeting metal contaminants, water soluble in its free form to allow easy soil application, and be able to form more lipophilic metal complexes, easily absorbed by plants, and thus decrease the risk of leaching. It should also be non-toxic and inexpensive. EDTA, the most widely tested chelating agent, is not ideal for a number of reasons. EDTA forms highly soluble complexes with Pb and other polluting metals, which can therefore be leached from the soil to the groundwater. EDTA is toxic, especially in its free form [41, 42], and poorly photo-, chemo- and biodegradable in soil [43]. A combined widespread use in fertilizers and slow decomposition has already led to background concentrations of EDTA in European surface waters in the range 10-50 mg L^{-1} [44].

The increase of soluble metal complexes in pore water following the application of chelating agents raises health, safety and environmental concerns. In a soil column study, we examined the effect of EDTA addition and different watering regimes on the uptake of Pb, Zn and Cd by *B. rapa*, the leaching of polluting metals and the toxicity effect of EDTA additions on plants and soil microfauna. The most effective was a single dose of 10 mmol EDTA kg^{-1} soil, whereby 0.06% of soil Pb, 0.03% of Zn and 0.13% of Cd were extracted into shoots, whereas up to 38% of initial total Pb in the soil, 10.5% of Zn and 56% of Cd were leached down the soil profile. EDTA addition had a strong phytotoxic effect on red clover (*Trifolium pratense*) and inhibited the development of arbuscular mycorrhiza on plants. The results of phospholipid fatty acid analyses indicated a toxic effect of EDTA addition on soil fungi and increased stress of soil microfauna in general, indicated by increased fatty acids trans/cis ratio [34]. Soil microorganisms depend directly on the soil solution for uptake of food, and elevated heavy metal concentrations might be responsible for the toxic effects. Soil microorganisms are largely responsible for important soil processes, such as mineralization and synthesis of soil organic matter, soil micro-aggregate formation, nitrification and denitrification etc., and disturbed microbial activity could thus effect soil functioning. Soil microorganisms are also at the base of the soil food chain. Römkens *et al.* [45] reported that the number of microbivorous nematodes was greatly reduced during EGTA-enhanced phytoextraction, presumably due to a smaller availability of food. Another hazard is the potential contamination of the food chain if animals graze on the metal contaminated phytoextracting vegetation.

Several other studies investigating phytoextraction enhanced by chelating agents have highlighted the risk of possible mobilization of Pb and other toxic metals from soil to groundwater [9, 46]. Wenzel *et al.* [47] used canola (*B. napus*) in a pot and out-door lysimeter experiment and also reported that leaching losses of Cu, Pb and Zn far exceeded the amounts of metal taken up by plants after EDTA was applied. These results indicate that the application of EDTA and other biologically persistent chelating agents may be limited to field conditions in which soil containment and hydrological control of the metal–enriched leachates can be safely achieved, for example to sites where the connection to receiving water has been "broken". In soils that are not hydrologically isolated, control over metal leaching could be possible by maintaining a neutral or negative soil water balance. If a chelating agent is applied and soil irrigation adjusted to natural precipitation in such a way that losses of soil water due to plant transpiration and evaporation are higher or equal to soil water gains, then leaching of metal complexes should in principle not occur. In practice, however, the control of natural factors affecting the soil water balance (precipitation, evapotranspiration) would be very difficult to achieve. To maximize Pb accumulation by plants and reduce the environmental risk of leaching, Epstein *et al.* [48] proposed that the chelating agent application rate should be selected that maximizes the concentration of complexed Pb, based on the extractability of Pb by the chelating agent. For example; Shen *et al.* [38] reported that application of EDTA in three separate doses was not just more effective in enhancing the accumulation of Pb in *B. rapa*, but also reduced mobility and the potential risk of soluble Pb movement into the groundwater.

To extend the time available for plants to accumulate heavy metals and in this way to reduce the leaching of heavy metals from the soil, we tested controlled release as an alternative method of chelating agent application [49]. Controlled release pesticide formulation was first proposed in agriculture to reduce pesticide leaching while maintaining control of pests in soil. We entrapped EDTA in different hydrogel carriers and used it in a column experiment with *B. rapa*. EDTA in acrylamide, starch, and carrageenan hydrogel granules increased Pb accumulation in a test plant by up to 9.4-times compared to the control. However, the addition of the corresponding amount of EDTA in water solution (5 mmol kg^{-1} in four separate doses) was more effective and increased the Pb plant concentration by 26.7-times. Controlled release of EDTA was not particularly successful in reducing the Pb leaching, either; 5.4 to 23.6% of initial total Pb was leached through the soil profile. EDTA applied in water solution leached 49.6 % of the total initial Pb.

6. Use of the biodegradable chelating agents for enhanced phytoextraction

The use of biodegradable instead of biologically persistent chelating agents for enhanced phytoextraction could curb off-site migration of their heavy metal complexes. For example, NTA is biodegradable in both aerobic (where it degrades as fast as glucose and citric acid) and anaerobic soil conditions [50]. Kulli *et al.* [46] used NTA for enhanced phytoextraction of soil contaminated with Cd (2 mg kg^{-1}), Cu (530 mg kg^{-1}) and Zn (700 mg kg^{-1}), with lettuce and Italian ryegrass as test plants. At the highest NTA dose (5.3 mol m^{-2} soil), the metal concentration in the aboveground plant biomass was 4 to 24 times greater than in the control plants. At this NTA dose, plant growth was almost completely inhibited. Severe visual symptoms indicated metal toxicity as the likely cause. Citric acid used for enhanced U phytoextraction could also represent a useful alternative to more persistent chelating agents. It induces very rapid accumulation of U in plants. Huang *et al.* [28] reported that shoot U concentrations of *B. juncea* and *Brassica chinensis* grown in a U-contaminated soil (750 mg kg^{-1}) increased from less than 5 mg kg^{-1} to more than 5000 mg kg^{-1} in citric acid-treated soils. Citric acid complex with U is degraded in soil within a few days. The resulting readsorption of residual U, not extracted by the plants, in the soil reduces the environmental hazard related to potential leaching of U to groundwater.

The [S,S]-isomer of EDDS is a particularly interesting chelating agent because it combines high biodegradability with high chelating strength. [S,S]-EDDS was first isolated as a metabolite of the soil actinomycete *Amycolatopsis orientalis* [51] and is naturally present in the soil. In addition, the environmental risk of its use in detergent application has already been assessed [52]. The toxicity to fish and daphnia was low (EC_{50} > 1000 mg/L). The [S,S]-EDDS is readily biodegradable using the criteria stipulated by the Organization for Economic Co-operation and Development (OECD). The OECD criteria state that 60% of the compound must biodegrade within 28 days. For [S,S]-EDDS, the final CO_2 yield exceeded 80% after 20 days, assessed by the modified Sturm test [53, 54]. [S,S]-EDDS biodegradation in a concentration of 0.0034 mmol kg^{-1} occurred rapidly in various environmental compartments. In unacclimated sludge amended soil, the half-life of [S,S]-EDDS was ca. 2.5 days and mineralization was completed in 28 days. [S,S]-EDDS was transformed into benign degradation products, first into N-(2-aminoethyl) aspartic acid, which significantly decreased the chelating capacity [54, 55].

To test the feasibility of using [S,S]-EDDS in enhanced phytoextraction, we compared its efficiency against a benchmark chelating agent EDTA [56]. Applied in a single 10 mmol kg^{-1} dose, both EDTA and [S,S]-EDDS were almost equally effective in increasing the concentrations of Pb (94.2 and 102.3-fold) and, to a lesser extent, also of Zn and Cd in the leaves of the test plant *B. rapa*. In separate doses, EDTA was more effective than [S,S]-EDDS, but caused leaching of approx. 22% of Pb, while [S,S]-EDDS leached only 0.8% of initial total Pb concentrations. A biotest with red clover (*Trifolium pratense*) indicated a greater phytotoxic effect of EDTA than [S,S]-EDDS addition. EDDS was also less toxic to soil fungi, as determined by PLFA analysis, and caused less stress to soil microorganisms, as indicated by the trans/cis PLFA ratio. [S,S]-EDDS and EDTA (5 mmol kg^{-1}, single dose) equally effectively promoted Zn and Cd uptake by oilseed rape (*B. napus*), amaranth (*Amaranthus spp.*), Chinese cabbage (*B. rapa*) and hemp (*C. sativa*) [37]. Generally, [S,S]-EDDS was also less efficient for Pb plant uptake, except for *C. sativa,* where the Pb biomass concentration was higher than in the EDTA treatment by 42%. The stability constant for complexes with Pb is substantially higher for EDTA than for [S,S]-EDDS (Table 2). These data therefore illustrate that data on logK does not provide sufficient information on the potency of specific chelating agents for enhanced phytoextraction. The efficiency of chelating agents for phytoextraction seems to be plant-specific, as well as being controlled by the stability constant and soil conditions, as already described above.

To retain the chelating agent solution in the topsoil and thus (i) improve chelating agent availability for Pb mobilisation and plant uptake, and (ii) reduce leaching of Pb-chelate complexes by prolonging the retention time available for [S,S]-EDDS complexes biodegradation in soil, we modified the soil water holding capacity by using synthetic acrylamide hydrogel [57]. Gel-forming soil conditioners are known from agronomy and forestry practice to be effective in increasing the soil water holding capacity, decreasing deep percolation, and minimising losses of water solution through leaching. The addition of 0.2% (w/w) of hydrogel amendment increased soil field water capacity from an initial 24.6% to 31.3%. The use of 5 mmol kg^{-1} [S,S]-EDDS in hydrogel amended soil increased Pb uptake by *B. rapa* by 18 times while only 0.2% of total initial Pb was leached through the soil profile. In the control, soil without hydrogel [S,S]-EDDS leached 1.2% of Pb. EDTA was more effective for phytoextraction but caused much higher Pb leaching in all treatments, under any of the soil water sorption condition tested. Using a higher 10 mmol kg^{-1} [S,S]-EDDS dose in hydrogel amended soil significantly reduced plant Pb uptake and increased Pb leaching to up to 44.2% of initial soil content. This was presumably caused by the toxic effect of high concentrations of [S,S]-EDDS on soil microorganisms (slower biodegradation) and plants (lower Pb uptake). In all treatments in which 10 mmol kg^{-1} EDTA or [S,S]-EDDS was applied, visual symptoms (necrotic lesions on the leaves of *B. rapa*) of toxicity were observed. The effect was much less pronounced in treatments with lower amounts of chelating agents.

A serious limitation of [S,S]-EDDS based enhanced phytoextraction is the high price of the chelating agent. The current price for 1 ton of [S,S]-EDDS is approx. 5000 GBP. As [S,S]-EDDS has been substituted for traditional chelating agents in a number of commercial products, e.g. industrial detergents, the price is expected to decrease. In future, the biosynthesis of EDDS by *A. orientalis*, which produces a chelating agent exclusively in the biodegradable S,S-configuration, instead of the current chemical synthesis, could significantly reduce the production costs [58].

7. Horizontal permeable reactive barriers in enhanced phytoextraction

To further reduce the hazard of heavy metal leaching and off-site migration from treated soil to the environment, we proposed the use of biodegradable chelating agents together with horizontal permeable reactive barriers placed below the contaminated topsoil [59]. A flowsheet of the method is shown in Figure 4. The construction and reactive materials in the barriers are a layer of nutrient enriched substrate with a high water sorption capacity and with extensive surfaces on which microbial films could form, and a layer of absorbent (apatite) for precipitation of metals. The main purpose of the substrate is to enhance the microbial degradation of the metal complexes. Once metal ions are released from the complex they are bound by absorbent, become immobilized and are no longer subject to leaching. After completion of the soil remediation process, the barrier can be excavated and removed.

We borrowed the concept of horizontal permeable reactive barriers from polluted groundwater remediation. Here, a vertical permeable reactive barrier is constructed below ground as a vertical underground wall, filled with reactive materials. The barrier is built by digging a long, narrow trench in the path of the polluted groundwater. Clean groundwater flows out of the other side of the wall. Reactive materials in the barrier trap harmful chemicals or change the chemicals into harmless ones. For example, zero-valent iron can be used for the reduction of toxic Cr^{6+} to harmless Cr^{3+}, and limestone for Pb precipitation.

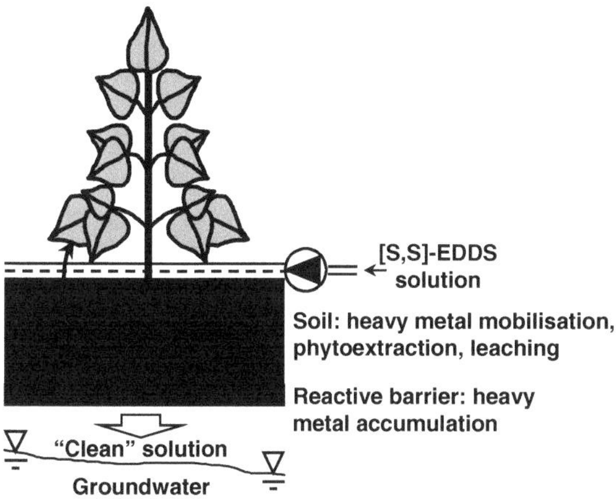

Figure 4. Conceptual representation of combined enhanced phytoextraction of heavy metals with biodegradable chelating agent and in situ soil washing using horizontal permeable reactive barriers.

In our first study, barriers were placed 20 cm deep in soil columns and tested for their ability to prevent leaching of Pb during enhanced phytoextraction using [S,S]-EDDS, EDTA and *B. rapa* [59]. The reactive materials in the barriers were nutrient enriched vermiculite, peat or agricultural hydrogel, and apatite as absorbent. EDTA and [S,S]-EDDS addition (10 mmol kg^{-1}) increased Pb concentrations in the test plant by 158 and 89 times compared to the control. In EDTA treatments, approx. 25% of total initial soil Pb was leached in a single cycle of chelating agent addition. This was expected, since EDTA is nonbiodegradable, so none of the reactive materials in the barriers were effective. The biodegradability of the chelating agent and metal complexes and binding capacity of adsorptive materials in the barrier for released heavy metals are essential for this type of barrier to function. In [S,S]-EDDS treatments, 20% of the initial Pb was leached from columns with no barrier, while barriers with vermiculite or hydrogel and apatite decreased leaching by more than 60 times, to 0.35%. In total 11.6% of initial Pb was removed from the soil above the barrier with vermiculite and apatite. However, almost all removed Pb was simply washed from the soil and accumulated in the barrier, only a fraction (0.03%) of Pb was phytoextracted. These results pointed toward the possibility of enhanced phytoextraction combined with *in situ* soil washing. With current plants, however, phytoextraction can be expected to make only a minor contribution.

Most soils are multi-contaminated, so we tested the feasibility of using reactive barriers for remediation of soil contaminated simultaneously with Pb, Zn and Cd. Hemp (*C. sativa*) was used as the phytoextracting plant [60]. The addition of [S,S]-EDDS (10 mmol kg^{-1} dry soil) yielded concentrations of 1026 ± 442, 330 ± 114 and 3.8 ± 1.5 mg kg^{-1} of Pb, Zn and Cd in the dry above-ground plant biomass, respectively. These concentrations were 1926-, 7.5-, and 11- times higher, respectively, compared to treatments with no chelating agent addition. Horizontal permeable barriers, composed of a 3 cm high layer of nutrient enriched sawdust and vermiculite and a 3 cm layer of soil mixed with vermiculite and apatite, were positioned at different soil depths. The barrier placed 30 cm deep reduced leaching of Pb, Zn and Cd by 435-, 4- and 53-times, respectively, compared to columns with no barrier. The lower positioned barrier did not prevent leaching of Zn. In total, 2.5% of initial Pb, 7.3% of Zn and 2.8% of Cd was removed from the contaminated soil in a single [S,S]-EDDS treatment and mostly accumulated in the barrier. The contribution of *in situ* soil washing prevailed over enhanced phytoextraction, despite a marked increase in Pb plant uptake induced by [S,S]-EDDS. The relative inefficiency of the barrier to prevent leaching of Zn indicates either a lower biodegradability of the Zn-[S,S]-EDDS complex or poor adsorption of released Zn ions. Indeed, Van Devivere *et al.* (2001) reported that the microbial degradation of Pb-[S,S]-EDDS was much faster than that of Zn-[S,S]-EDDS, and while apatite is an effective absorption material for Pb by conversion into pyromorphite, a poorly soluble Pb phosphate mineral [61], the sorption mechanisms of apatite for metals other than Pb are less clear.

In the follow-up study, the treatment of a vineyard soil with 5 mmol kg^{-1} [S,S]-EDDS increased the accumulation of Cu in *B. rapa* by 3.3-times over the control. The reactive barriers composed of nutrient enriched sawdust and apatite were effective and only 0.53±0.32% of the initial Cu was leached, while 36.7% of Cu was washed from the layer of contaminated soil and accumulated in the barriers [29]. Again, these results indicate that with current plants, the use of reactive barriers could be more justified for controlled and environmentally safe *in situ* soil washing of metals rather than for enhanced phytoextraction.

8. Conclusions

The use of hyperaccumulating plants and continuous phytoextraction is limited to situations in which all of the pollutants present are bioavailable and can be tolerated by plants. At present, it is difficult to judge whether chelating agent assisted phytoextraction can be economically feasible and environmentally sustainable. On the other hand, it is the only phytoextraction option for soils contaminated with highly insoluble metals such as Pb and U. The use of both phytoextraction methods is limited by the climatic and geologic conditions of the site to be cleaned.

The risk of metal leaching is one of the most important limitations of enhanced phytoextraction. We showed that it could be efficiently prevented by the use of a biodegradable chelating agent and reactive barriers. Further research might provide a new environmentally safe mobilizing agent, more effective in enhancing metal accumulation in plants than currently known agents. New methods of chelating agent application, such as alternative formulations allowing the controlled release, could be developed. Another possible option is the use of plants genetically modified to exude strong, metal-selective mobilizing compounds in their rhizosphere. Indeed, environmentally safe methods of enhanced heavy metal phytoextraction must be fully developed and tested before steps towards further commercialization of this remediation technology are attempted.

Since it is the bioavailable fraction and not the total concentration of metals in soil that interact with biological targets and pose an environmental and health threat, phytoextraction research should focus on stripping the bioavailable metal fractions, instead of trying to reduce the concentration of metals in soil below limits set by legislation. Nevertheless, to allow remediation within a reasonable period, the plant heavy metal uptake must be enhanced dramatically. This may be achieved by engineering high-biomass producing plants with as yet unidentified hyper-accumulating genes. One possible strategy to increase heavy metal plant uptake is to increase translocation of chelate complexes through the plant transpiration stream. This could be achieved using high-transpiration plants. For example, Gleba *et al.* [62] reported that a high-transpiration species of *B. juncea* phytoextracted 104 % more Pb than the wild-type plant.

Plants exhibiting a high capacity for heavy metal uptake when assisted by chelating agents, i.e. *B. juncea,* are also generally not very resistant to high levels of Pb and other heavy metals in their foliage. This requires harvesting soon after chelating agent application. There are some indications that the ability to hyperaccumulate heavy metals is the result of high resistance to the metals rather than greater rates of metal uptake [63]. Vacuolar sequestration is likely to be a key component of metal tolerance and hyperaccumulation. To increase resistance to metal by genetic manipulation is therefore another potential approach to improve the efficiency of phytoextraction. Freeman *et al.* [64] reported that glutathione biosynthesis might play a role in Ni tolerance in hyperaccumulator *Thlaspi goesingense.* Meagher [65] reported that bacterial genes encoding Hg reductase and organomercurial lyase were successfully transferred into the genome of several plants, including poplar, and that transformed plants showed increased tolerance to Hg.

Although fast progress is being made and studies have demonstrated the power of genetic engineering, more knowledge of the molecular mechanisms responsible for plant metal accumulation, root to shoot transfer and vacuolar sequestration, is necessary before the genetic traits can be transferred into high biomass plants. Practical aspects of enhanced heavy metal phytoextraction also need further research. These might include optimisation of agronomic practices (fertilization, crop protection, harvesting, irrigation) and development of economically feasible techniques for the disposal of metal-enriched plants or, when practical, for metal recovery. Finally, apart from deeper basic research, more pilot-scale outdoor studies and field trials are needed objectively to evaluate the real potential and feasibility of enhanced heavy metal phytoextraction.

References

[1] Alkorta, I; Hernandez-Allica, J; Becerril, JM; Amezaga, I; Albizu, I; Onaindia, M and Garbisu, C (2004) Chelate-enhanced phytoremediation of soils polluted with heavy metals. Rev Environ Sci Biotech 3: 55-70

[2] Salt, DE; Smith, RD and Raskin, I (1998) Phytoremediation. Annu Rev Plant Physiol Plant Mol Biol 49: 643-668

[3] Watanabe, ME (1997) Phytoremediation on the brink of commercialization. Environ Sci Technol 31: 182-186

[4] Cunningham, SC; Berti, WR and Huang, JW (1995) Phytoremediation of contaminated soils. TIBtech 13: 393-397

[5] Puschenreiter, M; Stoger, G; Lombi, E; Horak, O and Wenzel, WW (2001) Phytoextraction of heavy metal contaminated soils with *Thlaspi goesingense* and *Amaranthus hybridus*: Rhizosphere manuipulation using EDTA and ammonium sulfate. J Plant Nut Soil Sci 164: 615-621
[6] Finžgar, N; Kos, B and Leštan, D (2005) Heap leaching of lead contaminated soil using biodegradable chelator [S,S]-ethylenediamine disuccinate. Environ Technol 26: 553-560
[7] Huang, JW and Cunningham, SD (1996) Lead phytoextraction: species variation in lead uptake and translocation. New Phytol 134: 75-84
[8] Blaylock, MJ; Salt, DE; Dushenkov, S; Zakharova, O; Gussman, C; Kapulnik, Y; Ensley, BD and Raskin, I (1997) Enhanced accumulation of Pb in Indian mustard by soil-applied chelating agents. Environ Sci Technol 31: 860-865
[9] Cooper, EM; Sims, JT; Cunningham, SD; Huang, JW and Berti, WR (1999) Chelate-assisted phytoextraction of lead from contaminated soils. J Environ Qual 28: 1709-1719
[10] Pichtel, J; Kuroiwa, K and Sawyer, HT (2000) Distribution of Pb, Cd and Ba in soils and plants of two contaminated sites. Environ Pollut 110: 171-178
[11] Rieuwerts, JS; Thornton, I; Farago, ME and Ashmore, MR (1998) Factors influencing metal bioavailability in soils: preliminary investigations for the development of a critical loads approach for metals. Chem Spec Bioavail 10: 61-75
[12] Levy, DB; Barbarick, KA; Siemer, EG and Sommers LE (1992) Distribution and partitioning of trace metals in contaminated soil near Leadville, Colorado. J Environ Qual 21: 185-195
[13] Sauve, S; McBride, MB and Hedershot, WH (1997) Speciation of lead in contaminated soils. Environ Pollut 98: 149-155
[14] Janssen, RPT; Peijnenburg, WJGM; Posthuma, L and Van Den Hoop, MAGT (1997) Equilibrium partitioning of heavy metals in Dutch field soils. 1. Relationship between metal partitioning coefficient and soil characteristics. Environ Toxicol Chem 16: 2470-2478
[15] Ramos, L; Hernandez, LM and Gonzales, MJ (1994) Sequential fractionation of copper, lead, cadmium and zinc in soil from or near Donana national park. J Environ Qual 23; 50-57
[16] Chlopecka, A; Bacon, JR; Wilson, MJ and Kay, J (1996) Heavy metals in the environment. J Environ Qual 25: 69-79
[17] Leštan, D and Grčman, H (2001) Speciation of lead, zinc and cadmium in contaminated soils from Mežica Valley. Zb Bioteh Fak 77: 205-214
[18] Leštan, D; Grčman, H; Zupan, M and Bačac, N (2003) Relationsip of soil properties to fractionation of Pb and Zn in soil and their uptake into *Plantago lanceolata*. Soil Sediment Contam 12: 507-522
[19] Kabata-Pendias, A (2000) Trace Elements in Soils and Plants. CRC Press: Boca Raton. 432
[20] Li, X and Thornton, I (2001) Chemical partitioning of trace and major elements in soils contaminated by mining and smelting activities. Appl Geochem 16: 1693-1706
[21] Maiz, I; Arambarri, I; Garcia, R and Millan, E (2000) Evaluation of heavy metal availability in polluted soils by two sequential extraction procedures using factor analysis. Environ Pollut 110: 3-9
[22] Tessier, A; Campbell, PGC and Bisson, M (1979) Sequential extraction procedure for the speciation of particulate trace metals. Anal Chem 51: 844-851
[23] Alloway, BJ (1995) Cadmium, in Heavy Metals in Soils, BJ Alloway, Ed., Blackie Academic and Professional: Glasgow. 368
[24] Martell, AE and Smith, RM (2003) NIST critically selected stability constants of metal complexes; Version 7.0. NIST: Gaithersburg.
[25] Sun, B; Zhao, FJ; Lombi, E and McGrath, SP (2001) Leaching of heavy metals from contaminated soils using EDTA. Environ Pollut 113: 111-120
[26] Theodoratos, P; Papassiopi, N; Georgoudis, T and Kontopoulos, A (2000) Selective removal of lead from calcareous polluted soils using the Ca-EDTA salt. Water Air Soil Poll 122: 351-368
[27] Nowack, B (2002) Environmental chemistry of aminopolycarboxylate chelating agents. Environ Sci Technol 36: 4009-4016
[28] Huang, JW; Blaylock, MJ; Kapulnik, Y and Ensley, BD (1998) Phytoremediation of uranium-contaminated soils: role of organic acids in triggering uranium hyperaccumulation in plants. Environ Sci Technol 32: 2004-2008
[29] Kos, B and Leštan, D (2003) Chelator induced phytoextraction and in situ soil washing of Cu. Environ Pollut 132: 333-339
[30] McGrath, SP; Zhao, FJ and Lombi, E (2002) Phytoremediation of metals, metalloids and radionuclides. Advances in Agronomy 75: 1-56

[31] Lasat, MM; Norvell, WA and Kochian, LV (1997) Potential of phytoextraction of ^{137}Cs from a contaminated soil. Plant Soil 195: 99-106
[32] Hammer, D; Kayser, A and Keller, C (2003) Phytoextraction of Cd and Zn with *Salix viminalis* in field trials. Soil Use Manag 19: 187-192
[33] Vasssil, AD; Kapulnik, Y; Raskin, I and Salt, DE (1998) The role of EDTA in lead transport and accumulation by Indian mustard. Plant Physiol 117: 447-453
[34] Grčman, H; Velikonja-Bolta, Š; Vodnik, D; Kos, B and Leštan, D (2001) EDTA enhanced heavy metal phytoextraction: metal accumulation, leaching and toxicity. Plant Soil 235: 105-114
[35] Jarvis, MD and Leung, DWM (2001) Chelated lead transport in *Chamaecytisus proliferus* (L. f.) ling ssp. *proliferus* var. *palmensis* (H. Chris): an ultrastructural study. Plant Sci 161: 433-441
[36] Huang, JW; Chen, J; Berti, WR and Cunningham, SD (1997) Phytoremediation of lead contaminated soils: Role of synthetic chelates in lead phytoextraction. Environ Sci Technol 31: 800-805
[37] Kos, B; Grčman, H and Leštan, D (2003) Phytoextraction of lead zinc and cadmium from soil by selected plants. Plant Soil Environ 49: 548-553
[38] Shen, Z-G; Li, X-D; Wang, C-C; Chen, H-M and Chua, H (2002) Lead phytoextraction from contaminated soil with high-biomass plant species. J Environ Qual 31: 1893-1900
[39] Barocsi, A; Csintalan, Z; Kocsanyi, L; Duschenkov, S; Kuperberg, JM; Kucharski, R and Richter, PI (2003) Optimizing phytoremediation of heavy-metal contaminated soil by exploring plants stress adaptation. Int J Phytoremediation 5: 13-23
[40] Blaylock, MJ (2000) Field demonstration of phytoremediation of lead contaminated soils, in Phytoremediation of Contaminated Soil and Water, N Terry and TN Banuelos, Eds., Lewis Publishers, Boca Raton. 389
[41] Dirilgen, N (1998) Effects of pH and chelator EDTA on Cr toxicity and accumulation in *Lemma minor*. Chemosphere 37: 771-783
[42] Sillanpaa, M and Oikari, A (1996) Assessing the impact of complexation by EDTA and DTPA on heavy metal toxicity using Microtox bioassay. Chemosphere 32: 1485-1497
[43] Nörtemann, B (1999) Biodegradation of EDTA. Appl Microbiol Biotechnol 51: 751-759
[44] Kari, FG; Hilger, S and Canonica, S (1995) Determination of the reaction quantum yield for the photochemical degradation of Fe(III)-EDTA: Implications for the environmental fate of EDTA in surface waters. Environ Sci Technol 29: 1008-1017
[45] Römkens, P; Bouwman, L; Japenga, J and Draisma, C (2002) Potentials and drawbacks of chelate-enhanced phytoremediation of soils. Environ Pollut 116: 109-121
[46] Kulli, B; Balmer, M; Krebs, R; Lothenbach, B; Geiger, G and Schulin, R (1999) The influence of nitrilotriacetate on heavy metal uptake of lettuce and ryegrass. J Environ Qual 28: 1699-1705
[47] Wenzel, WW; Unterbrunner, R; Sommer, P and Sacco, P (2003) Chelate-assisted phytoextraction using canola (*Brassica napus* L) in outdoors pot and lysismeter experiments. Plant Soil 249: 83-96
[48] Epstein, AL; Gussman, CD; Blaylock, MJ; Yermiyahu, U; Huang, JW; Kapulnik, Y and Orser, CS (1999) EDTA and Pb-EDTA accumulation in *Brassica juncea* grown on Pb-amended soil. Plant Physiol 208: 87-94
[49] Grčman, H and Leštan, D (2003) Use of hydrogels in EDTA induced Pb phytoextraction. Fresen Environ Bull 12, 1044-1049
[50] Tiedje, JM and Mason, BB (1974) Biodegradation of nitrilotriacetate (NTA) in soil. Soil Sci Soc Am Proc 38: 278-283
[51] Nishikiori, T; Okuyama, A; Naganawa, T; Takita, T; Hamida, M; Takeuchi, T; Aoyagi, T and Umezawa, H (1984) Production of actinomycetes (S,S)-N,N'-ethylenediamine-dissuccinic acid, an inhibitor of phospholipase. C J Antibiot 37: 426-427
[52] Jaworska, JS; Schowanek, D and Feijtel, TCJ (1999) Environmental risk assessment for trisodium [S,S]-ethylene diamine disuccinate, a biodegradable chelator used in detergent applications. Chemosphere 38: 3597-3625
[53] Jones, PW and Williams, DR (2001) Chemical speciation used to assess [S,S]-ethylendiamine-disuccinic acid (EDDS) as a readily-biodegradable replacement for EDTA in radiochemical decontamination formulations. Appl Radiat Isotopes 54: 587-593
[54] Schowanek, D; Feijtel, TCJ; Perkins, CM; Hartman, FA; Federlen, TW and Larson, RJ (1997) Biodegradation of [S,S], [R,R] and mixed stereoisomers of ethylene diamine disuccinic acid (EDDS), a transition metal chelator. Chemosphere 34: 2375-2391

[55] Bucheli-Witschel, M and Egli, T (2001) Environmental fate and microbial degradation of aminopolycarboxylic acids. FEMS Microb Reviews 25: 69-106

[56] Grčman, H; Vodnik, D; Velikonja-Bolta, Š and Leštan D (2003) Ethylenediamine-disuccinate as a new chelate for environmentally safe enhanced lead phytoextraction. J Environ Qual 32: 500-506

[57] Kos, B and Leštan, D (2003) Influence of a biodegradable ([S,S]-EEDS) and nondegradable (EDTA) chelate and hydrogel modified soil water sorption capacity on Pb phytoextraction and leaching. Plant Soil 253: 403-411

[58] Zwicker, N; Theobald, U; Zähner, H and Fiedler, HP (1997) Optimization of fermentation conditions for the production of ethylene-diamine-disuccinic acid by *Amycolatopsis arientalis*. J Ind Microbiol Biotechnol 19:280-285

[59] Kos, B and Leštan D (2003) Induced phytoextraction/soil washing of lead using biodegradable chelate and permeable barriers. Environ Sci Technol 37: 624-629

[60] Kos, B and Leštan D (2004) Soil washing of Pb, Zn and Cd using biodegradable chelator and permeable barriers and induced phytoextraction by *Cannabis sativa*. Plant Soil 263: 43-51

[61] Laperche, V; Traina, SJ; Gaddam, P and Logan TJ (1996) Chemical and mineralogical characterization of Pb in a contaminated soil: reactions with synthetic apatite. Environ Sci Technol 30: 3321-3326

[62] Gleba, D; Borisjuk, NV; Borisjuk, LG; Kneer, R; Poulev, A; Skarzhinskaja, M; Dushenkov, S; Logendra, S; Gleba, YY and Raskin, I (1999) Use of plant roots for phytoremediation and molecular framing. Proc. Natl Acad Sci USA 96: 5973-5977

[63] Kramer, U; Smith, RD; Wenzel, WW; Raskin, I and Salt, DE (1997) The role of metal transport and tolerance in nickel hyperaccumulation by *Thlaspi goesingense* Halacsy. Plant Physiol 115: 1641-1650

[64] Freeman, JL; Persans, MW; Nieman K; Albrecht, C; Peer, W; Pickering, IJ and Salt, DE (2004) Increased glutathione biosynthesis plays a role in nickel tolerance in *Thlaspi* nickel hyperaccumulators. Plant Cell 16: 2176-2191

[65] Meagher, RB (2000) Phytoremediation of toxic elemental and organic pollutants. Curr Opin Plant Biol 3: 153-162

ENZYMES TRANSFERRING BIOMOLECULES TO ORGANIC FOREIGN COMPOUNDS: A ROLE FOR GLUCOSYLTRANSFERASE AND GLUTATHIONE S-TRANSFERASE IN PHYTOREMEDIATION

PETER SCHRÖDER
Institute of Soil Ecology, Department Rhizosphere Biology, GSF National Research Center for Environment and Health, D-85758 Neuherberg, FRG. E-mail: peter.schroeder@gsf.de

1. Introduction

In the general enzyme list of the Nomenclature Committee of the International Union of Biochemistry and Molecular Biology (NC-IUBMB, first published in 1961 and with the last printed edition in 1992) EC 2 is reserved for the enzyme family of transferases. Generally, transferases are enzymes transferring a functional group, for example, methyl- or glycosyl-groups, from one substrate (regarded as donor) to another substrate (regarded as acceptor). Hence, the classification is based on the scheme "donor:acceptor-group transferase". The common names of the enzymes belonging to this group are normally derived from acceptor group-transferase or donor group-transferase. In many cases the donor is a cofactor (coenzyme) carrying the group to be finally transferred.

Whereas most members and subclasses of EC2 are confined to the metabolism of biogenic and natural compounds two subgroups, the glycosyltransferases of EC 2.4 and the aryl/alkyl transferases of EC 2.5, have been recognized as having crucial functions in the metabolism of foreign compounds, xenobiotics, in both animals and plants.

This role is very important, as all organisms are frequently exposed to an array of potentially toxic substances. Organic chemicals are particularly threatening. They may have natural sources e.g. fires, volcano eruptions or processes of biodegradation. They may also be the products of microbial or animal metabolism, or from the secondary metabolism of plants [1]. These organic substances may play a role in defence or in allelopathic reactions. Furthermore, increasing industrialization has provided two novel sources of foreign compounds: (1) through the invention and use of agrichemicals for the protection of crops from pests and weeds, and (2) through the emission of organic xenobiotics in chemical manufacturing processes or the use of synthetic chemicals. The latter compounds of solely anthropogenic origin represent a threat to our environment as these synthetic chemicals are emitted without any control. For plants, the situation is especially difficult as they are rooted in the ground and are dependent on that site for survival. Plants therefore, have to rely on effective detoxification mechanisms.

M. Mackova et al. (eds.), Phytoremediation Rhizoremediation, 133–142.

The uptake of xenobiotics from polluted media, i.e. air, water or soil, follows the laws of phase distribution and diffusion. Plants therefore, have only limited possibilities to avoid accumulation of xenobiotics in their tissue and the associated detrimental consequences. In recent years, some plant species have been recognized as potent accumulators or detoxifiers of such compounds. These plants are capable of removing these dangerous chemicals from the environment. Hence, they are to be utilized in the green technology of phytoremediation, helping to solve some of our environmental problems in aninexpensive, reliable and natural manner. However, information on the underlying biochemical principles involved in these processes is generally scarce.

2. EC 2.4 Glycosyltransferases

All enzymes transferring glycosyl groups to acceptor molecules belong to this class. Some of these enzymes also catalyse hydrolysis, which can be regarded as transfer of a glycosyl group from a donor molecule to water. Also, inorganic phosphate can act as an acceptor in the case of phosphorylases; phosphorolysis of glycogen is regarded as transfer of one sugar residue from glycogen to phosphate. However, the more usual scenario is the transfer of a sugar from an oligosaccharide or a high-energy compound to another carbohydrate molecule as acceptor. This subclass, EC2.4 is further subdivided, according to the nature of the sugar residue being transferred, into hexosyltransferases (EC 2.4.1), pentosyltransferases (EC 2.4.2) and those transferring other glycosyl groups (EC 2.4.99). This mechanism is widespread in the plant kingdom, and the resulting glycosides represent the largest group of natural substances in plants, contributing factors to whether plants are colourful, tasty, or poisonous. The mechanism of sugar transfer in plants was discovered early in plant biochemistry [2]. The earliest reports of plant glucosides were associated with the metabolism of secondary compounds, such as flavonoids, anthocyanins and phenylpropanoids. Intermediates as well as storage forms of these compounds are frequently glucosylated. Glucosides possess lower reactivity than aglyca [3], they have a high hydrophilicity [4], they are used in detoxification of endogenous products and xenobiotics [5], and they may be compartmentalized in the plant [3]. A physiological role for glucosides is seen in pathogen defence, allelopathy and plant inherent signals [6, 7].

The transfer of glucose to a xenobiotic molecule requires the presence of an acceptor group on the target. Such an acceptor functional group might be an –OH, -NH or –SH function, and correspondingly, the plant glucosyltransferases are named O-glucosyl-, N-glucosyl-, and S-glucosyltransferases.

Molecules that do not bear these functional groups may be conjugated with sugars after chemical activation, i.e. hydroxylation by one of the plant P450 monooxygenases. For a significant number of herbicides, activation by P450 prior to detoxification by O-glycosyltransferases has been reported. Recently the formation of plant cyanoglucosides via stepwise activation by P450 and final glucosylation (e.g. dhurrin or triglochin) has attracted considerable interest [8, 9].

The second requirement for glucosyltransfer reactions is the availability of activated sugars, such as UDP-glucose (uridine[5′]diphospho-[1]-α-D-glucose, Fig. 1A). This metabolite is formed via phosphorylation of glucose in an ATP-driven reaction, to yield glucose-6-phosphate, followed by conversion to glucose-1-phosphate. Glucose-1-phosphate reacts with UTP (uridine triphosphate), yielding UDP-glucose plus pyrophosphate. The UDP-glucose is the final activated intermediate that donates its glucose residue to the xenobiotic in an energetically favourable reaction. It has to be noted that the reaction of UDP-glucose on hydroxylated ring systems is the most frequently described reaction of sugar transfer in plants. Usually ß-1-D bonds are formed between the sugar moiety and the second substrate, but ß-5 and ß-3-conjugates have also been reported [3].

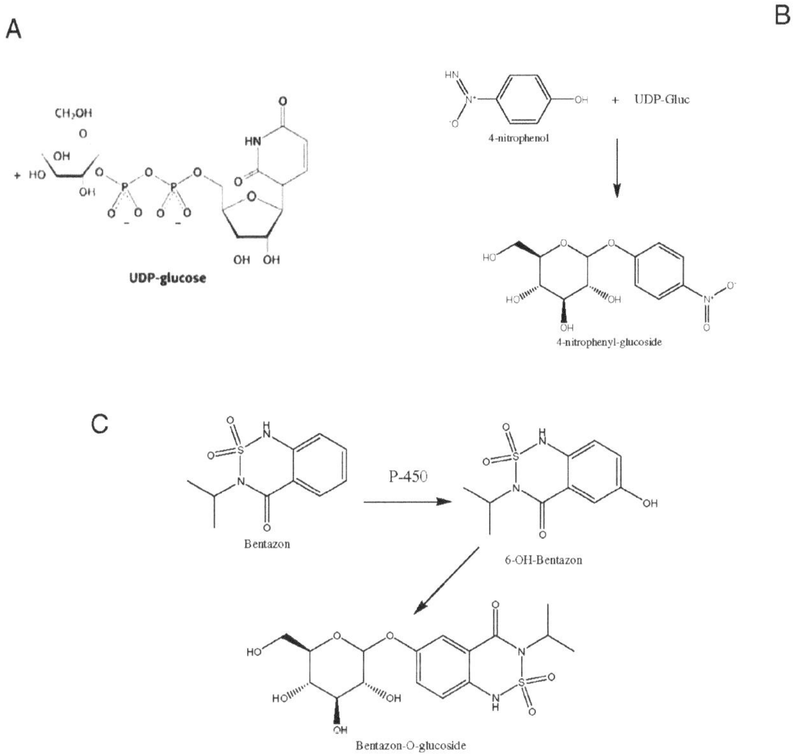

Figure 1. A) UDP-glucose, B) glucosyl transfer to a target molecule; C) conjugation of the herbicide, Bentazon, after activation by P450.

Numerous herbicides are conjugated to sugars via O-glucosyl-transfer or N-glucosyl-transfer in tolerant plants. The non-identity of the responsible enzymes has been demonstrated several times, although overlapping activities have been found in some cases. Conjugation may occur either at OH-groups of the molecule to form O-glucosides or at carboxy-groups to form acylglucosides. For N-glucosyltransfer, coupling to NH_2-groups of the molecule is crucial. From a practical point of view, predominantly phenolic pollutants, as well as components of ammunition (TNT and metabolites) or pesticide spills, might be candidates for detoxification via glucose transfer in plants. A possible exploitation of these enzymatic mechanisms has recently been reviewed [10].

It has been shown that glucosyl-transferase activities are hardly inducible and might thus represent a class of housekeeping enzymes. Attempts to increase their activity using herbicide safeners are rare. Recently Brazier and co-workers [11] demonstrated the increase of O-GT in black grass after dichlormid or cloquintocet mexyl treatment. In each of the treatments increased activity was found for the conjugation of quercetin but not for xenobiotic compounds. On the other hand, evidence has been found to suggest that the individual enzymes responsible for these reactions might well be under developmental control and that the conjugation of single xenobiotics can not be expected to proceed throughout the plant's life and in every plant part [12, 13]. Of course, this fact has consequences for the practical use of plants in the detoxification of foreign compounds, and it is especially important when considering plants for use in phytoremediation, because it has to be ensured that the detoxification capacity meets with the xenobiotic burden of the system.

Table 1: Examples for xenobiotic substrates of glucosyl transferases in plants (adapted from [14])

O-glucosyl transfer		N-glucosyl transfer	
Direct conjugation	after activation	direct conjugation	after activation
1,2,5-Trichlorophenol	2,4-D	Chloramben	Dinoben
2,4-dichloroanilin	Chlorpropham	Metribuzin	Propanil
4-nitrophenol	Cisanilide		Pyridate
4-nitrophenol	DDT / DDE		
Chloramben	Dicamba		
Clopyralid	Bentazon		
Dimethenamid	Diclofop		
Fenoxaprop ethyl	Diphenamid		
Maleic hydrazide	Methylphenylureas		
MCPA	Perfluridone		
Pentachlorophenol	Sulfonylureas		
Picloram	Terbacil		
Quinclorac			

It is important to note, that glucosyl conjugates may be cleaved by glucosidases and, in the case of acylglucosides (-COOH substitution) by esterases. Both enzyme activities are abundant in plant cells. However, these activities might be compartmentalized or under developmental control. The action of these enzymes will yield the respective aglyca that are spontaneously reprotonated under the conditions of the cytosol. Thus, the original xenobiotic substrate may be regenerated. This reversibility represents a great disadvantage of glucosylation for its practical consideration in phytoremediation, because previously detoxified compounds may regain their toxicity under certain conditions. In oats, the formation of an acylglucoside from Diclofop in the presence of an esterase explains this plant's susceptibility to this herbicide. In wheat, an O-glucoside is formed from Diclofop that is not readily cleaved [15].

This reaction chain of activating, conjugating and releasing a certain compound makes sense in the course of natural compound formation. It has also been shown that homeostasis of salicylic acid and indole acetic acid (auxin) is maintained in plant cells by this mechanism. For practical application, it has to be ensured that the cleavage of the glycosyl-conjugate happens only under conditions where the aglycon can be inserted into the cell wall and covalently bound to lignin or other polymerous structures.

3. EC 2.8 Glutathione S-transferases

Various xenobiotics possess electrophilic centres, i.e. centres of low electron density that can accept an electron pair to form a covalent bond. This feature makes them dangerous because they can to react spontaneously with corresponding nucleophilic sites of proteins and genetic material, i.e. DNA and RNA and thereby disturb metabolic networks.

The action of such electrophilic xenobiotics appears to be dependent on particular cellular enzymes called glutathione S-transferases (GSTs) [16, 14]. Electrophilic centres necessary for GSH conjugation are found in arene-oxides, aliphatic and arylic halides, in α–β-unsaturated carbonyls, organonitro-esters and organic thiocyanates. Industrial substrates for GSTs are haloalkanes, chlorobenzenes, thiocarbamates, diphenylethers, triazines, chloracetanilides [see 17, 18]. In animals the oxidants acrolein, propenals, lipid hydroperoxides, chlorambucil and fosfomycin are additional substrates [19].

Such compounds will not be conjugated by glucosyl transferases. Instead, the reactions are performed by a somewhat heterogeneous class of enzymes, GSTs, which catalyze the transfer of aliphatic, aromatic, or heterocyclic radicals as well as epoxides and arene oxides to glutathione. The transfer reaction takes place at the sulphur atom and has been annotated as the enzyme class coding EC 2.5.1.18. GST enzymes occur ubiquitously [20]. The binding of the foreign compound and the transfer of glutathione follows two mechanisms catalyzed by glutathione S-transferases [21, 22]:

- (a) Nucleophilic displacement of an alkyl or aryl halogen or a nitro group is the most frequently observed step. Conjugation of many pesticides like atrazine, propachlor or pentachloronitrobenzene (PCNB) are examples of this type of reaction. Halogens or nitrogroups of these molecules are soft electrophiles and react readily with GSH. Further substitution reactions are found in the detoxification of diphenylether herbicides (e.g. fluorodifen, fenoxaprop-ethyl). Here, an ether bond is cleaved and substituted by the thiolate. Moreover, the standard enzyme assays for GST activity use 1-chloro-2,4-dinitrobenzene (CDNB) or 1,2-dinitro-4-chlorobenzene (DCNB) as substrates (see Fig. 2, B).
- (b) Nucleophilic addition (Michael reaction): Addition of the thiolate to carbon-carbon-double bonds is a special type of reactions on compounds with reactive carbon-carbon double bonds neighboured by an electron withdrawing group [23]. Conjugation of tridiphane or cinnamic acid may be examples for this type of reaction [24, 25]. The conjugation on these bonds leads to a labile conjugate that may be sensitive to pH changes.

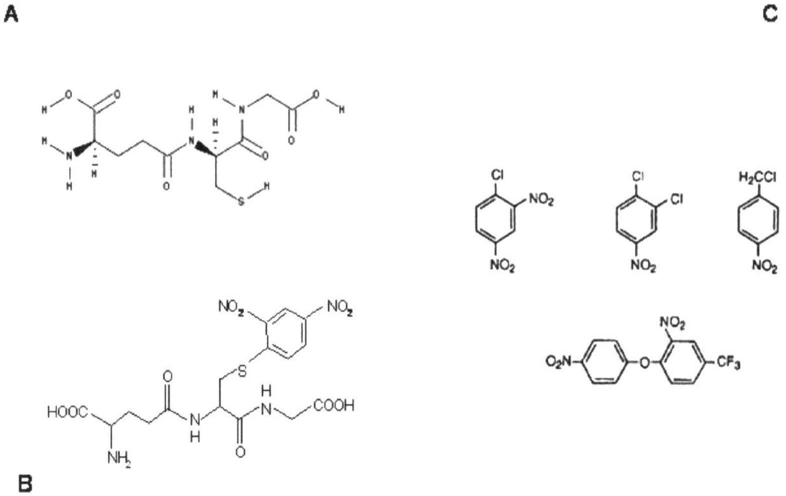

Figure 2. (A) reduced glutathione (γ-glutamyl-cysteinyl-glycin), (B) S-dinitrobenzyl-glutathione; (C) four typical plant GST substrates: chlorodinitrobenzene, dichloronitrobenzene, nitrobenzylchloride, and fluorodifen.

In contrast to the metabolism of glucosyl conjugates, and with the exception of the cleavage of "Michael-reaction"-type conjugates formed at double bonds of molecules (see above), the dissolution of glutathione conjugates does not lead to the liberation of the original toxic or lipophilic foreign compound. This is due to the fact that nucleophilic substitution or displacement removes the significant electrophilic centre from the target molecule to introduce the glutathione thio-function. When cleaved, the electrophilic moiety is lost, and the toxicity of any released parent residue is significantly lowered.

The unravelling of the *Arabidopsis* genome has confirmed the observation that the multiple reactions of GST enzyme activity are attributes to a large number of isoenzymes encoded for by more than 50 GST genes. These genes, many of them formed by gene reduplication in the simple *Arabidopsis* genome, can be clustered in four distinct groups. According to the mammalian GST system, which names the GSTs with Greek letters, a comprehensive nomenclature of plant GSTs has been proposed by Edwards and coworkers [26].

To date, five distinct classes of GSTs, Lambda (and DHAR, dehydroascorbate reductases), Phi, Tau, Theta and Zeta have been identified in plants. According to Edwards & Dixon [27], class Tau, the predominant class, catalyzes the detoxification of xenobiotics by nucleophilic substitution reactions. Class Phi and Tau GSTs, the next most abundant, appear to catalyze reactions with endogenous toxic metabolites or are involved in the metabolism of reactive oxygen species. The two remaining, classes Lambda and Zeta, are unusual, Lambda GSTs occur as monomeric enzymes or enzymes with transmembrane regions, functioning as redox mediators. Zeta GSTs are isomerases.

The multitude of GSTs in plants reflects their ability to conjugate to GSH a large array of different substrates, many of them of anthropogeneous origin. Furthermore, GSTs are inducible by different forms of stress, including xenobiotics [28]. This property also confers herbicide tolerance or resistance in many plant species. Whereas herbicides are designed to kill weeds and leave crops unaffected, organic pollutants might have properties that stress plants in a similar manner. GST activities have been shown to be crucial for the detoxification of a number of these compounds.

4. Physiological roles of conjugates

Summarizing the above results, we must conclude that glutathione and sugar conjugation lead to the formation of detoxified and more or less stable products in the plant cytosol. Upon chemical contact or under conditions of pesticide application, plants may encounter relatively high concentrations of the respective foreign compounds, and, provided the defence enzymes work properly, they will form significant amounts of conjugates from them. Besides eventual chemical lability this accumulation of conjugates will lead to an unfavourable situation as conjugates might inhibit enzymes by feedback mechanisms or may affect specific binding.

Glucosyl- and glutathione-conjugates do not accumulate in the cytosol of plant cell, but are translocated into the vacuole through ABC-transporters [29, 30, 15]. The presence of a conjugate is mandatory for this translocation. An interesting observation is that glucosylconjugates build up in pools inside the plant, whereas glutathione conjugates do not. They undergo rapid and complete metabolism [31]. As well as the ABC-transporters on the tonoplast, ABC-transporters have recently been found in the plasmalemma. They accept the same conjugates and allow for long range transport of these metabolites. Some hints exist that a physiological role for conjugates might be the signalling of pollutant stress, and induction of detoxification reactions has been described after application of xenobiotics to plant cells [32]. In the context of phytoremediation, such conjugates might also play a role in enhancing the plants capacity for detoxification.

5. Conjugation reactions and options for phytoremediation

The use of plants in phytoremediation of organic pollutants has been reviewed thoroughly in the frame of the COST action 837 [33, 34, 10, 14). Literature on the degradation of herbicides in crops demonstrates clearly that xenobiotic conjugates are usually further processed to more complex conjugates [21, 31], or cleaved to form reactive molecules that are excreted from the living cell and reside in the cell wall or the apoplast [35]. In our context of phytoremediation it is, of course, crucial to know which type of primary conjugation occurred, because this determines the final fate of the compound [36, 37].

Further breakdown steps include incorporation of metabolites into the cell wall in the pectin, lignin, hemicellulose and cellulose fraction [35, 38, 39]. This has been demonstrated with numerous cereals, soybean and the respective cell cultures. Few metabolites have also been found in the rhizosphere, where they might disappear in microbial and mycorrhizal metabolism, and single findings point to the volatilization of metabolites after the action of methyl transferases [31]. One crucial question remaining is how and why the conjugating enzymes detect and meet the foreign compounds attacking the cells. Given the large number of isoforms of detoxification enzymes on the one hand, and the diverse group of xenobiotics on the other, it is hard to imagine how the different reactions occur in the cytosol in an orderly manner. Even if the respective enzymes would occupy distinct positions in a metabolic network, the question would remain how the detoxification is channelled in a cell or a tissue. Our present knowledge is that the conjugating enzymes have low specificity, i.e. that they are able to accept a larger number of endogenous and xenobiotic substrates. Only very specifically designed poisons at higher concentrations will represent a severe threat.

Historically, our present state of knowledge on foreign compound metabolism is focused on crops and a small number of ornamental plants. Only a few reports exist on plants that might be interesting with respect to their potential in phytoremediation. Conjugative metabolism in the outstanding candidates *Arundo donax*, *Brassica juncea*, *Phragmites sp..*, *Typha sp..*, *Plantago majus*, *Populus sp..*, *Salix sp.*, to name but a few, has not been investigated in any depth. This situation is awkward as there are already numerous existing field sites that seem to be very successful in the removal of xenobiotics from soil and water. Knowing about the mechanisms involved, the efficiency of these systems could probably be improved when methods to increase metabolism rates would be applied. One option could be to utilize gene manipulation methods to over express the desired enzymes. Another option could be to add inducers of herbicide resistance to the plants. Finally it is possible that xenobiotic uptake and transport to tissues with high degradative activity would be enhanced. Each of these attempts would improve commercial and public acceptance of the use of plants to improve the environment with biological methods. Plants have a large potential to cope with their environment [40], and they will also be able to help us solve some of our pollution problems in an environmentally friendly way.

References

[1] Naumann, K (1993) Chlorchemie der Natur. Chemie in unserer Zeit 27(1): 33-41

[2] Ciamician, GL and Ravenna, C (1916), without title, RC Acad Linei 18: 419

[3] Hösel, W (1981) Glycosylation and glycosidases, in The Biochemistry of Plants, pp. 725-753, Conn, EE, Ed., New York, Academic Press

[4] Bartz, W; Köster, J; Weltring, KM and Strack, D (1985) Recent advances in the metabolism of phenolic compounds in plants and animals, in The Biochemistry of Plant Phenolics. Ann Proc Phytochem Soc Europe, 25, pp. 307-347, VanSumere, CF and Lea, PL, Eds., Oxford, Clarendon

[5] Gallant, ER and Balke, NE (1995) Xenobiotic glucosyltransferase activity from suspension cultured *Glycine max* cells. Pestic Sci 43: 31-40

[6] Matile, P (1976) Biochemistry and function of vacuoles. Ann Rev Plant Phys 29: 193-213

[7] Balke, NE and Schulz, M (1987) Potential impact of enzyme glucosylation of allelopathic phenolic compounds, in Invited Lectures, Section 4: Industrial Chemistry, 31st International Congress on Pure and Applied Chemistry, Sophia, Bulgarian Academy of Sciences, pp. 17-23

[8] Bak, S; Olsen, CE; Halkier, BA and Moller, BL (2000) Transgenic tobacco and Arabidopsis plants expressing the two multifunctional Sorghum cytochrome P450 enzymes, CYP79A1 and CYP71E1, are cyanogenic and accumulate metabolites derived from intermediates in dhurrin biosynthesis. Plant Physiol 123: 1437-1448

[9] Nielsen, JS and Moller, BL (1999) Biosynthesis of cyanogenic glucosides in *Triglochina maritima* and the involvement of cytochrome P450 enzymes. Arch Biochem Biophys 368: 121-130

[10] Harvey, P; Campanella, B; Castro, PML; Harms, H; Lichtfouse, E; Schäffner, A; Smrzek, S and Werck-Reichhart, D (2001) Phytoremediation of Polyaromatic Hydrocarbons, Anilines and Phenols. ESPR Environmental Sciences and Pollution Research 9(1): 29-41

[11] Brazier, M; Cole, JD and Edwards, R (2002) O-Glucosyltransferase activities toward phenolic natural products and xenobiotics in wheat and herbicide-resistant and herbicide-susceptible black-grass (Alopecurus myosuroides). Phytochem 59: 149-156

[12] Haas, M (1997) Metabolisierung von Xenobiotika durch pflanzliche Zellkulturen und Enzyme. Dissertation, University Weihenstephan

[13] Schröder, P (2001) The role of glutathione and glutathione S-transferases in the adaptations of plants to xenobiotics, in Significance of Glutathione in Plant Adaptation to the Environment. Handbook Series of Plant Ecophysiology, pp. 157-182, Grill, D, Tausz, M and DeKok, LJ, Eds., Boston, Dordrecht, London, Kluwer Acad Publishers

[14] Schröder, P and Collins, CJ (2002) Conjugating enzymes involved in xenobiotic metabolism of organic xenobiotics in plants, Int J Phytorem 4(4): 247-265

[15] Shimabukuro, RH; Walsh, WC and Hoerauf, RA (1979) Metabolism and selectivity of Diclofop-methyl in wild oat and wheat. J Agric Food Chem 27: 615-623

[16] Coleman JOD; Randall RA and Blake-Kalff, MMA (1997) Detoxification of xenobiotics by plants: chemical modification and vacuolar compartimentation. TIPS 2: 144-151

[17] Schröder, P (1997) Fate of glutathione S-conjugates in plants: Cleavage of the glutathione moiety, in Hatzios, KK, Ed., Regulation of enzymatic systems detoxifying xenobiotics in plants. NATO ASI Series Vol. 37, pp. 233-244, Kluwer, NL

[18] Sheehan, D; Meade, G; Foley, VM and Dowd, CA (2001) Structure, function and evolution of glutathione transferases: implications for classification of non-mammalian members of an ancient enzyme superfamily. Biochem J 360: 1-16

[19] Hayes, JD and Pulford, DJ (1995) The glutathione S-transferase supergene family: Regulation of GST and the contribution of the isoenzymes to cancer chemoprotection and drug resistance. Crit Rev Biochem Mol Biol 30: 445-600

[20] Pflugmacher, S; Sandermann, H and Schröder, P (2000) Taxonomic distribution of plant glutathione S-transferases acting on xenobiotics. Phytochemistry 54: 267-273

[21] Lamoureux, GL and Rusness, DG (1989) The role of glutathione and glutathione S-transferases in pesticide metabolism, selectivity and mode of action in plants and insects, in Dolphin, D, Poulson, R and Avramovic, O, Eds., Glutathione: Chemical Biochemical and Medical Aspects, Vol IIIB, Ser: Enzyme and Cofactors, pp. 153-196, New York, Wiley & Sons

[22] Lamoureux, GL, Rusness, DG and Schröder, P (1993) Metabolism of a diphenylether herbicide to a volatile thioanisole and a polar sulphonic acid metabolite in spruce (Picea). Pestic Biochem Physiol 47: 8-20

[23] Talalay, P; de Long, MJ and Prochaska, HJ (1988) Identification of a common chemical signal regulating the induction of enzymes that protect against chemical carcinogenesis. PNAS 85: 8261-8265

[24] Lamoureux, GL and Rusness, DG (1986) Xenobiotic conjugation in higher plants, in Paulson, GD, Caldwell, J, Hutson, DH, and Menn, JJ, Eds., Xenobiotic Conjugation Chemistry, pp. 62-105, Washington, Am Chem Soc

[25] Dean, JV; Devarenne, TP; Lee, IS and Orlofsky, LE (1995) Properties of a maize glutathione S-transferase that conjugates coumaric acid and other phenylpropanoids. Plant Physiol 108: 985-994

[26] Dixon, DP; Cole, JD and Edwards, R (2001) Cloning and characterization of plant theta and zeta class GSTs: implications for plant GST classification. Chem Biol Interact 133: 33-36

[27] Edwards, R and Dixon, DP (2004) Metabolism of natural and xenobiotic substrates by the plant glutathione S-transferase superfamily, in Sandermann H, Ed., Molecular ecotoxicology of plants, Ecol Studies Vol. 170, Springer, Berlin, Heidelberg, pp. 17-52

[28] Dean, JV; Gronwald, JW and Eberlein, CV (1991) Induction of glutathione S-transferase isozymes in sorghum by herbicide antidotes. Plant Physiol 92: 467-473

[29] Marrs, KA (1996) The functions and regulation of glutathione S-transferases in plants. Annu. Rev. Plant Physiol. 47: 127-158

[30] Wolf, A.E; Dietz, K.J and Schröder, P (1996) A carboxypeptidase degrades glutathione conjugates in the vacuoles of higher plants. FEBS Lett 384: 31-34

[31] Lamoureux, GL and Rusness, DG (1993) Glutathione in the metabolism and detoxification of the xenobiotics in plants, in De Kok, LJ, Stulen, I, Rennenberg, H, Brunold, C and Rauser, W, Eds., Sulphur Nutrition and Assimilation in Higher Plants, pp. 221-239. The Hague, SPB Acad. Press

[32] Diekmann, F; Nepovim, A and Schröder, P (2004) Influence of *Serratia liquefaciens* and a xenobiotic glutathione conjugate on the detoxification enzymes in a hairy root culture of horse radish (*Armoratia rusticana*). J Appl Bot 78: 64-67

[33] Chaudhry, Q; Schröder, P; Werck-Reichhart, D; Grajek, W and Marecik R (2001) Prospects and Limitations of Phytoremediation for the Removal of Persistent Pesticides in the Environment. ESPR - Environmental Sciences & Pollution Research 9(1): 4-17

[34] Coleman, JOD; Frova, C; Schröder, P and Tissut, M (2002) Exploiting plant metabolism for the phytoremediation of persistent herbicides. ESPR, Environmental Sciences and Pollution Research 9(1): 18-28

[35] Schmidt, B; Ebing, W and Schupahn, I (1988) Einsatz eines Pflanzenzellkultur-Tests zur Ermittlung der Metabolisierbarkeit von Pflanzenschutzmitteln. Gesunde Pflanzen 40: 245-249

[36] Frear, DS (1976) Pesticide conjugates – glycosides, in Kaufman, DD, Still, GG, Paulson, GD, and Bandal, SK, Eds., Bound and conjugated pesticide residues, ACS Symposium 29, pp. 35-54. Washington DC, American Chemical Society.

[37] Kreuz, K; Tommasini, R and Martinoia, E (1996) Old enzymes for a new job. Herbicide detoxification in plants. Plant Physiol 111: 349-353

[38] Langebartels, C and Harms, H (1986) Plant cell suspension cultures as test systems for an ecotoxicological evaluation of chemicals. Angew Bot 60: 113-123

[39] Sandermann, H; Haas, M; Meßner, B; Pflugmacher, S; Schröder, P and Wetzel, A (1997) The role of glucosyl and malonyl conjugation in herbicide selectivity, in Hatzios, KK, Ed., Regulation of Enzymatic Systems Detoxifying Xenobiotics in Plants. NATO ASI Series Vol. 37 pp. 211-232 Kluwer, Dordrecht

[40] Schnoor, JL; Licht, LA; McCutcheon, SC; Wolfe, NL and Carreira, LH (1995) Phytoremediation of organic and nutrient contaminants, Env Sci Tech 29: 318-323

PHYTOREMEDIATION OF POLYCHLORINATED BIPHENYLS

MARTINA MACKOVA[1,2], DIANE BARRIAULT[3], KATERINA FRANCOVA[1], MICHEL SYLVESTRE[3], MONIKA MÖDER[4], BLANKA VRCHOTOVA[1,2], PETRA LOVECKA[1], JITKA NAJMANOVA[1], KATERINA DEMNEROVA[1], MARTINA NOVAKOVA[1,2], JAN REZEK[1,2] AND TOMAS MACEK[2]

[1] Dept. Biochemistry and Microbiology., Faculty of Food and Biochemical Technology, ICT Prague, Technicka 3, Prague, 166 28 Czech Republic, [2] Dept. of Natural Products, Institute of Organic Chemistry and Biochemistry, Academy of Science of the Czech Republic, Flemingovo n. 2, 166 10 Prague, Czech Republic, E-mail: tom.macek@uochb.cas.cz [3]Institut National de la Recherche Scientifique, INRS-IAF, 245 Boul. Hymus, Pointe-Claire, Québec, Canada, H9R 1G6, [4] Laboratory of Analytical Chemistry, UFZ Leipzig-Halle, Permoserstrasse 17, Leipzig, Germany, [4]

1. Introduction

Many hydrophobic organic compounds including polychlorinated biphenyls are priority soil contaminants because of their toxicity and tendency to persist in soils, sediments and to escape biological degradation. The fate of contaminants in the environment is controlled by a combination of interacting processes. They can be classified as physical, chemical or biological. Physical processes are responsible for the movement of contaminants through the soil and subsurface away from their source. Chemical and biological processes determine the extent to which compounds will be transformed. In some cases the most important transformations occur abiotically, while in other cases they are mediated by microorganisms with the help of other living organisms. Knowing the physicochemical properties of environmental contaminants is a prerequisite for predicting the fate of organic contaminants or new products in the environment, predicting the efficiency of treatment systems, and assessing ecotoxicological and human health risks. Unfortunately such data are available only for a relatively small number of compounds. At this time, several reviews described the pathway used by bacteria and fungi to transform PCBs in the environment as well as the various alternate pathways created by engineered organisms. However, the contribution of plants to the ecodeposition of PCBs has not

M. Mackova et al. (eds.), Phytoremediation Rhizoremediation, 143–167.

yet been well investigated. In this review, we will place emphasis on studies regarding the pathways used by plants to transform PCB in the environment.

2. Chemistry and use of PCBs

PCBs were manufactured from 1929 until their use begins to be restricted in the 1970s and their production banned in most countries in 1979. Because of their excellent chemical and thermal stabilities they were used on a large scale as dielectric fluids in transformers and capacitors, hydraulic fluids and lubricants, organic diluents, paints, carbon-less copy paper, etc but also in small household electrical appliance and domestic products. The extensive application of the PCBs resulted in their wide distribution in the environment. PCBs are actually a large family of 209 molecules that all share the biphenyl backbone but differ by the number and the position of the chlorine atoms on the biphenyl ring. We call isomer the compounds having the same number of chlorine atoms and congeners those, which bear different number of chlorine. The congeners are designated by describing the position of the chlorine atoms on the biphenyl ring or, more simply, by the IUPAC numbering system. The congeners differ in their physical properties according to the number and the position of chlorine atoms. The high-chlorinated biphenyls are less water-soluble and less volatile than the low-chlorinated ones. The degree of chlorine substitution influences their biodegradability that decreases with increasing chlorination. The toxicity for the biota is also related to the number of chlorines but their position on the biphenyl ring is of prime importance. The congeners that take a co-planar configuration such as 3,4,3',4'-TCB (IUPAC #77) are the more toxic ones.

The PCBs are prepared by direct chlorination of the biphenyl ring. The commercial formulations display various overall percentages of chlorine and different congener distribution. For example, Aroclor 1242 contains 42% of chlorine with a predominance of congeners bearing three and four chlorine atoms; Aroclor 1260 has 60% chlorine content with a predominance of six- and seven-chlorinated congeners. These mixtures typically contain more than 70 different congeners [1] and were sold under different names (Aroclor, Phenoclor, Clophen, Delor and Kanechlor), depending on the manufacturer.

3. PCBs in the environment

The PCBs have been detected in the environment for the first time in 1966 by Jensen [2] along with DDT contamination. Since then, PCBs were found everywhere, including polar regions, both Artic and Antarctic [3]. They are now considered as the most widespread pollutant on the planet. In industrial countries, the contamination originates from inadequate disposal and leaks from equipments. In remote areas where PCBs were not used, the contamination resulted from atmospheric transport [4]. The high chemical stability of PCBs explains their persistence in the environment. Their low water solubility results in their accumulation in the biota (in fats) and their concentration in the food chain [5]. The marine mammals at the top of the food chain concentrate high level of PCBs [6, 7]. Human are contaminated as well and breast milk contains measurable content of PCBs [8]. In the industrialized countries the PCB blood level of an individual increases with age [9, 10].

The distribution of the PCBs in the different environmental compartments in the United Kingdom has been extensively studied by Harrad *et al.* [11]. They found that the PCB concentration peaked in the early to mid 1960s, declined rapidly following restrictions of their use in the 1970s and continued to decrease since then, but at a lesser rate. According to their estimates, most of the PCBs reside in soil (92.5%), sea water (3.5%) and marine sediments (2%). Up to 0.2% might be in humans. The amount in soil represents about 370 tons for UK only. The pattern for other industrialized countries should be similar, adjusted upscale or downscale depending of the amount of PCBs used and the duration of their use. Today, even if the PCB production has ceased, there is still a flux of PCBs originating from leaks of transformers and capacitors remaining in use (0.2-0.3 tons/year for UK) and from volatilization from soil (8,1 to 40 tons/year for UK) [11].

The composition of PCB mixtures found in the environment differs from the original commercial preparations. At or near contamination points there may be a shift to the more highly chlorinated congeners because the lowest chlorinated congeners being more volatile and more easily degradable have either migrated or have been transformed by microorganisms. So the pattern observed for long term contaminated soil displays a higher proportion of high chlorinated congeners than the mixture at the origin of the contamination. In remote areas where PCBs were not used, the shift is toward low-chlorinated congeners which are more volatile and more easily translocated through the atmospheric currents. Contaminated sediments also present a higher proportion of low-chlorinated congeners that is attributable to reductive dechlorination resulting from microbiological activity.

Although toxicological studies of PCBs are complex, because they should take into account numerous different molecules with possible synergic effects, studies suggest toxic and carcinogenic properties, particularly for co-planar congeners [12, 13]. Extensive studies *in vivo* and *in vitro* have been conducted on the toxicological effects of PCBs in different mammalian systems. One area that has received less attention is the uptake, translocation and transformation of PCBs in terrestrial plants. To investigate the mobility of PCBs in any phase of the environment, it is very essential to understand their physical and chemical properties, transport, translocation and accumulation in the food chain.

Proof of biotransformation is not itself sufficient to guarantee protection, the biotransformation must not result in the production of harmful intermediates with long lifetimes. For this reason, biotransformation pathways must also be understood. Additionally, good mass balances are difficult to make. Alternative approaches are often necessary to provide proof that the desired transformation process is occurring. Evidence is provided by detection of intermediates and products or of the consumption of other chemicals that are associated with biotransformation.

4. Bacterial degradation of PCBs

Despite the high chemical stability of PCBs, the finding that microorganisms are able to degrade some congeners opened the door to implement bioremediation technologies. Essentially, bacteria can degrade PCBs by two processes: aerobic degradation via the biphenyl pathway and anaerobic dechlorination. The former being the most documented.

4.1. AEROBIC PATHWAY

The first report on the isolation of PCB degrading bacteria dates back to 1973 when Ahmed and Focht found two species of *Achromobacter* capable of degrading PCBs [14]. Rapidly, other strains were isolated from PCB contaminated sites, some with better degradation efficiencies and broader spectrum of activity [15, 16, 17, 18, 19, 20]. These strains were extensively studied and the biphenyl degradation pathway was elucidated. The genes encoding the different enzymes of the pathway were sequenced and in some cases cloned in high expression vectors to obtain large amount of enzymes. The enzymes produced and purified from clones or from native strains were used to assess *in vitro* activity against specific PCB congeners or mixtures of them [21, 22].

The biphenyl pathway (Fig. 1) is identical for all the aerobic bacterial strains. Biphenyl dioxygenase is a multi-component enzyme constituted by an iron-sulfur protein, (ISP_{bph}), which interacts directly with the substrate to introduce two oxygen atoms on adjacent positions on the biphenyl ring. Electrons are transferred from NADH to ISP_{bph} by the two other components of the dioxygenase, a ferredoxine and a reductase. The biphenyl dioxygenase components are encoded by four genes clustered in a single operon in most of the strains: *bphA* (α-subunit of ISP_{bph}), *bphE* (β-subunit of ISP_{bph}), *bphF* (FER_{bph}) and *bphG* (RED_{bph}). The product of the first enzymatic step is *cis*-(2R,3S)-dihydroxy-1-phenylcyclohexa-4,6-diene. A dehydrogenase (encoded by *bphB* gene) further transforms the first metabolite to a dihydroxybiphenyl that is transformed by a second dioxygenase (encoded by *bphC* gene) in an open-ring compound. Finally a hydrolase (encoded by *bphD* gene) cleaves the molecule to form benzoic acid. This is the so-called higher pathway. The mineralization of benzoic acids is performed by another group of enzymes constituting the lower pathway.

Chlorinated biphenyls are processed by the same pathway. However not all congeners can be degraded and biphenyl dioxygenases isolated from different bacterial strains have different specificity patterns. Biphenyl dioxygenases generally have a good activity towards congeners containing three or less chlorine atoms, but lower or no activity towards more substituted biphenyl. *Burkholderia xenovorans* LB400 is the most potent natural occurring strain known to date. It can degrade most of chlorinated biphenyls containing 3 chlorine atoms and some containing 4 and 5 chlorine atoms [23, 24]. Several evolved biphenyl dioxygenases have been reported which exhibit extended catalytic activity toward PCBs [25, 26, 27, 28]. However, in some case the metabolites produced by these modified enzymes or even by native

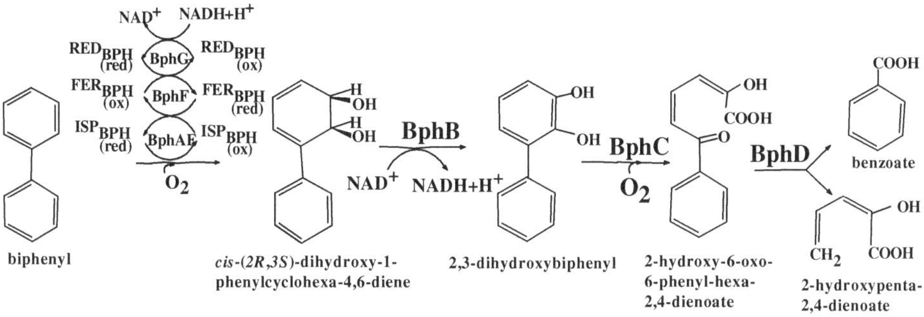

Figure 1. Aerobic bacterial biphenyl degradation pathway.

large spectrum dioxygenases cannot be further processed downstream, the first bottleneck being at the third step of the pathway (2,3-dihydroxybiphenyl 1,2-dioxygenase) [29]. Thus the development of more efficient PCB degrading strains must address the improvement of every enzyme of the pathway.

4.2. ANAEROBIC DECHLORINATION

The observation that PCB contaminated sediments display a higher content of low chlorinated congeners than the mixture at the origin of the contamination was the first clue that anaerobic bacteria are able to degrade PCBs by a reductive dechlorination process. This process has been well documented for several aquatic sediments [30, 31, 32]. Anaerobic enrichment cultures able to dechlorinate PCBs were maintained in laboratory conditions, with and without sediment [33, 34, 35, 36, 37]. The dechlorination release preferentially *meta* and *para* chlorines [35, 38] but there are some reports of *ortho* chlorine release [39, 40, 41]. Brown *et* al described eight different microbiological processes responsible for the different dechlorination patterns and congener specificities [30, 42]. It is thought that these processes are performed by different strains and different enzymes. Until recently, all attempts to obtain a pure culture able to dechlorinate PCBs had failed. Now, one pure anaerobic strain and one consortium have been reported [43, 44]. However, isolation of more strains and studies of the enzymes responsible for the dechlorination are still to come. The effect of anaerobic dechlorination is double: first the removal of *meta* and *para* chlorines reduces the toxicity of the contaminated sediments by reducing the occurrence of co-planar congeners [45]. Second, the reduction of the number of chlorine atoms on the biphenyl ring potentially facilitates the aerobic degradation through the biphenyl pathway.

5. Degradation of PCBs in higher organisms

The first reports of PCBs degradation by white rot fungi date back to 1985 [46, 47]. The enzymes that are potential candidates for this activity are peroxidases and laccase. These enzymes are involved in lignin degradation and are also active against a wide range of aromatic compounds (PAH, substituted phenols, chlorinated pesticides [48, 49] etc.). The mechanism of degradation by lignin-peroxidase, manganese-peroxidases and laccases is described in degradation of many environmental pollutants. The ligninolytic enzyme system is non-specific, extracellular and free radical based that allows them to degrade structurally diverse range of xenobiotic compounds. Lignin peroxidase and manganese peroxidase carry out direct and indirect oxidation as well as reduction of xenobiotic compounds. Indirect reactions involved redox mediators such as veratrylalcohol and Mn^{2+}. Reduction reactions are carried out by carboxyl, superoxide and semiquinone radicals, etc. Methylation is used as detoxification mechanism by WRF (white rot fungi). Highly oxidized chemicals are reduced by transmembrane redox potential. However, the mechanism of PCB degradation has not yet been fully understood. Several authors documented transformation of PCBs determined by fungi and some studies concluded that neither lignin peroxidases nor Mn-dependent peroxidases are involved in PCB degradation [50, 51]. A major problem is the adsorption of the different PCB congeners to the biomass which introduces a bias when the degradation is measured by substrate depletion. One alternative is to measure the complete mineralisation by using radiolabeled substrates but it is likely that many congeners cannot go through the entire process. The other alternative is the identification of the metabolites resulting from PCB degradation. Dietrich *et al.* [52] reported the presence of two metabolites produced from 4,4'-dichlorobiphenyl by *P. chrysosporum* confirming that this fungi can actually degrade this congener. Schultz *et al.* [53] described dehalogenation and metabolisation of chlorinated hydroxybiphenyls by laccase from white rot fungi *Pycnoporus cinnabarinus*, Romero *et al.* [54] documented biotransformation of biphenyl by ascomyceteous fungi *Talaromyces helicus* giving hydroxylated biphenyls and 4-phenyl-2-pyrone-6-carboxylic acid. Several studies showed capacity of fungal cultures on di, tri, tetra- and penta-chlorinated phenols when first step is characterized by oxidative dehalogenation mediated by extracellular peroxidases to form benzoquinones. The level of degradation of halogenated aromatics, reported in scientific literature, is very different from one study to the other. More extensive investigation will be necessary to elucidate the PCB degradation pathway in these eukaryotic microorganisms.

5.1. TRANSFORMATION OF PCBS BY PLANT CELLS

5.1.1 Metabolism of PCBs by plant cells and products formed

Plants have shown the capacity to withstand relatively high concentrations of metals or organic chemicals without significant toxicity symptoms. Also in some cases they can uptake and transform organic compounds to less phytotoxic metabolites. Several investigations have shown that PCBs can be translocated from soil to various parts of the plants and can accumulate in particular tissues in higher concentrations than in others.

The metabolism of PCBs varies between the plant species and is affected by the substitution pattern and the degree of chlorination [55]. Wilken *et al.* [55] studied metabolism of 10 different congeners of PCBs in cell cultures of 12 plant species. They stated observed that the metabolism of defined PCB congeners was dependent of the plant species. They detected various monohydroxylated and dihydroxylated compounds that were likely PCB metabolites. Macková *et al.* [56, 57, 58] studied the ability of the callus, root and shoot cultures of various species cultivated *in vitro* to degrade PCBs. Root cultures derived from single plant cells genetically transformed by the Ri-plasmid of *Agrobacterium rhizogenes*, the so-called hairy roots, proved to be a very useful tool to investigate the metabolic routes used by plants for PCB conversion [57, 58, 59, 60]. About 40 axenic cell cultures of different plant species [62, 63, 64, 65] were screened for the ability to transform PCBs. Delor 103 - a mixture of PCBs commercially produced in the former Czechoslovakia until mid eighties (similar to Aroclor 1242) [61, 62] was used in the screening as model mixture of PCBs in these investigations. This PCB mixture contains about 59 congeners with an average of 3 chlorine substitution per biphenyl molecule. Degradative abilities in relation to the origin and morphology of the cultures have been evaluated. The PCB transformation capacities varied considerably within different cultures of different but even the same species [58]. The growth of plant biomass in the presence of individual congeners was lower compared to the controls without PCBs and correlated with the metabolic activities toward PCBs. In general, lesser-chlorinated congeners were metabolised more rapidly than those with higher number of chlorine atoms. However, there are some exceptions, which indicate that not only the number of chlorine atoms, but also the position of chlorine substitution and molecular structure are also important factors controlling the PCB metabolism by plants. Previous studies [65, 55] have documented that as the degree of chlorination of the biphenyl ring increases, the rate of their conversion by plant cells of different species decreases, mostly in agreement with the increasing phytotoxicity of the tested compound. Comparison of different clones of one plant species proved, that clones can significantly differ in their ability to metabolise PCBs within one species and thus the ability to transform PCBs should always be checked and compared with metabolism of the normal plants. Differentiated and transformed plant cell cultures metabolised xenobiotics with higher efficiency than amorphous and non-transformed ones of the same species. This ability was comparable with that of intact plants.

Differentiated cultures of horseradish and black nightshade cells exhibited the best transformation abilities as well as growth characteristics [65, 66]. Transformed roots are excellent model systems for screening higher plants that are tolerant to various inorganic and organic pollutants, and for determining the role of the root matrix in the uptake and further metabolism of contaminants [67].

The results of the removal of PCB congeners by plant cell cultures exposed to Delor 103 are shown in Table 1 and Table 2. Both plant species were chosen for further studies to investigate not only the capacity of cell cultures to remove PCBs but also the PCB degrading ability of the whole plants and those grown under natural conditions in real contaminated soil. Plant tissue cultures of two other species, plant tissue cultures of *Nicotiana tabacum* (tobacco) and *Medicago sativa* (alfalfa), grew well in the presence of PCBs. Other reasons motivated the choice of these two species: tobacco is a well studied general plant model, especially for molecular biology and genetic studies; alfalfa is a plant species with high metabolic activities within rhizosphere and thus could potentially serve as model for investigations addressing the interaction between the plant of rhizosphere interactions of plants and indigenous bacteria.

Table 1. Assignment of individual PCB congeners to peaks of Delor 103 analysed by gas chromatography with electron capture detector and their conversion shown as the residual amount of each congener after 14 days incubation, expressed as % of residual PCB content in the dead cell control flasks (hairy root clone SNC-9O at initial PCB (Delor 103) content 5 mg per flask and 2.5 mg per flask with 100 ml of media). Modified according to [65].

Peak No.	IUPAC No.	Cl-substitution of congeners in the GC peak	Residual content of PCB (%)	
			Initial Delor 103 conc.	
			*+D 5mg/10ml	* + D 2.5mg/100ml
1	5 + 8	2,3 + 2,4	91	33
2	15 + 18	4,4' + 2,2',5	90	36
3	17	2,2',4	90	29
4	16 + 32	2,2',3 + 2,4',6	80	35
5	26	2,3',5	82	35
6	31	2,4',5'	84	36
7	28	2,4,4'	80	34
8	20 + 33+53	2,3,3' + 2',3,4 + 2,2',5,6'	86	33
9	45	2,2',3,6	80	28
10	52	2,2',5,5'	78	26
11	49	2,2',4,5'	77	26
12	47 + 75	2,2',4,4' + 2,4,4',6	81	26
13	48	2,2',4,5	80	29
14	44	2,2',3,5'	76	27
15	37 + 42+ 59	3,4,4' + 2,2',3,4' + 2,3,3',6	79	39
16	41 + 64	2,2',3,4 + 2,3,4',6	77	43
17	96	2,2',3,6,6'	79	35
18	74	2,4,4',5	75	35
19	70	2,3',4',5	75	38
20	66 + 88+ 95	2,3,4,4' + 2,2'3,4.6 + 2,2',3,5',6	75	40
21	101	2, 2',4,5,5'	75	37
22	77 + 110	3,3',4,4' + 2,3,3',4',6	69	45

+D = mixture of PCBs (Delor 103) was added in described amounts. The results were calculated from 3 independent experiments, each GC value was measured in 2 parallels.

Studies were focused to investigate the fate of PCB congeners in plant cells, to identify the enzymes involved in their metabolism and the products formed. These experiments were performed at laboratory under aseptic conditions and in real soil conditions. The study was performed first with PCB mixture – Delor 103 and then with individual monochlorobiphenyls and dichlorobiphenyls in concentration of 0.3 mg/100ml [65, 66, 67].

The results showed slight transformation of all mono and dichlorobiphenyl congeners except 4,4'-dichlorobiphenyl [67]. 3- and 4-Chlorobiphenyl, PCB 4 – 2,2'-dichlorobiphenyl and PCB 10 – 2,6-dichlorobiphenyl were the most easily degraded congeners [68]. All these three congeners are relatively well soluble in water their log K_{ow} is rather low. These features are likely to increase their bioavailability for plants. Similar results were reported by the group of prof. Harms [68]. The authors showed that 12 plant species were able to metabolise 10 different PCBs. Except 4,4'-dichlorobiphenyl, all other PCBs congeners tested were at least slightly metabolised. They hypothesised that 4,4'- dichlorobiphenyl is sterically protected from the enzyme attack in spite of being more soluble in water and having lower K_{ow}.

Table 2. Assignment of PCB congeners to peaks of Delor 103 analyzed by gas chromatography and their conversion by horse radish cultured cells K54 after 2 weeks of incubation with PCB content 2.5, 5mg/100 ml of medium, expressed as the residual amount of each congener in % of the control. Modified according to [65].

Peak No.	IUPAC No.	Cl-substitution of congeners in the GC peak	Residual amount of PCBs after transformation (%)	
			+D* 5mg/100 ml	+D* 2.5mg/100 ml
1	5+8	2,3+2,4'	54	7
2	15+18	4,4'+2,2',5	55	10
3	17	2,2',4	50	8
4	16+32	2,2',3+2,4',6	54	10
5	26	2,3',5	46	8
6	31	2,4',5'	41	10
7	28	2,4,4'	36	12
8	20+33+53	2,3,3'+2',3,4 +2,2',5,6'	47	10
9	45	2,2',3,6	50	8
10	52+69	2,2',5,5'+2,4,6,3'	43	12
11	49	2,2',4,5'	35	11
12	47+75	2,2',4,4'+2,4,4',6	38	9
13	48	2,2',4,5	41	8
14	44	2,2',3,5'	42	10
15	37+42+59	3,4,4'+2,2',3,4'+2,3,3'	47	9
16	41+64	2,2',3,4 + 2,3,4',6	38	12
17	96	2,2',3,6,6'	42	8
18	74	2,4,4',5	39	8
19	70	2,3',4',5	43	9
20	66+88+95	2,3,4,4'+2,2'3,4.6 +2,2',3,5',6	31	9
21	101	2, 2',4,5,5'	38	8
22	77+110	3,3',4,4'+2,3,3',4',6	35	22

+D = mixture of PCBs (Delor 103) was added in described amounts. The results were calculated from 3 independent experiments, each GC value was measured in 2 parallels.

In our studies we further paid attention to the metabolites generated by the plant cells. In order to identify the structure of the metabolites and the position of attack to investigate number and structure of the products, using improved and more sensitive analytical procedures and increasing the biomass used for the assays, we showed that although 4,4'-dichlorobiphenyl was metabolised poorly, one hydroxylated product generated from this congener (Table 3) was formed.

Analyses of the transformation products of plant cultures exposed to dichloro-, trichloro- and tetrachlorobiphenyls showed that even the most persistent 3,3',4,4'-tetrachlorobiphenyl can be metabolised [69].

Table 3. Hydroxylated derivatives of monochlorobiphenyls formed from individual monochlorobiphenyls during 14 days incubation with plant cells of tobacco, black-nightshade horseradish and alfalfa.

PCB	Plant species	Number of hydoxy products	Identified structure
PCB 1, 2-chlorobiphenyl	Tobacco	4	No product was precisely identified
	Black nightshade	5	One product identified as 2Cl-5OH biphenyl
	Horseradish	5	Two products were identified as 3Cl-2OH biphenyl and 2Cl-5OH biphenyl
	Alfalfa	0	No product was identified
PCB 2, 3-chlorobiphenyl	Tobacco	3	No product was precisely identified
	Black nightshade	4	Two products identified 3Cl4-OH biphenyl, 3Cl-6OH biphenyl
	Horseradish	4	One product identified as 3Cl-4OH biphenyl
	Alfalfa	0	No product was identified
PCB 3, 4-chlorobiphenyl	Tobacco	3	4Cl-4'OH biphenyl
	Black nightshade	2	4Cl-4'OH biphenyl
	Horseradish	4	4Cl-4'OH biphenyl
	Alfalfa	1	4Cl-4'OH biphenyl
PCB 4 2,2'dichlorobiphenyl	Tobacco	3	Dichlorohydroxybiphenyls No product was precisely identified
	Black nightshade	3	Dichlorohydroxybiphenyls No product was precisely identified
PCB 9 2,5 -dichlorobiphenyl	Tobacco	2	2,5diChl-3OH biphenyl 2,5diChl-4OH biphenyl
	Black nightshade	6	2,5diChl-2OH biphenyl 2,5diChl-3OH biphenyl 2,5diChl-4OH biphenyl

To identify chemical structure, position of chlorines and hydroxyl groups plant cells of four different species were incubated with monochlorobiphenyls. GC-MS analysis revealed that metabolites identified in plant tissue were mono- and dihydroxy-compounds. Some of them could be identified by GC-MS by comparison with the structures of the standards. The results are summarized in Table 3. Altogether, data showed that although different plant species metabolise PCB congeners following a similar pattern for any PCB congener, the different plant species form the same types of hydroxychloroderivatives but number of products and positions of hydroxyl group(s) differ with the plant species. Alfalfa has quite low potential to metabolise PCBs, and its ability is limited only to transformation of 4-chlorobiphenyl, which is metabolised to 4Cl-4'OH biphenyl, the same product found in cells of other plant.

5.1.2 Enzymes responsible for PCB transformation in plants

As was previously documented in case of many environmental xenobiotics, phytotransformation refers a process by which organic contaminants are uptake by plants of organic contaminants from soil and groundwater and subsequently the xenobiotics are metabolised or transformed. Various plant xenobiotics-converting enzymes, cytochrome P450, peroxidases, glutathione–S-transferases, carboxyesterases, O-glucosyl transferases, O-malonyl transferases, N-glucosyl transferases and N-malonyl transferases, were isolated and purified. They were proved to metabolise efficiently various xenobiotics such as PCBs, organic solvents and chlorinated pesticides [70, 71].

Organic pollutants undergo in plants a three -stage metabolic processes similar to mammalian metabolism [72]. First phase which includes oxidation, reduction or hydrolysis of xenobiotics often generates metabolites with increased polarity [67, 71, 72, 73]. These reactions are catalyzed by mixed functional oxidases (cytochrome P450) and/or peroxidases to introduce an hydroxyl group or to substitute common functional groups (nitro, carboxyl, alkyl or halogens) [68]. Plants hydroxylate PCBs to form more soluble hydroxyderivatives [67, 74, 75, 76]. This reaction suggests that polychlorinated biphenyls oxidases, cytochrome P450 and/or peroxidases present in plant cells are involved in PCB conversion. The identity of the enzymes responsible for plant transformation of xenobiotics was already a matter of discussion in the early seventieth, when the first reports about PCB metabolisation were published [77]. In 1992 it was discussed by Lee and Fletcher [74] who studied involvement of cytochrome P450 and peroxidase in PCB transformation in presence of specific inhibitors for cytochrome P450 and peroxidases. Their study provided evidence that cytochrome rather than peroxidases were the major contributors to plant metabolism of PCB. More recently, we got evidence that peroxidases are also likely [62, 63] involved in PCB transformation. It was shown that plant tissue cultures of various species, exhibiting higher peroxidase activities in PCB presence, possess higher capabilities to transform them. Chroma *et al.* [63] used different specific inhibitors for both enzymic systems and measured not only PCB removal but also peroxidase activities. Presence of all inhibitors resulted in a decrease of PCB degradation efficiency compared to the controls without any inhibitor. Nevertheless, further following analyses showed a decrease of peroxidase activity not only when peroxidase's but also cytochrome P450 inhibitors were added (see Table 4). This phenomenon supported idea that both these enzymic systems are involved in the metabolism of PCBs in plant cells.

Tab. 4. The effect of inhibitors on peroxidase of SNC-9O and its PCB transformation. Modified according [63].

Inhibitor (Enzyme inhibited)	**PCB**	**Removal of PCB [%]**	**POX activity [RLUd × 107/mg protein]**
None	–	–	390
None	+	45	318
Propylgallate (POX)	+	10	48
Sodium benzoate (POX)	+	15	84
Aminobenztriazol.(cyt P450)	+	20	49
Metyrapone (cyt P450)	+	12	37

Transformation and degradation of organic pollutants by peroxidases has been already described in many reports [49, 50, 51, 52, 78, 79, 80, 81, 82, 83, 84]. Most of these addressed the peroxidases of the so-called white rot fungi (WRF), basidiomycetes strains involved in wood decay. Degradation of a number of environmental pollutants by ligninolytic system of white rot fungi was described many times [49, 84, 85].

Plant peroxidases, namely horseradish peroxidase [86] and their mechanism of catalytic reaction on various xenobiotics [87] (including phenolic compounds [88, 89]), were described in detail. The classical reaction catalysed by heme-peroxidases is oxidative dehydrogenation, although they also catalyze a variety of related reactions, including oxygen transfer, hydrogen peroxide cleavage, and peroxidative halogenations.

The plant enzyme-mediated decomposition and biodegradation of PCB were recently investigated [76]. 5-dichlorobiphenyl (PCB 9) and 2,2',5,5'- tetrachlorobiphenyl (PCB 52) were used as model compounds to study efficiency and mechanism of the degradation processes. It was found that the application of commercial horseradish peroxidase (HRP) together with defined amounts of hydrogen peroxide removed 90% of the initial concentration of PCB 9 and 55% of the initial concentration of PCB 52 from an aqueous solution after a reaction period of 220 min. Dechlorination was observed as the initial step. Although the metabolites identified were mainly chlorinated hydroxybiphenyls, benzoic acids and non-substituted 1,10-biphenyl, some higher chlorinated biphenyl isomers were also detected. As a comparison can be added that the biodegradation of PCB 9 using the white rot fungus *Trametes multicolor* took about four weeks and reduction was about 80% of the initial concentration. The metabolites produced (dichlorobenzenes, chlorophenols and alkylated benzenes) were not quite the same as those observed upon incubation with HRP.

To investigate the capacity of peroxidases of different plant species to transform PCBs, plant peroxidases from the cells of tobacco, black nightshade and alfalfa were isolated, partially purified and concentrated by ammonium sulphate precipitation.

Then transformation experiments with various PCB congeners (mono and dichlorobiphenyls) were performed. All cultures (hairy root of black nightshade, callus of tobacco and callus of alfalfa) synthesized both intracellular and extracellular peroxidases. Peroxidases of tobacco and black nightshade changed in the presence of PCBs their total activities and some changes in isoenzyme patterns were also visible after native electrophoresis. Total peroxidase activity of alfalfa exhibited the highest values when measured in absence of PCB presence. Its activity decreased and isoenzyme pattern was not influenced by the presence of PCBs and this plant was generally less active in PCB transformation (data not shown) [63]. Peroxidases isolated from tobacco and black nightshade transformed various PCBs with different efficiency. Significant differences were detected when congeners with chlorines at different positions were used for the reaction with peroxidases. In several cases peroxidases from tobacco or nightshade exhibited different potential to transform the same congeners. The highest yield was obtained after reaction with PCB 3 and PCB 9, the lowest transformation was detected in cases of PCB 13 and 15. The results are summarized in Table 5. These experiments confirmed that peroxidases of different plant species can actively metabolise polychlorinated biphenyls, but reaction specificity is not clear yet. Efficiency of degradation of structurally different PCB is related to their chemical structure, position and number of chlorines, physico-chemical properties but also plant species, and PCB toxicity to plant cells of diverse species.

Small amounts of numerous products formed from PCBs during reactions of peroxidases, isolated from black nightshade, with were detected by GC-MS after silylation. Reactions with tobacco peroxidase did not give any obvious products, only traces of compounds that are possibly di-OH-dichlorobiphenyl occurred after reaction of tobacco peroxidase with PCB 9. After derivatization the concentration of products was under detection limit and original hypothesis could not be confirmed. In reaction mixtures containing peroxidase from black nightshade, dechlorination was observed as the initial step. As metabolites less chlorinated biphenyls than the original ones or even the non-substituted biphenyl were detected. Reaction of horseradish peroxidase with PCBs gave very low concentrations of hydroxychlorobiphenyls [76]. In our study surprisingly no hydroxylated chlorobiphenyls were found. This is most likely explained by further fast reactions following the first dechlorination step without any accumulation of intermediates. Regarding oxidative degradation, the cleavage of the ring system and subsequent reactions gave benzoic acid and hydroxybenzoic acids as the products (see Figure 2). As a result of POX radical mechanism, higher chlorinated isomers were formed. Traces of phenylacylchlorides were also found in reaction mixtures [63].

Table 5. Transformation of mono- and dichlorobiphenyls by intracellular plant peroxidases isolated from cells of black nightshade and tobacco.

IUPAC No.	PCB	Degradation of PCB by POX from black nightshade (%)	Degradation of PCB by POX from tobacco (%)
Delor 103	Mix	45 ± 5	20 ± 3
PCB 1	2-Cl	35 ± 5	–*
PCB 2	3-Cl	35 ± 7	–*
PCB 3	4-Cl	100 ± 10	79 ± 10
PCB 4	2,2'- diCl	56 ± 9	60 ± 8
PCB 5	2,3 - diCl	63 ± 8	17 ± 3
PCB 7	2,4 - diCl	59 ± 7	47 ± 6
PCB 8	2,4'- diCl	16 ± 2	44 ± 5
PCB 9	2,5 - diCl	66 ± 4	90 ± 8
PCB10	2,6 - diCl	14 ± 2	48 ± 5
PCB 11	3,3'- diCl	28 ± 4	66 ± 5
PCB 12	3,4 - diCl	–*	59 ± 8
PCB 13	3,4'- diCl	– *	0
PCB 14	3,5 - diCl	16 ± 2	12 ± 2
PCB15	4,4'- diCl	8 ± 1	–*

* Transformation was not measured

Residual amount of individual congeners were analysed after 20 hours of the reaction of peroxidases and 5-10μmol/l of PCB congeners. Reaction was started with 5mmol/l of hydrogen peroxide

Our results have proved that an important role in PCB transformation can be attributed also to plant peroxidases. In plant biochemistry peroxidases are mostly studied as functional enzymes in plant cell growth and development, ethylene biosynthesis, lignification, important also in protection of plants during stress as protective enzymes against reactive oxygen species generated during biotic and abiotic stress [90]. Our results confirmed that peroxidases from different plant sources (tobacco, black-nightshade) than horseradish (the most studied model peroxidase) can be involved not only in antioxidative defense mechanisms, but they even could play important role in direct transformation and degradation of xenobiotics [91].

5.1.3 Cooperation and interactions between PCB metabolising bacteria and plants can affect their removal from the environment

Microbial activity has been deemed the most influential and significant cause of PCB removal. Numerous studies have been conducted on microbial consortia and enrichment, and several diverse genera of bacteria have been isolated (for a review, see reference [92]). Recent work has indicated that the stimulation of microbial activity in the rhizosphere of plants can also stimulate biodegradation of various toxic

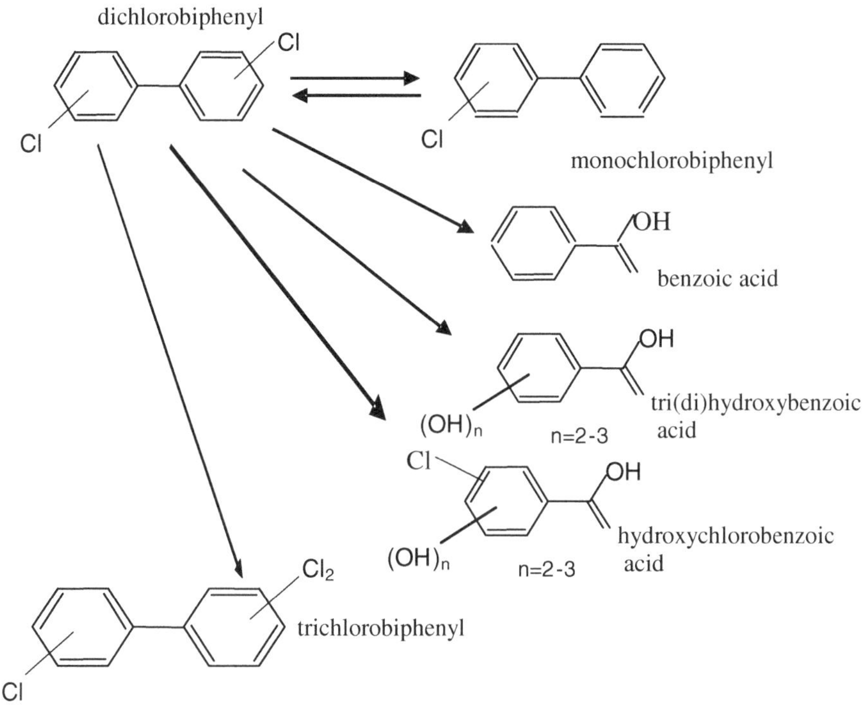

Figure 2. Products identified after reaction of peroxidase isolated from black nightshade with individual congeners of PCBs (modified according [91]).

organic compounds [93]. The rhizosphere soil has been described as the zone of soil under the direct influence of plant roots and usually extends a few millimetres from the root surface and is a dynamic environment for microorganisms. The rhizosphere microbial community is comprised of microorganisms with different types of metabolism and adaptive responses to variation in environmental conditions. The production of mucilaginous material and the exudation of a variety of soluble organic compounds by the plant root play an important part in root colonization and maintenance of microbial growth in the rhizosphere. Microbial activity is thus generally higher in the rhizosphere due to readily biodegradable substrates exuded from the plants [94, 95, 96, 97, 98]

Recent studies indicate that degradation of PAHs and polychlorinated biphenyls (PCBs) in soils can be enhanced by plants. Published results support the hypothesis and provide mechanistic understanding of rhizosphere degradation {98, 99]: (1) purified natural plant compounds (i.e. flavanoids) stimulate the growth and activity of PCB degrading bacteria [99]; (2) Plant roots release phenolic compounds that support the growth of PCB degrading bacteria, but all plant species are not effective [100]; (3) Flavanoid compounds that support the growth of PAH-degrading bacteria accumulate in aging/dying fine roots of mulberry [100]. The combined interpretation of these data is that the roots of some plant species are capable of growing to immobile soil contaminants (PCBs, PAHs) and deliver cometabolites (i.e. flavanoids) upon fine root death. These natural cometabolites foster the growth and activity of degradative microbes. The dead/decayed roots also create soil cavities that facilitate soil aeration.

Plant-microbial interactions in contaminated soil taking place on the level of growth and support of degradation abilities were already studied and a lot of information about cooperative mechanism was described. Less information is available about possible linking up the metabolism of bacteria and plants as a result of followed metabolisation of metabolic products formed during primary degradation of PCBs in both types of organisms. Particularly our interest was directed to ability of plants originally transforming PCB, to metabolise also bacterial PCB degradation products, i.e. chlorobenzoates. On the contrary if bacteria degrading PCBs are able to transform plant primary metabolites – hydroxychlorobiphenyls [101]. Similarly Sondossi *et al.* [102, 103] studied transformation of nonchlorinated hydroxybiphenyls. Commercially available hydroxychlorobiphenyls were transformed *in vitro* by bacterial enzymes isolated from two PCB degrading bacteria *Pseudomonas xenovorans* LB400 and *Comamonas testosteroni* B356 [101]. Products of each step of bacterial PCB pathway were detected after derivatization by GC-MS. Data obtained after the first reaction step of three hydroxychlorobiphenyls, 3Cl-2OH; 5Cl2OH and 3,5Cl2OH with biphenyldioxygenase (BPDO) confirmed, that both B-356 BPDO and LB400 BPDO oxygenate mono-substituted-hydroxy- and hydroxychlorobiphenyls on the non-substituted ring. A *meta-para* oxygenation of the *ortho*-hydroxylated ring would have required a dehalogenation of 2-hydroxy-3,5-dichlorobiphenyl, but no dehalogenated metabolites were detected when this compound was used as substrate. This confirms that neither LB400 BPDO nor B-356 BPDO can oxygenate the substituted ring of the *ortho*-hydroxylated biphenyl. Furthermore, the fact that the metabolites generated by LB400 BPDO were identical to those generated by B-356 BPDO for all three hydroxychlorobiphenyls tested exclude any other mode of attack of these compounds by LB400 BPDOs than the *ortho-meta* oxygenation. When the reaction was done with 3Cl-2OH; 5Cl-2OH and 3,5Cl-2OH, and BphAEFGBC, metacleavage compound was produced. Thus, these compounds could be converted, probably efficiently to HOPDA. However, it seemed that BphD did not work on the HOPDAs carrying an hydroxyl group on the phenyl ring. When 2Cl-4OH and 3Cl-4OH were submitted to BphAEFGBC, a lots of HOPDA was produced, thus the triOH was a good substrate for BphC. Similarly, when submitted to BphAEFGBCD, hydroxybenzoic acid was detected. This observation indicated that with respect to the catalytic activity of BPDO, the mode of oxygenation of substrate analogs as well as the stability of the metabolites produced can greatly differ depending on the type of substituents present on the biphenyl ring. Thus caution should be taken when transposing data reported for the bacterial degradation of PCB congeners to the bacterial degradation of hydroxylated PCB metabolites derived from plants. From above mentioned results can be concluded that some intermediates of plant transformation of polychlorinated biphenyls can be further degraded by the same bacteria which degrade original PCBs. More soluble hydroxychlorobiphenyls are more available than original PCBs and can be released during decaying of plants and root turnover. On the other hand they are more toxic [104], especially for non-degrading bacteria. In context of this observation we suggest that PCB degrading bacteria possess unique ability and the action of microbial associations can be responsible for generally wider degrading capacity of biological systems important when evaluating final degradation potential in contaminated environment.

Chlorobenzoic acids (CBAs) are one of the end products of so-called upper aerobic bacterial degradation pathway (see Figure 1) of PCBs and they can be available to plants growing in consortia with PCB degrading microbes. To explain the fate of xenobiotics and their intermediates we also intended to find the answer, if plants were able to metabolise CBAs. As a model we chose four plant tissue cultures used before to examine their PCB transformation potential and peroxidases activities, namely black nightshade (*Solanum nigrum*), tobacco (*Nicotiana tabacum*), and horseradish (*Armoracia rusticana*) alfalfa (*Medicago sativa*) [105]. Plant cells of horseradish and black nightshade showed to be the most active cultures in respect of CBAs metabolism. Both species metabolised about 70% of 2-chloro and 2,5-dichlorobenzoic acids within one week (see Table 6). 2CBA is probably the most suitable substrate due to its best solubility and bioavailability. Cells of alfalfa and tobacco plants did not degrade significant amounts of CBAs and all chosen CBAs proved much higher toxic effect on vitality and growth of these two species in comparison with horseradish and black nightshade cultures. When evaluating the results of CBAs transformation by plants it can be concluded that ortho-position in molecule of chlorobenzoate supports the transformation potency. As well as in previous cases [55, 63], chlorine in ortho-position is crucial for transformation not only of PCBs [55, 75], but also in metabolism of chlorobenzoic acids.

Our study proved that certain plant species are able to degrade some chlorobenzoic acids entering the environment due to microbial PCB degradation but also industrial and agricultural activities. From the literature it is known, that plants help the rhizospheric microflora to remove several organic and inorganic pollutants by supporting its activities and living conditions. Our experiments showed that plant cells could help to remove CBAs also by their direct metabolisation and removal from the environment.

Direct and indirect interactions and mutual cooperation of plants and growth promoting rhizosphere bacteria were already studied and the aspects, advantages and pitfalls of the processes were described for many types of inorganic and organic chemicals [106, 107, 108, 109]. Many laboratories are studying processes of biological degradation by different systems to increase the existing knowledge about fate, toxicity and effect of contaminants. Such approach needs not only detailed information about uptake, transport, metabolism and toxicity of pollutants and their products, but also physiology of individual organisms, enzymes involved and complex interactions of organisms included within the environment [110, 111, 112, 113].

The storage of organic compounds, the presence of transformed and conjugated derivatives are also causes for concern [114] to explain unknown metabolic and toxicological impacts. Our study shows further possibility of interactions between bacteria and plants in contaminated environment and on example of model contaminant – PCB, gives more detailed information about abilities of biological systems to metabolised original xenobiotics and also their intermediates and products.

TABLE 6. Degradation of chlorobenzoic acids during 14-days cultivation with plant tissue cultures of Armoracia rusticana K54. Solanum nigrum SNC 9O and Nicotiana tabacum WSC 38.

	Residual CBA [%]		
CBA	***Armoracia rusticana***	***Solanum nigrum***	***Nicotiana tabacum***
	Day 14	Day 14	Day 14
2-CBA	5 ± 3	13 ± 2	88 ± 1
3-CBA	90 ± 3	91 ± 4	95 ± 4
4-CBA	82 ± 5	99 ± 2	94 ± 3
2,3-di CBA	77 ± 5	58 ± 10	99 ± 6
2,4-di CBA	65 ± 2	61 ± 10	79 ± 9
2,5-di CBA	77 ± 5	65 ± 3	85 ± 4
2,6-di CBA	66 ± 10	80 ± 5	98 ± 2
3,4-di CBA	93 ± 3	89 ± 12	98 ± 6
3,5-di CBA	92 ± 6	95 ± 12	100 ± 5
2,3,5-tri CBA	95 ± 7	88 ± 11	96 ± 9
2,4,6-tri CBA	90 ± 2	95 ± 2	99 ± 3

Initial concentration of CBAs was 200 mg/l. As controls cells killed by boiling for 20 minutes were used.

6. Conclusions

During last 15 years many examples describing various plant species accumulating and/or metabolising organic xenobiotics were described [55, 56, 57, 63, 75, 105, 107, 108, 109, 110, 111, 112, 113]. First interest was directed to uptake and translocation of organics within plants [114], later the approach was shifted to experiments investigating the metabolic routes leading to detoxification of xenobiotics in individual organisms. Sandermann [115], Burken and Schnoor [114] described general so-called "green liver model" showing common steps in plant and mammalian metabolism, others studied particular metabolisation of xenobiotics. Many laboratories have been studying plant metabolism on level of plant tissue cultures (*in vitro* systems) or hydroponic systems that provide information on the metabolism, phytotoxicity, products formed and enzymes involved. Many investigations showed that qualitatively the metabolic pathways in plant tissue cultures and intact plants are the same [75]. As a result of more comprehensive information about natural systems, more sophisticated analytical methods and development of molecular biology and genetic engineering, also metabolic pathways and interactions between all involved natural organisms and complexity of their action became important in phytoremediation studies [71, 74, 92, 94, 95, 96, 97, 98,

99, 100, 105, 106]. Nowadays, laboratories are focused on processes for the removal of the contaminants based on cooperation of different systems, optimizing agronomic and agricultural practises, post-harvesting processes, ecological risk assessments or construction of genetically modified plants combining autotrophic energy independence of plants with heterotrophic enzymatic processes needed for complete mineralization [114, 116]. The metabolic degradation routes of microorganisms (namely bacteria) and higher organisms are different [91, 117, 118, 119] and lead to different products from the point of structure and toxicity [110, 111, 116]. From this point of view transgenic plants have great potential for field applications.

It seems that in spite we have available a lot of information to use the plants for phytoremediation purposes, still much work is necessary to forecast all aspects of the beneficial application of plants in the field [107, 108, 112, 114]. In this context needs for bench and field-testing are still actual to extend theoretical knowledge and practical experience.

Acknowledgement

The authors thank to grant agencies of Ministry of Education of the Czech Republic which sponsored the experiments and work described above by the grant No. MSM 6046137305, No. 1PO5ME 745 and Z40550506, Grant Agency of the Czech Republic – grants No. GACR 521/05/0500, and 203/06/0563

References

[1] Schulz, DE; Petrick, G and Duinker, JC (1989) Complete characterization of polychlorinated biphenyl congeners in commercial Aroclor and Clophen mixtures by multidimensional gas chromatography –electron capture detection. Environ Sci Technol 23: 852-859

[2] Jensen, S (1966) Report of a new chemical hazard. New Science 32: 612

[3] Risebrough, RW; Walker, W; Schmidt, TT; deLappe, BW and Connors, CW (1976) Transfer of chlorinated biphenyls to Antarctica. Nature 264(5588): 738-739

[4] Montone, RC; Taniguchi, S and Weber, RR (2001) Polychlorinated biphenyls in marine sediments of Admiralty Bay, King George Island, Antarctica. Mar Pollut Bull 42(7): 611-614

[5] Safe, S (1980) Polyhalogenated aromatic compounds, in R Kimbrough, Ed., Halogenated biphenyls, naphtalenes, dibenzodioxins and related products. Elsevier North Holland Pub, Amsterdam

[6] Ross, PS; Jeffries, SJ; Yunker, MB; Addison, RF; Ikonomou, MG and Calambokidis, JC (2004) Harbour seals (*Phoca vitulina*) in British Columbia, Canada, and Washington State, USA, reveal a combination of local and global polychlorinated biphenyls, dioxin and furan signals. Environ Toxicol Chem 23(1): 157-165

[7] Kannan, K; Kajiwara, N; Watanabe, M; Nakata, H; Thomas, NJ; Stephenson, M; Jessup, DA and Tanabe, S (2004) Profiles of polychlorinated biphenyls congeners, organochlorine pesticides and butyltins in southern sea otters and their prey. Environ Toxicol Chem 23(1): 49-56

[8] Solomon, GM and Weiss, PM (2002) Chemical contaminants in breast milk: time trends and regional variability. Environ Health Perspect 110(6): A339-347

[9] Bates, MN; Buckland, SJ; Garrett, N; Ellis, H; Needham, LL; Patterson, DG Jr; Turner, WE and Russell, DG (2004) Persistent organochlorines in the serum of non-occupationally exposed New Zeland population. Chemosphere 54(10): 1431-1443

[10] Orloff, KG; Dearwent, S; Metcalf, S; Kathman, S and Turner, W (2003) Human exposure to polychlorinated biphenyls in a residential community. Arch Environ Contam Toxicol 44(1): 125-131

[11] Harrad, JS; Sewart, AP; Alcock, R; Boumphrey, R; Burnett, V; Duarte-Davidson, R; Halsall, C; Sanders, G; Waterhouse, K; Wild, SR and Jones, KC (1994) Polychlorinated biphenyls (PCBs) in the British environment: sink, sources and temporal trends. Environ Pollut 85(2): 131-146

[12] Golden, R; Doull, J; Waddell, W and Mandel, J (2003) Potential human cancer risks from exposure to PCBs: a tale of two evaluations. Crit Rev Toxicol 33(5): 543-580

[13] Faroon, OM; Keith, S; Jones, D and De Rosa, C (2001) Carcinogenic effects of polychlorinated biphenyls. Toxicol Ind Health Mar 17(2): 41-62

[14] Ahmed, M and Focht, DD (1973) Degradation of polychlorinated biphenyls by two species of *Achromobacter*. Can J Microbiol 19: 47-52

[15] Furukawa, K; Matsumura, F and Tonomura, K (1978) *Alcaligenes* and *Acinetobacter* capable of degrading polychlorinated biphenyls. Agr Biol Chem 42: 543-548

[16] Sylvestre, M and Fauteux, J (1982) A new facultative anaerobe capable of growth on 4-chlorobiphenyl. J Gen Appl Microbiol 28: 61-72

[17] Bopp, LH (1986) Degradation of highly chlorinated PCBs by *Pseudomonas* strain LB400. J Ind Microbiol 1: 23-29

[18] Bedard, DL; Wagner, RE; Brennan, MJ; Haberl, ML and Brown, JF Jr (1987) Extensive degradation of Aroclors and environmentally transformed polychlorinated biphenyls by *Alcaligenes eutrophus* H850. Appl Environ Microbiol 53(5): 1094-1102

[19] Asturias, JA and Timmis KN (1993) Three different 2,3-dihydroxybiphenyl-1,2-dioxygenase genes in the Gram-positive polychlorobiphenyl-degrading bacterium *Rhodococcus globerulus* P6. J Bacteriol 175(15): 4631-4640

[20] Sierra, I; Valera, JL; Marina, ML and Laborda, F (2003) Study of the biodegradation process of polychlorinated biphenyls in liquid medium and soil by a new isolated aerobic bacterium (*Janibacter* sp). Chemosphere 53(6): 609-618

[21] Haddock, JD and Gibson, DT (1995) Purification and characterization of the oxygenase component of biphenyl 2,3-dioxygenase from *Pseudomonas* strain LB400. J Bacteriol 177: 5834-5839

[22] Hurtubise, Y; Barriault, D; Powlowski, J and Sylvestre, M (1995) Purification and characterization of the *Comamonas testosteroni* B-356 biphenyl dioxygenase components. J Bacteriol 177: 6610-6618

[23] Haddock, JD; Horton, JR and Gibson, DT (1995) Dihydroxylation and dechlorination of chlorinated biphenyls by purified biphenyl 2,3-dioxygenase from *Pseudomonas* sp. Strain LB400. J Bacteriol 177: 20-26

[24] Seeger, M; Zielinski, M; Timmis, KN and Hofer, B (1999) Regiospecificity of dioxygenation of di- to penta-chlorobiphenyls and their degradation to chlorobenzoates by the *bph*-encoded catabolic pathway of *Burkholderia* sp strain LB400. Appl Environ Microbiol 65: 3614-3621

[25] Kumamaru, T; Suenaga, H; Mitsuoka, M; Watanabe, T and Furukawa, K (1998) Enhanced degradation of polychlorinated biphenyls by directed evolution of biphenyl dioxygenase. Nat Biotechnol 16(7): 663-666

[26] Brühlmann, F and Chen, W (1999) Tuning biphenyl dioxygenase for extended substrate specificity. Biotechnol Bioeng 63(5): 544-551

[27] Suenaga, H; Watanabe, T; Sato, M; Ngadiman, T and Furukawa, K (2002) Alteration of regiospecificity in biphenyl dioxygenase by active-site engineering. J Bacteriol 184(13): 3682-3688

[28] Barriault, D and Sylvestre, M (2004) Evolution of the biphenyl dioxygenase from *Burkholderia xenovorans* LB400 by random mutagenesis of multiple sites in region III. J Biol Chem 279 (46): 47480-47488

[29] Dai, S; Vaillancourt, FH; Maaroufi, H; Drouin, NM; Neau, DB; Snieckus, V; Bolin, JT and Eltis, LD (2002) Identification and analysis of a bottleneck in PCB biodegradation. Nat Struct Biol 9(12): 934-939

[30] Brown, JF and Wagner, RE (1990) PCB movement, dechlorination and detoxification in the Acushnet estuary. Environ Toxicol Chem 9: 1215-1233

[31] Lake, JL; Pruell, RJ and Osterman, FA (1992) An examination of dechlorination processes and pathways in New Bedford harbor sediments. Marine Environ Res 33: 1

[32] Abramowicz, DA; Brennan, MJ; Van Dort, HM and Gallagher, EL (1993) Factors influencing the rate of PCB dechlorination in Hudson River sediments. Environ Microbiol 58: 1057

[33] Hartcamp-Commandeur, LCM (1996) Reductive dehalogenation of polychlorinated biphenyls by anaerobic microorganisms enriched from Dutch sediments. Chemosphere 32: 1275-1286

[34] Williams, WA (1997) Stimulation and enrichment of two microbial polychlorinated biphenyl reductive dechlorination activities. Chemosphere 34: 665-669

[35] Bedard, DL; Van Dort, HM; May, RJ and Smullen, LA (1997) Enrichment of microorganisms that sequentially *meta, para*-dechlorinate the residue of Aroclor 1260 in Housatonic River sediments. Environ Sci Technol 31: 3300-3307

[36] Kuipers, B; Cullen, WR and Mohn, WW (1999) Reductive dechlorination of nonachlorobiphenyls and selected octachlorobiphenyls by microbial enrichment cultures. Environ Sci Technol 33: 3577-3583

[37] Cuttel, L; Sowers, KR and May, HD (1998) Microbial dechlorination of 2,3,5,6,- tetrachlorobiphenyl undr anaerobic conditiond in the absence of soil or sediment. Appl Environ Microbiol 64(8): 2966-2969

[38] Quensen, JF; Tiedje, JM and Boyd, SA (1988) Reductive dechlorination of PCBs by anaerobic microorganisms from sediments. Science 242: 752-754

[39] Berkaw, M; Sowers, KR and May, HD (1996) Anaerobic ortho dechlorination of polychlorinated biphenyls by estuarine sediments from Baltimore Harbor. Appl Environ Microbiol 62: 2534-2539

[40] Van Dort, HM and Bedard, DL (1991) Reductive ortho and meta dechlorination of a polychlorinated biphenyl congener by anaerobic microorganisms. Appl Environ Microbiol 57: 1576-1578

[41] Wu, Q; Bedard, DL and Weigel, J (1997) Effect of incubation temperature on the route of microbial reductive dechlorination of 2,3,4,6-tetrachlorobiphenyl in polychlorinated biphenyl (PCB)-contaminated and PCB-free freshwater sediments. Appl Environ Microbiol 63: 2836-2843

[42] Brown, JF; Wagner, RE; Feng, H; Bedard, DL; Brennan, MJ; Carnahan, JC and May, RJ (1987) Environmental dechlorination of PCBs. Environ Toxicol Chem 6: 579-593

[43] Cutter, LA; Watts, JE; Sower, KR and May, HD (2001) Identification of a microorganism that links its growth to the reductive dechlorination of 2,3,5,6-chlorobiphenyl. Environ Microbiol 3(11): 699-709

[44] Wu, Q; Watts, JE; Sower, KR and May, HD (2002) Identification of a bacterium that specifically catalyzes the reductive dechlorination of polychlorinated biphenyls with doubly flanked chlorines. Appl Environ Microbiol 68(2): 807-812

[45] Tiedje, JM; Quensen, JF; Chee-Sanford, J; Schimel, JP and Boyd, SA (1993-1994) Microbial reductive dechlorination of PCBs. Biodegradation 4(4): 231-240

[46] Bumpus, JA; Tien, M; Wright, D and Aust, SD (1985) Oxidation of persistent environmental pollutants by a white rot fungus. Science 228: 1434-1436

[47] Eaton, D (1985) Mineralization of polychlorinated biphenyls by *Phanerochaete chrysosporium*: a lignolytic fungus. Enzyme Microb Technol 7: 194-196

[48] Hundt, K; Jonas, U; Hammer, E and Schauer, F (1999), Transformation of diphenyl ethers by *Trametes versicolor* and characterization of ring cleavage products. Biodegradation 10: 279-286

[49] Davila-Vazquez, G; Tinoco, R; Pickard, MA and Vazquez-Duhalt, R (2005) Transformation of halogenated pesticides by versatile peroxidase from *Bjerkandera adusta*. Enzyme Microb Technol 36: 223-231

[50] Thomas, DR; Carswell,, KS and Georgiou, G (1992) Mineralization of biphenyl and PCBs by the white rot fungus *Phanerochaete chrysosporiun*. Biotechnol Bioeng 40: 1395-1402

[51] Beaudette, LA; Davies, S; Fedorak, PM; Ward, OP and Pickard, MA (1998) Comparison of gas chromatography and mineralization experiments for measuring loss of selected polychlorinated biphenyl congeners in cultures of white rot fungi. Appl Environ Microbiol 64(6): 2020-2025

[52] Dietrich, D; Hickey, WJ and Lanar, R (1995) Degradation of 4,4'-dichlorobiphenyl, 3,3',4,4'-tetrachlorobiphenyl and 2,2',4,4',5,5'-hexachlorobiphenyl by white rot fungus *Phanerochaete chrysosporium* Appl Environ Microbiol 61: 3904-3909

[53] Schultz, A; Jonas, U; Hammer, E and Schauer F (2001) Dehalogenation of chlorinated hydroxybiphenyls by fungal laccase. Appl Environ Microbiol 67: 4377-4381
[54] Romero, MC; Hammer, E; Hanschke, R; Arambarri, AM and Schauer, F (2005) Biotransformation of biphenyl by the filamentous fungus *Talaromyces helicus*. World J Microbiol Biotechnol 21: 101-106
[55] Wilken, A; Bock, C; Bokern, M and Harms, H (1995) Metabolism of different PCB congeners in plant cell cultures. Environ Chem Toxicol 14: 2017-2022
[56] Mackova, M; Macek, T; Kucerova, P; Burkhard, J; Pazlarova, J and Demnerova, K (1997) Degradation of polychlorinated biphenyls by hairy root culture of *Solanum nigrum*. Biotechnol Let 19: 787-790
[57] Mackova, M; Macek, T; Burkhard, J; Ocenaskova, J; Demnerova, K and Pazlarova, J (1997) Biodegradation of polychlorinated biphenyls by plant cells. Int Biodeterior Biodegrad 39: 317-325
[58] Macek, T; Kotrba, P; Suchová, M; Skácel, F; Demnerová, K and Ruml, T (1994) Accumulation of cadmium by hairy-root cultures of *Solanum nigrum*. Biotechnol Let 16 (6): 621-624
[59] Macek, T; Kotrba, P; Ruml, T; Skácel, F and Macková, M (1997) Accumulation of cadmium by hairy root cultures, in PM Doran, Ed., Hairy Roots, pp. 133-138, Gordon and Breach/Harwood Academic, London, Sydney, UK
[60] Demnerová, K; Burkhard, J; Kostál, J; Macek, T; Macková, M; Pazlarová, J, Kastánek, F and Kuncová, G (1997) Biodegradation of alkanes and PCBs. Experience in the Czech Republic, in FW Holm, Ed., Mobile Alternative Technologies, 53-70, Kluwer Academic Publishers, Dordrecht, Boston, London
[61] Kas, J; Burkhard, J; Demnerová, K; Kostál J; Macek, T; Macková, M and Pazlarová, J (1997) Perspectives in biodegradation of alkanes and PCBs. Pure Appl. Chem. 69, (11): 2257-2369
[62] Kucerová, P; Macková, M; Poláchová, L; Burkhard, J; Demnerová, K; Pazlarová, J and Macek, T (1999) Correlation of PCB transformation by plant tissue cultures with their morphology and peroxidase activity changes. Coll Czech Chem Commun 64 (9): 1497-1509
[63] Chromá, L; Macková, M; Kucerova, P; in der Wiesche, C; Burkhard, J and Macek, T (2002) Enzymes in plant metabolism of PCBs and PAHs. Acta Biotechnol 22: 35-41
[64] Groeger, AW and Fletcher, JS (1988) The influence of increasing chlorine content on the accumulation and metabolism of polychlorinated biphenyls by Paul's Scarlet rose cells. Plant Cell Rep 7: 329-332
[65] Macková, M; Chromá, L; Kucerová, P; Burkhard, J; Demnerová, K and Macek, T (2001) Some aspects of PCBs metabolism by horseradish cells. Int J Phytorem 7 (7): 401-414
[66] Pletsch, M; de Araujo, BS and Charlwood, BV (1999) Novel biotechnological approaches in environmental remediation. Biotechnol Advances 17: 679-687
[67] Kucerová, P; Macková, M; Chromá, L; Burkhard, J; Tríska, J; Demnerová, K and Macek, T (2000) Metabolism of polychlorinated biphenyls by *Solanum nigrum* hairy root clone SNC-9O and analysis of transformation products. Plant Soil 225: 109-115
[68] Bock, C; Kolb, M; Bokern, M; Harms, H; Mackova M; Chroma, L; Macek, T; Hughes, J; Just, C and Schnoor, J (2002) PCB – approaches to possible removal from the environment, in D Reible and K Demnerova, Eds., Advances in phytoremediation, phytotransformation, NATO ASI Series, pp. 115-140. Kluwer Academic Publishers, Dordrecht
[69] Rezek, J; Macek, T and Macková, M (2004) The effect of vegetation on decrease of PAH and PCB content in long-term contaminated soil, in Proceedings ESEB 2004, (W Verstraete, Ed.), pp. 833-837, Taylor and Francis Group, London
[70] Suresh, B and Ravishankar, GA (2004) Phytoremediation – A novel and promising approach for environmental clean-up. Crit Rev Biotechnik 24: 97-124
[71] Macek, T; Macková, M and Kás, J (2000) Exploitation of plants for the removal of organics in environmental remediation. Biotechnol Adv 18 (1), 23-35
[72] Schnoor, JL; Licht, A; McCutcheon, SC; Wolfe, NL and Carreira, LH (1995) Phytoremediation of organic and nutrient contaminants. Environ Sci and Technol 29, 318-323
[73] Susarla, S; Medina, VF and McCutcheon, SC (2002) Phytoremediation: An ecological solution to organic chemical contamination. Ecol Engineer 18, 647-658
[74] Lee, I and Fletcher, JS (1992) Involvement of mixed function oxidase systems in PCB metabolism by plant cells. Plant Cell Rep 11: 97-100

[75] Harms, H; Bokern, M; Kolb, M and Bock, C (2003) Transformation of organic contaminants by different plant systems, in SC McCutcheon and JL Schnoor, Eds., Phytoremediation. Transformation and control of contaminants: 285-316, Wiley Interscience, John Wiley and Sons, Inc., Hoboken, New Jersey

[76] Koeller, G; Moeder, M and Czihal, K (2000) Peroxidative degradation of selected PCB: a mechanistic study. Chemosphere 41: 1827-1834

[77] Moza, P; Weisberger, I; Klein, W and Korte, F (1973) Distribution and metabolism of carbon-labelled 2,2'- dichlorobiphenyl in the higher marsh plants *Veronica beccabunga*. Chemosphere 5: 217-222

[78] Van Aken, B; Hofrichter, M; Scheibner, K; Hatakka, A; Naveau, H and Agathos, S (1999) Transformation and mineralization of 2,4,6-trinitrotoluene (TNT) by manganese peroxidase from the white-rot basidiomycete *Phlebia radiata*. Biodegradation 10: 83-91

[79] Ikehata, K; Buchanan, I and Smith, D (2004) Recent developments in the production of extracellular fungal peroxidases and laccases for waste treatment. J Environ Engin Science 3: 1-19

[80] Wesenberg, D; Kyriakides, I and Agathos, SN (2003) White-rot fungi and thein enzymes for the treatment of industrial dye effluents. Biotechnol Adv 22: 161-187

[81] Cameron, S; Timofeevski, G and Aust, SD (2000) Mini-review: Enzymology of *Phanerochaete chrysosporium* with respect to the degradation of recalcitrant compounds and xenobiotics. Appl Microbiol Biotechnol 54: 751-758

[82] Novotny, C; Erbanova, P; Cajthaml, T; Rothschild, N.; Dosoretz, C and Sasek, V (2000) *Irpex lacteus*, a white rot fungus applicable to water and soil bioremediation. Appl Microbiol Biotechnol 54: 850-853

[83] Reddy GVB and Gold MH (2000) Degradation of pentachlorophenol by *Phanerochaete chrysosporium*: intermediates and reactions involved, Microbiology 146: 405-413

[84] Cajthaml, T; Moeder, M; Kacer, P; Sasek, V and Popp, P (2002) Study of fungal degradation products of polycyclic aromatic hydrocarbons using gas chromatography with ion trap mass spectrometry detection. J Chromatog A 974: 213-22

[85] Shrivastava, CHV; Shukla, R; Modi, HA and Vyas, BR (2005) Degradation of xenobiotic compounds by lignin-degrading white-rot fungi: enzymology and mechanisms involved. Indian J Exp Biol 43: 301-12

[86] Veitch, NC (2004) Horseradish peroxidases: A modern view of a classic enzyme, Phytochemistry 65: 249-259

[87] Moeder, M; Martin, CM and Koeller, G (2004) Degradation of hydroxylated compounds using laccase and horseradish peroxidases immobilized on microporous propylene hollow fiber membranes, J Membrane Sci 245: 183-190

[88] Klibanov, AM; Tu, TM and Scott, KL (1983) Peroxidase catalyzed removal from coal-conversion waste waters. Science 221: 259-260

[89] Cooper, VA and Nicell, JA (1996) Removal of phenols from foundry wastewater using HRP. Wat Res 30, 954-959

[90] Inze, D and Van Montagu, M, Eds., (2002) Oxidative Stress in Plants. Taylor & Francis, London.

[91] Chromá, L; Moeder, M; Kucerova, P; Macek, T; and Mackova, M (2003) Plant enzymes in metabolism of PCBs. Fres Environ Bull 12: 291-295

[92] Pieper, DH (2005) Aerobic degradation of polychlorinated biphenyls. Appl Microbiol Biotechnol 67: 170-191

[93] Gilbert, ES and Crowley, DE (1997) Plant compounds that induce polychlorinated biphenyl biodegradation by *Arthrobacter* sp. strain B1B. Appl Environ Microbiol 63: 1933-1938

[94] Gilbert, ES and Crowley, DE (1998) Repeated application of carvone-induced bacteria to enhance biodegradation of polychlorinated biphenyls in soil. Appl Microbiol Biotechnol 50: 489-494

[95] Singer, AC; Smith, D; Jury, WA; Hathuc, K and Crowley, DE (2003) Impact of the plant rhizosphere and augmentation on remediation of polychlorinated biphenyl contaminated soil. Environ Toxicol Chem 22(9): 1998-2004

[96] Singer, AC; Crowley, DE and Thompson, IP (2003) Secondary plant metabolites in phytoremediation and biotransformation. Review. Trends Biotechnol 21(3): 123-30

[97] Leigh, MB; Fletcher, JS; Nagle, DP; Mackova, M and Macek, T (2001) Vegetation and fungi at Czech PCB-contaminated sites as bioremediation candidates, in A Leeson, EA Foote, MK Banks, V Magar, Eds., Phytoremediation, Wetlands and Sediments, the Sixth International In Situ and On-Site Bioremediation Symposium, San Diego, 61-68, Battelle Press, Columbus, OH.
[98] Donnelly, PK; Hedge, RS and Fletcher, JS (1994) Growth of PCB-degrading bacteria on compounds from photosynthetic plants. Chemosphere 28: 981-988
[99] Fletcher, JS and Hegde, RS (1995) Release of phenols by perennial plant roots and their potential importance in bioremediation. Chemosphere 31: 3009-16
[100] Leigh, MB; Fletcher, JS; Fu, X and Schmitz, FJ (2002) Root turnover: an important source of microbial substances in rhizosphere remediation of recalcitrant contaminants. Environ Sci Technol 36: 1579-83
[101] Francova, K; Mackova, M; Macek, T and Sylvestre, M (2004) Ability of bacterial biphenyl dioxygenases from *Burkholderia* sp. LB400 and *Comamonas testosteroni* B-356 to catalyse oxygenation of ortho-hydroxybiphenyls formed from PCBs by plants. Environ Pollution 127/1: 41-48
[102] Sondossi, M; Sylvestre, M; Ahmad, D and Massé, R (1991) Metabolism of hydroxybiphenyl and chlorohydroxybiphenyl by biphenyl/chlorobiphenyl degrading *Pseudomonas testosteroni*, strain B-356. J Ind Microbiol 7: 77-88
[103] Sondossi, MD; Barriault, D and Sylvestre, M (2004) Metabolism of 2,2'- and 3,3'-Dihydroxybiphenyl by the Biphenyl Catabolic Pathway of *Comamonas testosteroni* B-356. Appl Environ Microbiol 70: 174-181
[104] Lovecká, P; Macková, M; Zlamaliková, J; Kochanková, L; Ryslavá, E and Demnerová, K (2005) Use of biological systems for measurement of ecotoxicity during biological degradation of PCBs, in N. Kalogerakis, Ed., Proceedings of 3rd European Bioremediation Conference, P139, Chania, Greece, CD-ROM, Tech University of Crete
[105] Mackova, M; Francova, K; Vrchotová, B; Najmanová, J; Kochánková, L; Sylvestre M; Zídková, J; Demnerova, K and Macek, T (2005) Biotransformation of PCBs by plants and bacteria – consequences of plant microbe interactions, in N Kalogerakis, Ed., Proceedings of 3rd European Bioremediation Conference, P63, Chania, Greece, CD-ROM, Tech University of Crete
[106] Vilacieros, M; Whelan, CM; Mackova, M; Molgaard, J; Sanchez-Contreras, M; Lloret, J; Aguirre de Carcer, D; Bolanos, L; Oruezabal, RI; Macek, T; Karlson, U; Dowling, DN; Martin, M and Rivilla, R (2005) PCB rhizoremediation by *Pseudomonas fluorescens* F113 derivatives using a *Sinorhizobium meliloti* nod system to drive bph gene expression. Appl Environ Microbiol 71 (4): 2687-2694
[107] McCutcheon, SC and Schnoor, JL (2003) Phytoremediation. Transformation and control of contaminants. Wiley Interscience, John Wiley and Sons, Inc., Hoboken, New Jersey
[108] Macek, T; Francova, K; Kochankova, L; Lovecka, P; Ryslava, E; Rezek, J; Sura, M; Triska, J; Demnerova, K and Mackova, M (2004) Phytoremediation: Biological cleaning of a polluted environment. Rev Environ Health 19 (1): 63-82
[109] Trapp, S and Karlson, U (2001) Aspects of phytoremediation of organic pollutants. J Soil Sediments 1(1): 37-43
[110] Meagher, RB (2000) Phytoremediation of toxic elemental and organic pollutants. Cur Opin Plant Biol 3: 153-162
[111] Newman, LA and Reynolds, CM (2004) Phytodegradation of organic compounds. Curr Opin Biotechnik 14: 225-230
[112] Glick, BR (2003) Phytoremediation: synergistic use of plants and bacteria to clean up the environment. Biotech Advances 21: 383-393
[113] Winter, AC; Crowley, DE and Thompson, IP (2003) Secondary plant metabolites in phytoremediation and biotransformation. Trends Biotechnol 21 (3): 123-130
[114] Burken, JG (2003) Uptake and metabolism of organic compounds: Green liver model, in SC McCutcheon and JL Schnoor, Eds., Phytoremediation. Transformation and control of contaminants. Wiley Interscience, John Wiley Sons, Inc., Hoboken, New Jersey
[115] Sandermann, H Jr (1994) Higher plant metabolism of xenobiotics: the "green liver concept" Pharmacogenetics 4: 225-241

[116] Macek, T; Surá, M; Pavlíková, D; Francová, K; Scouten, WH; Szekeres, M; Sylvestre, M and Macková, M (2005) Can Tobacco Have Potentially Beneficial Effect to our Health? Zeitschrift fuer Naturforschung C, 60: 292-299
[117] Furukawa, K; Suenaga, H and Goto, M (2004) Biphenyl dioxygenases: functional versatilities and directed evolution. J Bacteriol 186: 5189-5196
[118] Sylvestre, M (2004) Genetically Modified Organisms to Remediate Polychlorinated Biphenyls. Where Do We Stand? Intl Biodeterioration Biodegradat 54: 153-162
[119] Unterman, R (1996) A history of PCB degradation, in Biotechnology Research Series: Bioremediation Principles and Applications. RL Crawford and DL Crawford, Eds., Cambridge, U.K.: Cambridge University Press, 209-253

METABOLISM AND GENETIC ENGINEERING STUDIES FOR HERBICIDE PHYTOREMEDIATION

MELISSA P. MEZZARI[1] AND JERALD L. SCHNOOR[2]
[1]*Biochemistry and Cell Biology, Rice University, Houston, Texas, USA*
[2]*Civil and Environmental Engineering, University of Iowa, Iowa City, IA, 52242, USA – Fax 01 319 3355660*
E-mail: jerald-schnoor@uiowa.edu

1. Introduction

Herbicides are extensively used for the purpose of crop cultivation and still remain as an important strategy for the commercial agriculture. However, the widespread use of herbicides has resulted in both point source and non-point source contamination of shallow groundwater and surface waters, which became a major issue and a serious environmental problem [1]. Phytoremediation is an alternative technique to effectively cleanup herbicides from soil and groundwater, since plants possess highly efficient systems for the removal and transformation of these compounds. Plant-based remediation can accelerate natural attenuation processes from contaminated sites by taking up significant quantities of water, and herbicides may be transformed into less toxic forms.

Some of the most commonly used herbicides in the world can be removed from the environment using phytoremediation techniques with several plant species. The structure of the major herbicides are shown in Figure 1.

The use of non-target and tolerant plants, such as trees, shrubs and grasses, have been proposed for phytoremediation of pesticide-contaminated soil and water [2]. Poplar trees are able to uptake, hydrolyze, and dealkylate atrazine to less toxic metabolites [3, 4]. The retention ability of pesticides, such as simazine, atrazine (triazines), isoproturon, linuron (phenylureas), carbaryl (carbamates), fenamiphos (organophosphorus) and permethrin (pyrethroids) was also evaluated in *Lupinus angustifolius* seeds, and the apparent degradation of these pesticides in seeds showed to be promissing in water contaminated sites [5]. In addition, mixed prairie grasses (big bluestem, yellow indiangrass, and switchgrass) have been considered as potential tools for phytoremediation of atrazine, alachlor, metolachlor, and pendimethalin [2, 6]. These studies have indicated three successful types of plant-based remediation techniques: the containment of contaminated groundwater plumes by trees; sequestration of pesticides from water and sediments using aquatic plants; and uptake of pesticides from contaminated soils by terrestrial plants.

M. Mackova et al. (eds.), Phytoremediation Rhizoremediation, 169–178.

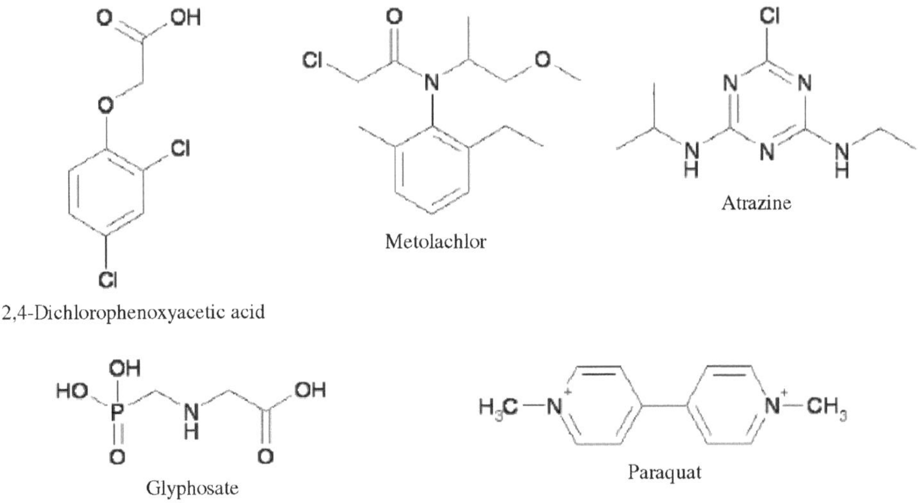

Figure 1. Structures of chemicals that are commonly used worldwide as herbicides.

The elucidation and understanding of degradation pathways in plants used for phytoremediation is of great importance to discern the end point transformation of herbicides. Mineralization of these chemicals by selected plants and the rhizosphere in degraded environments would be the final goal for successful phytoremediation processes.

2. Herbicide transformation pathways in plants

Plant herbicide metabolism is mediated by three groups of enzymes involved in phase I (conversion), phase II (conjugation) and phase III (compartmentation). Conversion and transformation reactions are often catalyzed by cytochrome P450 monooxygenases, where herbicides are converted to more hydrophilic metabolites [7]. Phase II of xenobiotic metabolism is usually mediated by a sugar, an amino acid, glutathione *S*-transferases (GST), or a cellular macromolecule, which generally form polar and water soluble products. The formation of insoluble, non-extractable bound components of a variety of toxic exogenous compounds, such as herbicides, is often performed by GSTs [8]. In Phase III, ATP-dependent membrane pumps recognize the conjugates and transfer them across membranes, where GST conjugates are either sequestered in the vacuole or transferred to the apoplast [9].

Numerous studies have contributed to the understanding of enzymes in plants and the overall plant metabolism towards herbicides. Evolution of GSTs, P450s, and other enzymes towards herbicide tolerance is mostly related to the complex and versatile chemistry developed by higher plants to synthesise a large number of natural products. A more complete picture of the main enzymes and pathways that can be explored for enhanced transformation processes of herbicides *in planta* are represented in Table 1.

Table 1. Plant enzymes involved on transformation processes of some herbicides commonly present in surface and ground waters of Europe and North America

Herbicide (Chemical Class)	Mode of Action	Metabolism in Resistant Plants	References
Atrazine, Simazine (triazines)	Photosynthetic inhibitors	Dechlorination followed by hydroxylation; N-dealkylation; GST conjugation	[10-13]
Acetochlor, Metolachlor (chloroacetanilides)	Protein and lipid synthesis inhibitors	GST conjugation followed by vacuolar compartmentation; O-demethylation by cytochrome P450s	[14-16]
2,4-Dichlorophenoxyacetic acid (phenoxyalkanoic acid)	Systemic hormone-type that causes abnormal growth and development	Hydroxylation; conjugation with sugars or amino acids	[17, 18]
Glyphosate (unique chemical class)	Inhibition of the biosynthesis of aromatic compounds; inhibition of 5-enolpyruvylshikimate-3-phosphate synthase enzyme (EPSPS)	Compartmentation to vacuole and organelles or insensitivity of the EPSPS enzyme	[19-22]
Chlorotoluron, chlorosulfuron, diuron, isoproturon, linuron (ureas)	Photosynthetic inhibitors	N-demethylation by P450s and C-hydroxylation followed by glucose conjugation	[23-25]

3. Exploitation of plant genes

Herbicides have important characteristics for their removal from the environment or detoxification during phytoremediation, since they are designed to kill unwanted plants by blocking vital biochemical and physiological processes [26]. However, weed and crop resistance to various herbicides have been reported, and the resistance could result from modification of the target site or by induced expression of genes that can lead to enhanced metabolism, and alteration in uptake, translocation, and compartmentation of the herbicide [26]. Plant herbicide tolerance can be achieved or improved by using metabolic engineering of enzymes. However, genetic engineering has been mostly focused and applied into crop plants in order to confer herbicide selectivity and enhance crop safety and production.

Genomic sequencing information from poplar (*Populus* sp.) has been progressively developing for gene discovery, expression profiling, genetic mapping, and exploitation in forest biotechnology and phytoremediation [27]. In addition, genomics of the model plant *Arabidopsis* and major agronomic crops, such as corn (*Zea mays* L.) and rice (*Oryza sativa* L.) have provided useful information about the genetic, physiological, and biochemical mechanisms involved in xenobiotic transformations in plants. The genomic sequence of *Arabidopsis*, with an estimate of 25,500 genes in 11,000 gene families, is widely accepted as a model for detailed functional characterization of plant genes, which comprises the gene families that could be used in phytoremediation applications, including regulatory networks (e.g. transcription factors) and tissue-specific transporters [28]. The functional definition of each gene of a gene family involved in xenobiotic metabolism is now the challenge for phytoremediation applications.

Studies on herbicide biochemical pathways now comprise modern biochemistry with metabolic processes and other mechanisms present at the cellular and molecular level. Plant molecular biology has rapidly evolved and several advanced methods are available for the isolation of a large number of genes that are thought to be involved in xenobiotic transformation. Genes that express the various enzymes involved in the metabolism of herbicides can be identified by several available methods such as screening of complementary DNA (cDNA) using protein-specific antibodies or a polymerase chain reaction (PCR) based approach [29]. The accelerating determination of genome sequences and their interpretation (genomics) in the last few years have provided new methods for phytoremediation approaches such as serial analysis of gene expression (SAGE), development of genetically modified organisms, and DNA microarrays. These efficient methods are available to isolate desired genes and to express them in relevant plant species or lines for phytoremediation, and to improve catalysts for specific functions. A combination of these tools and approaches provides an enormous resource for the development of new and more efficient detoxification and/or remediation by plant enzymes. Thus, the biggest challenge in modern bioscience is to determine the structure, function and expression of all the corresponding proteins that are encoded in these genes of interest.

3.1. C-DNA LIBRARIES

Numerous enzyme sequences have been identified in *Arabidopsis*, but little headway has been made in matching specific enzymes with their function on herbicide substrates. Molecular information on enzymes and function identification towards herbicides has been studied based on cDNA libraries. cDNA libraries are prepared from the mRNA of interest that is synthesized to DNA (cDNA) strand, which contains a specific coding region of a genome that is inserted into vectors and cloned [30].

Studies on P450 metabolism of herbicides have reported the use of two yeast-expressed plant enzymes, CYP73A1 [31] and CYP81B1 [32], in which ring-methyl hydroxylation of chlortoluron was observed. In both cases, however, the reaction was extremely slow if compared to a previous investigation, where a very fast *N*-demethylation catalysis was observed with CYP76B1 [33]. It was found that CYP76B1 catalyzes the mono- and di-*N*-demethylation of both chlortoluron and isoproturon in higher plants, and plays a significant role in the detoxification of most phenylureas, including the methoxylated forms [33]. Other studies showed that 2,4-D treated S401 tobacco cells could metabolize chlorotoluron using four different P450 enzymes identified as CYP71A11, CYP81B2, CYP81C1 and CYP81C2 [34]. Because only CYP71A11 and CYP81B2 were highly induced, both cDNA clones were expressed in yeast *S. cerevisiae*, which exhibited enhanced chlorotoluron metabolism [34].

Recent studies have investigated specific GST isoform function in response to herbicides (2,4-dichlorophenoxyacetic acid and metolachlor) and other chemicals by overexpressing specific GST cDNAs from plants in *E. coli* [35]. Results from this research showed that three GST isoform genes (*At*GSTF2, *At*GSTF6 and *At*GSTF8) are increased in the presence of metolachlor or 2,4-D [35]. Further literature studies with *Arabidopsis* plants exposed to chloroacetanilide herbicides have supported the information about *AtGST*F2 induction not only for metolachlor but also for acetochlor [36]. However, it was also observed that both herbicides induced other GST isoform genes, such as *AtGST*U24 and *AtGST*U1 [36]. The expression of *Arabidopsis* GSTs has also been studied in response to herbicide safeners – chemicals used to enhance herbicide tolerance of crop plants by a number of different mechanisms, specially by the induction of GST activity [37]. Studies from the proteomic analysis of *Arabidopsis* GSTs reported an effective strategy for examining the *Arabidopsis* GST family in response to the benoxacor safener, where it was observed a 6-fold induction of the gene *AtGST*F9 [38].

Phase III transformation pathway in plants rely on the transport of xenobiotics to the vacuole and storage in this compartment. The transporters involved in this process are mainly the ATP-binding cassete (ABC), which contain a cytoplasmic domain (the ABC protein) that binds and hydrolyzes ATP to energize solute translocation accross the cytoplasmic membrane [39, 40]. Induction of transport activity in *Arabidopsis* treated with herbicides and herbicide safeners has been observed previously, indicating that several glutathione-conjugate transporters exist [40]. Recent studies have functionally analyzed a full length cDNA of *Arabidopsis* coding for a protein (*At*MRP3) that exhibits high homologies to MPR1 and YCF1, two gluthathione-conjugate transporters of humans and yeast, respectively [40]. It was observed that *At*MRP3 was not only a glutathione-conjugate but also a chlorophyll catabolite transporter [40]. This investigation confirmed the multifunctionality between ABC-transporters in plants and opens the need for further identification and functional analysis of other members of the MRP family. Additional studies have provided evidence that the oligopeptide transporter (OPT) family in *Arabidopsis* mediate the transport of glutathione derivatives and metal complexes, thus participation in stress resistance [41]. RNA blots on yeast cell suspensions and real-time reverse transcription-PCR on *Arabidopsis* plants indicated that *AtOPT6* expression is strongly induced by the herbicide primisulfuron [41].

3.2. SAGE

Serial analysis of gene expression (SAGE) is a method that allows the quantitative and simultaneous analysis of a large number of transcripts [42]. The first application of this technique in higher plants was used for profiling expressed genes in rice (*Oryza sativa* L.) seedlings [43]. Recent applications of SAGE technique in plants consist of wood formation in loblolly pine [44]; gene expression pattern in the crop plant cassava (*Manihot esculenta*) from viral disease resistance and susceptible genotypes; and a few studies of gene expression patterns in *Arabidopsis* plants [45-49]. Only one study was directly related to phytoremediation applications, in which genes involved in the metabolism of TNT were identified in *Arabidopsis* plants [46]. This study provided important information for our understanding of the mechanisms that might be involved in plant root TNT degradation. Therefore, the SAGE technique could strengthen some existing theories of plant tolerance and metabolism of xenobiotic compounds, including herbicides.

3.3. MICROARRAY

The cDNA microarray has been successfully applied to the simultaneous expression of many thousands of known and unknown genes, and to large-scale gene discovery, as well as polymorphism screening and mapping of genomic DNA clones. cDNA microarrays were first developed for *Arabidopsis*, in which differential gene expression was simultaneously assayed in roots and shoots using a microarray of 48 duplicate cDNA elements [50]. Microarray experiments have been analyzed for plant response to drought and cold stresses, mechanical wounding and insect feeding, herbivory, nitrate treatments, and others [51].

Phytoremediation studies using cDNA microarrays are very few and some interesting articles report on metals [52-54]. A few studies have provided further information on the expressed genes in plants towards herbicides. GST sequences from soybean (*Glycine max*) and maize (*Zea mays*) were identified, expressed in cDNA libraries and studied in microarray analysis [55]. These studies have discussed the diversity of GSTs in plants which has enabled herbicide tolerance in plants. Other studies using a DNA array (MetArray) investigated stress-induced *Arabidopsis* in response to herbicides, UV-B radiation, endogenous stress hormones, and pathogen infection [56]. *Arabidopsis* transcriptional reaction was investigated towards sulfonylurea herbicides and bromoxynil, and results from this study supported the induction of specific cytochrome P450s, glycosyltransferases, glutathione S-transferases and ABC transporters [56]. The MetArray is a tool to detect interactions and important overlaps between abiotic and biotic stress responses, as indicated by the potential mutual interactions of pathogen defense and response to herbicides.

Populus is an ideal model system among tree species and microarray studies have revealed the putative functions of several genes involved in wood formation, root development, and gene expression patterns under different environmental conditions [57]. A novel investigation has reported the use of *Populus* as a model tree species to examine the genes up-regulated by safeners and their role in herbicide metabolism and the molecular mechanisms of herbicide detoxification [58]. These studies also showed induction of the enzymes involved in phase I (oxidation), phase II (conjugation) and phase III (sequestration). A few safener-induced genes that were not previously reported to be induced by safeners, such as 12-Oxophytodienoate reductase, MtN19 and thioredoxin h, were significantly expressed [58]. The newly identified genes could have potential for application in genetic engineering of plants for herbicide detoxification and tolerance.

3.4. GENETICALLY MODIFIED ORGANISMS

Changes in gene expression can be informative and useful in developing transgenic plants that respond to contaminants for phytoremediation enhancement. Plant metabolism rarely mineralizes xenobiotic contaminants and genetic engineering is a necessary approach to achieve sustainable *in situ* recycling of contaminants [59]. Genetically modified crops which can tolerate or metabolise herbicides are well known [7, 60-62]. Engineering of herbicide tolerance in higher plants can result from the overexpression of an endogenous enzyme, modification of the metabolic activity through introduction and expression of novel genes, or improved physiological capabilities, such as alterations in uptake and translocation, using classical breeding [7, 59, 63].

A few recent investigations have been developed for the tolerance of herbicides using transgenic crops [7, 60, 64]. Thus, direct herbicide phytoremediation approaches using transgenic plants have been reported by overexpressing a bacterial gamma-glutamylcysteine synthetase in *Populus* [65]. This study showed that the transgenic poplar was more tolerant to chloroacetanilide herbicides than the wild-type.

Phytoremediation and genetic engineering are a recent combination of technologies that have shown, in laboratory experiments, a great potential for the decontamination of areas with little risk of biomagnification of xenobiotics in the food chain. At this point, a potential solution for the phytoremediation of contaminated areas could be achieved, however, engineering plants for herbicide resistance raises an environmental concern regarding the transfer of foreign genes via pollen dispersion, consequently resulting in resistant weeds or genetic pollution among other crops [61]. Agrobiotechnology companies such as Monsanto and Novartis have recognized the risk of horizontal gene spreading and new technologies are being developed to avoid major environment impact. Genetic engineering of the chloroplast genome offers a novel way to obtain high expression without the risk of spreading the transgene via pollen [61, 66, 67]. This technique relies on the engineering of the maternal chloroplast DNA, which is lost during pollen maturation and hence is not transmitted to the next generation [68].

4. Conclusions

Understanding transformation pathways and determining the enzymes involved in plant tolerance to xenobiotics is vital for feasible phytoremediation application. Herbicides are of major concern due to the worldwide contamination of soil and associated groundwater. There is a complex array of plant enzymes that are involved in the detoxification of these compounds, however, the exact number of plant enzymes potentially involved in the metabolism of herbicides remains unknown. New biotechnological approaches are now tools for the identification of missing xenobiotic pathways. The large amount of genomic sequence information presents a challenge not only to the plant biology community but also to environmental engineers that are exploring new phytoremediation approaches to reduce human exposure to toxic pollutants.

References

[1] Suri CR; Raje M and Varshney GC (2002) Immunosensors for pesticide analysis: Antibody production and sensor development. Crit Rev Biotechnol 22: 15-32

[2] Karthikeyan R; Davis LC; Erickson LE; Al-Khatib K; Kulakow PA; Barnes PL; Hutchinson SL and Nurzhanova AA (2004) Potential for plant-based remediation of pesticide-contaminated soil and water using nontarget plants such as trees, shrubs, and grasses. Crit Rev Plant Sci 23: 91-101

[3] Burken JG and Schnoor JL (1997) Uptake and metabolism of atrazine by poplar trees. Environ Sci Technol 31: 1399-1406

[4] Schnoor JL; Licht LA; McCutcheon SC; Wolfe NL and Carreira LH (1995) Phytoremediation of Organic and Nutrient Contaminants. Environmental Science and Technology 29: A318-A323

[5] Garcinuno RM; Fernandez-Hernando P and Camara C (2003) Evaluation of pesticide uptake by Lupinus seeds. Water Res 37: 3481-3489

[6] Belden JB; Phillips TA and Coats JR (2004) Effect of prairie grass on the dissipation, movement, and bioavailability of selected herbicides in prepared soil columns. Environ Toxicol Chem 23: 125-132

[7] Kawahigashi H; Hirose S; Inui H; Ohkawa H and Ohkawa Y (2005) Enhanced herbicide cross-tolerance in transgenic rice plants co-expressing human CYP1A1, CYP2B6, and CYP2C19. Plant Science 168: 773-781

[8] Dixon DP; Cummins I; Cole DJ and Edwards R (1998) Glutathione-mediated detoxification systems in plants. Current Opinion in Plant Biology 1: 258-266

[9] Marrs KA (1996) The functions and regulation of glutathione S-transferases in plants. Annual Review of Plant Physiology and Plant Molecular Biology 47: 127-158

[10] Beynon KI; Stoydin G and Wright AN (1972) A comparison of the breakdown of the triazine herbicides cyanazine, atrazine and simazine in soils and in maize. Pestic Biochem Physiol 2: 153-161

[11] Deprado R; Romera E and Menendez J (1995) Atrazine Detoxification in *Panicum dichotomiflorum* and Target Site *Polygonum lapathifolium*. Pestic Biochem Physiol 52: 1-11

[12] Hatton PJ; Dixon D; Cole DJ and Edwards R (1996) Glutathione transferase activities and herbicide selectivity in maize and associated weed species. Pestic Sci 46: 267-275

[13] Schmidt B; Siever M; Thiede B; Breuer J; Malcherek K and Schuphan I (1997) Biotransformation of ring-U-C-14 atrazine to dealkylated and hydroxylated metabolites in cell-suspension cultures. Weed Research 37: 401-410

[14] Scarponi L; Perucci P and Martinetti L (1991) Conjugation of 2-Chloroacetanilide Herbicides With Glutathione - Role of Molecular-Structures and of Glutathione-S-Transferase Enzymes. J Agric Food Chem 39: 2010-2013

[15] Jablonkai I and Hatzios KK (1993) In-Vitro Conjugation of Chloroacetanilide Herbicides and Atrazine with Thiols and Contribution of Nonenzymatic Conjugation to Their Glutathione-Mediated Metabolism in Corn. J Agric Food Chem 41: 1736-1742

[16] Moreland DE; Corbin FT; Fleischmann TJ and McFarland JE (1995) Partial Characterization of Microsomes Isolated from Mung Bean Cotyledons. Pestic Biochem Physiol 52: 98-108

[17] Scheel D and Sandermann H (1981) Metabolism of 2,4-Dichlorophenoxyacetic Acid in Cell-Suspension Cultures of Soybean (*Glycine max* L.) and Wheat (*Triticum aestivum* L.) .1. General Results. Planta 152: 248-252

[18] Chkanikov DI; Makeyev AM; Pavlova NN; Grygoryeva LV; Dubovoi VP and Klimov OV (1976) Variety of 2,4-D Metabolic Pathways in Plants - Its Significance in Developing Analytical Methods for Herbicides Residues. Arch Environ Contam Toxicol 5: 97-103

[19] Hetherington PR; Marshall G; Kirkwood RC and Warner JM (1998) Absorption and efflux of glyphosate by cell suspensions. J Exp Bot 49: 527-533

[20] Simarmata M; Kaufmann JE and Penner D (2003) Potential basis of glyphosate resistance in California rigid ryegrass (*Lolium rigidum*). Weed Sci 51: 678-682

[21] Feng PCC; Tran M; Chiu T; Sammons RD; Heck GR and Jacob CA (2004) Investigations into glyphosate-resistant horseweed (*Conyza canadensis*): retention, uptake, translocation, and metabolism. Weed Sci 52: 498-505

[22] Koger CH and Reddy KN (2005) Role of absorption and translocation in the mechanism of glyphosate resistance in horseweed (*Conyza canadensis*). Weed Sci 53: 84-89

[23] Mougin C; Cabanne F and Scalla R (1992) Additional Observations on the Chlorotoluron Hydroxylase and N-Demethylase Activities in Wheat Microsomes. Plant Physiology and Biochemistry 30: 769-778

[24] Hall LM; Moss SR and Powles SB (1995) Mechanism of resistance to chlorotoluron in two biotypes of the grass weed *Alopecurus myosuroides*. Pestic Biochem Physiol 53: 180-192

[25] Glassgen WE; Komossa D; Bohnenkamper O; Haas M; Hertkorn N; May RG; Szymczak W and Sandermann H (1999) Metabolism of the herbicide isoproturon in wheat and soybean cell suspension cultures. Pestic Biochem Physiol 63: 97-113

[26] Coleman JOD; Frova C; Schroder P and Tissut M (2002) Exploiting plant metabolism for the phytoremediation of persistent herbicides. Environmental Science and Pollution Research 9: 18-28

[27] Bhalerao R; Nilsson O and Sandberg G (2003) Out of the woods: forest biotechnology enters the genomic era. Curr Opin Biotechnol 14: 206-213

[28] Kolukisaoglu HU; Bovet L; Klein M; Eggmann T; Geisler M; Wanke D; Martinoia E and Schulz B (2002) Family business: the multidrug-resistance related protein (MRP) ABC transporter genes in *Arabidopsis thaliana*. Planta 216: 107-119
[29] Zhen RG and Singh BK (2001) From inhibitors to target site genes and beyond-herbicidal inhibitors as powerful tools for functional genomics. Weed Sci 49: 266-272
[30] Lodish H; Berk A; Zipursky LS; Matsudaira P; Baltimore D and Darnell J (2000) Molecular Cell Biology, vol 2, New York: WH Freeman & Company
[31] Pierrel MA; Batard Y; Kazmaier M; Mignottevieux C; Durst F and Werck-Reichhart D (1994) Catalytic Properties of the Plant Cytochrome-P450 Cyp73 Expressed in Yeast - Substrate-Specificity of a Cinnamate Hydroxylase. Eur J Biochem 224: 835-844
[32] Cabello-Hurtado F; Batard Y; Salaun JP; Durst F; Pinot F and Werck-Reichhart D (1998) Cloning, expression in yeast, and functional characterization of CYP81B1, a plant cytochrome P450 that catalyzes in-chain hydroxylation of fatty acids. J Biol Chem 273: 7260-7267
[33] Robineau T; Batard Y; Nedelkina S; Cabello-Hurtado F; LeRet M; Sorokine O; Didierjean L and Werck-Reichhart D (1998) The chemically inducible plant cytochrome P450 CYP76B1 actively metabolizes phenylureas and other xenobiotics. Plant Physiology 118: 1049-1056
[34] Yamada T; Kambara Y; Imaishi H and Ohkawa H (2000) Molecular cloning of novel cytochrome P450 species induced by chemical treatments in cultured tobacco cells. Pestic Biochem Physiol 68: 11-25
[35] Wagner U; Edwards R; Dixon DP and Mauch F (2002) Probing the diversity of the arabidopsis glutathione S- transferase gene family. Plant Mol Biol 49: 515-532
[36] Mezzari MP; Walters K; Jelinkova M; Shih MC; Just CL and Schnoor JL (2005) Gene expression and microscopic analysis of *Arabidopsis thaliana* exposed to chloroacetanilide herbicides and explosive compounds: a phytoremediation approach. Plant Physiology. In Press
[37] Davies J and Caseley JC (1999) Herbicide safeners: a review. Pestic Sci 55: 1043-1058
[38] Smith AP; DeRidder BP; Guo WJ; Seeley EH; Regnier FE and Goldsbrough PB (2004) Proteomic analysis of Arabidopsis glutathione S-transferases from benoxacor- and copper-treated seedlings. J Biol Chem 279: 26098-26104
[39] Davies TGE and Coleman JOD (2000) The *Arabidopsis thaliana* ATP-binding cassette proteins: an emerging superfamily. Plant Cell Environ 23: 431-443
[40] Tommasini R; Vogt E; Schmid J; Fromentau M; Amrhein N and Martinoia E (1997) Differential expression of genes coding for ABC transporters after treatment of *Arabidopsis thaliana* with xenobiotics. FEBS Lett 411: 206-210
[41] Cagnac O; Bourbouloux A; Chakrabarty D; Zhang MY and Delrot S (2004) AtOPT6 transports glutathione derivatives and is induced by primisulfuron. Plant Physiology 135: 1378-1387
[42] Velculescu VE; Zhang L; Vogelstein B and Kinzler KW (1995) Serial Analysis of Gene-Expression. Science 270: 484-487
[43] Matsumura H; Nirasawa S and Terauchi R (1999) Transcript profiling in rice (*Oryza sativa* L.) seedlings using serial analysis of gene expression (SAGE). Plant J 20: 719-72
[44] Lorenz WW and Dean JFD (2002) SAGE Profiling and demonstration of differential gene expression along the axial developmental gradient of lignifying xylem in loblolly pine (*Pinus taeda*). Tree Physiol 22: 301-310
[45] Jung SH; Lee JY and Lee DH (2003). Use of SAGE technology to reveal changes in gene expression in Arabidopsis leaves undergoing cold stress. Plant Mol Biol 52: 553-567
[46] Ekman DR; Lorenz WW; Przybyla AE; Wolfe NL and Dean JFD (2003) SAGE analysis of transcriptome responses in Arabidopsis roots exposed to 2,4,6-trinitrotoluene. Plant Physiology 133: 1397-1406
[47] Chakravarthy S; Tuori RP; D'Ascenzo MD; Fobert PR; Despres C and Martin GB (2003) The tomato transcription factor Pti4 regulates defense-related gene expression via GCC box and non-GCC box cis elements. Plant Cell 15: 3033-3050
[48] Fizames C; Munos S; Cazettes C; Nacry P; Boucherez J; Gaymard F; Piquemal D; Delorme V; Commes TS; Doumas P; Cooke R; Marti J; Sentenac H and Gojon A (2004) The Arabidopsis root transcriptome by serial analysis of gene expression. Gene identification using the genome sequence. Plant Physiology 134: 67-80
[49] Munos S; Cazettes C; Fizames C; Gaymard F; Tillard P; Lepetit M; Lejay L and Gojon A (2004) Transcript profiling in the chl1-5 mutant of Arabidopsis reveals a role of the nitrate transporter NRT1.1 in the regulation of another nitrate transporter, NRT2.1. Plant Cell 16: 2433-2447

[50] Schena M; Shalon D; Davis RW and Brown PO (1995) Quantitative Monitoring of Gene-Expression Patterns with a Complementary-DNA Microarray. Science 270: 467-470
[51] Aharoni A and Vorst O (2002) DNA microarrays for functional plant genomics. Plant Mol Biol 48: 99-118
[52] Weber M; Harada E; Vess C; von Roepenack-Lahaye E and Clemens S (2004). Comparative microarray analysis of *Arabidopsis thaliana* and *Arabidopsis halleri* roots identifies nicotianamine synthase, a ZIP transporter and other genes as potential metal hyperaccumulation factors. Plant J 37: 269-281
[53] Bovet L; Feller U and Martinoia E (2005) Possible involvement of plant ABC transporters in cadmium detoxification: a cDNA sub-microarray approach. Environ Int 31: 263-267
[54] Kovalchuk I; Titov V; Hohn B and Kovalchuka O (2005) Transcriptome profiling reveals similarities and differences in plant responses to cadmium and lead. Mutation Research-Fundamental and Molecular Mechanisms of Mutagenesis 570: 149-161
[55] McGonigle B; Keeler SJ; Lan SMC; Koeppe MK and O'Keefe DP (2000) A genomics approach to the comprehensive analysis of the glutathione S-transferase gene family in soybean and maize. Plant Physiology 124: 1105-1120
[56] Glombitza S; Dubuis PH; Thulke O; Welzl G; Bovet L; Gotz M; Affenzeller M; Geist B; Hehn A; Asnaghi C; Ernst D; Seidlitz HK; Gundlach H; Mayer KF; Martinoia E; Werck-Reichhart D; Mauch F and Schaffner AR (2004) Crosstalk and differential response to abiotic and biotic stressors reflected at the transcriptional level of effector genes from secondary metabolism. Plant Mol Biol 54: 817-835
[57] Sterky F; Bhalerao RR; Unneberg P; Segerman B; Nilsson P; Brunner AM; Charbonnel-Campaa L; Lindvall JJ; Tandre K; Strauss SH; Sundberg B; Gustafsson P; Uhlen M; Bhalerao RP; Nilsson O; Sandberg G; Karlsson J; Lundeberg J and Jansson S (2004) A Populus EST resource for plant functional genomics. Proc Natl Acad Sci USA 101: 13951-13956
[58] Rishi AS; Munir S; Kapur V; Nelson ND and Goyal A (2004) Identification and analysis of safener-inducible expressed sequence tags in Populus using a cDNA microarray. Planta 220: 296-306
[59] McCutcheon SC; Medina VF and Larson SL (2003) Proof of phytoremediation for explosives in water and soil, in SC McCutcheon and JL Schnoor, Eds., Phytoremediation: transformation and control of contaminants. Hoboken, New Jersey, John Wiley & Sons, Inc, 429-480
[60] Inui H; Shiota N; Motoi Y; Ido Y; Inoue T; Kodama T; Ohkawa Y and Ohkawa H (2001) Metabolism of herbicides and other chemicals in human cytochrome P450 species and in transgenic potato plants co-expressing human CYP1A1, CYP2B6 and CYP2C19. J Pestic Sci 26: 28-40
[61] Daniell H; Khan MS and Allison L (2002) Milestones in chloroplast genetic engineering: an environmentally friendly era in biotechnology. Trends Plant Sci 7: 84-91
[62] Morant M; Bak S; Moller BL and Werck-Reichhart D (2003) Plant cytochromes P450: tools for pharmacology, plant protection and phytoremediation. Curr Opin Biotechnol 14: 151-162
[63] Didierjean L; Gondet L; Perkins R; Lau SMC; Schaller H; O'Keefe DP and Werck-Reichhart D (2002) Engineering herbicide metabolism in tobacco and Arabidopsis with CYP76B1, a cytochrome P450 enzyme from Jerusalem artichoke. Plant Physiology 130: 179-189
[64] Inui H and Ohkawa H (2005) Herbicide resistance in transgenic plants with mammalian P450 monooxygenase genes. Pest Manag Sci 61: 286-291
[65] Gullner G; Komives T and Rennenberg H (2001) Enhanced tolerance of transgenic poplar plants overexpressing gamma-glutamylcysteine synthetase towards chloroacetanilide herbicides. J Exp Bot 52: 971-979
[66] Daniell H and Dhingra A (2002) Multigene engineering: dawn of an exciting new era in biotechnology. Curr Opin Biotechnol 13: 136-141
[67] Daniell H (2002) Molecular strategies for gene containment in transgenic crops. Nat Biotechnol 20: 581-586
[68] Daniell H; Wiebe PO and Millan AFS (2001) Antibiotic-free chloroplast genetic engineering - an environmentally friendly approach. Trends Plant Sci 6: 237-239

PESTICIDES REMOVAL USING PLANTS: PHYTODEGRADATION VERSUS PHYTOSTIMULATION

JEAN-PAUL SCHWITZGUÉBEL[1], JOANA MEYER[1] AND PETRA KIDD[2]
[1]*Laboratory for Environmental Biotechnology, Station 6, EPFL, CH-1015 Lausanne, Switzerland, Fax+41216934722, E-mail: jean-paul.schwitzguebel@epfl.ch* [2]*Soil Biochemistry, IIAG, Consejo Superior de Investigaciones Científicas (CSIC), Aptdo. 122, E-15780 Santiago de Compostela, Spain, E-mail: edpetra@usc.es*

1. Introduction

Pesticides are chemicals used for crop protection and pest control, and are probably the most widely distributed contaminants in the environment over the last century. Although it is extremely difficult to obtain precise figures concerning their production and use per country [1], millions of tons of pesticides are produced and spread annually all over the world. Thousands of different synthetic molecules are used as pesticides: carbamates, thiocarbamates, dipyridyls, triazines, phenoxyacetates, coumarins, nitrophenols, pyrazoles, pyrethroids, etc. Most of these chemicals contain chlorine, phosphorus, tin, mercury, arsenic or copper atoms. Pesticides are divided into different groups according to their target, e.g. herbicides, against weeds and toxic vegetation; insecticides, against harmful insects; fungicides, against blights, mildews, mould and rusts; algaccides, for the sanitary control of lakes, channels, water pools, reservoirs; bactericides, against some pathogenic microbes; etc.

To be efficient, a pesticide must remain in the appropriate environmental compartment and be stable enough to act against the target pest for a certain period of time. However, less than 5% of these products are estimated to reach the target organism, the remainder being deposited on the soil and nontarget organisms, as well as moving into the atmosphere and water [2].

Once in the environment, the persistence of a pesticide depends on its chemical stability, degradability by microorganisms, climatic conditions (influencing pesticide degradation through soil genesis), soil physicochemical properties (especially amount and nature of organic matter) and uptake by terrestrial and aquatic species including plants. The degree of environmental contamination is thus dependent on many factors and on physicochemical properties of the pesticide: volatility, reactivity, absorption and adsorption, solubility in water, partition between polar and non-polar phases (log Kow)

M. Mackova et al. (eds.), Phytoremediation Rhizoremediation, 179–198.

and between soil and water (Kd). Depending upon their properties, many pesticides used in the field end up in surface and groundwater (Figure 1).

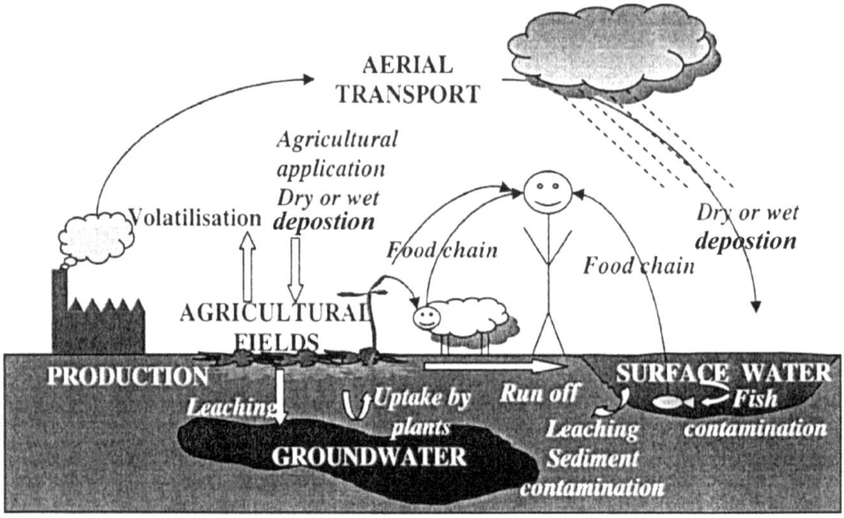

Figure 1. Fate of pesticides in the environment.

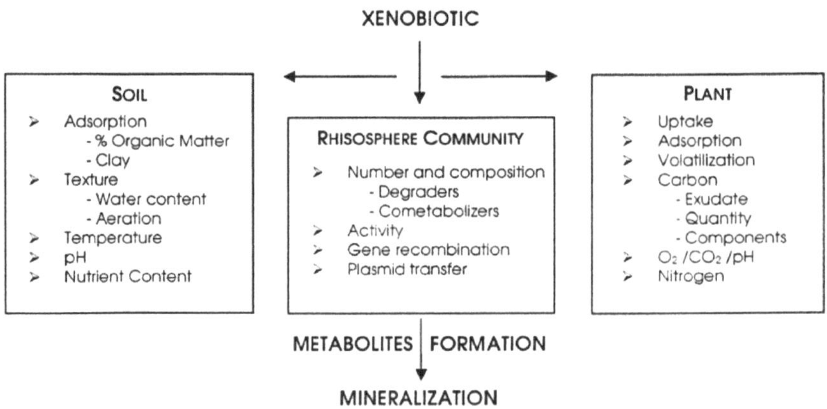

Figure 2. Factors affecting the formation of metabolites and mineralization of pesticides in soil, plant and rhizosphere.

Widespread and large-scale use of pesticides during the last decades has led to a global problem of pollution of soils and water resources. Heavy environmental contamination by pesticides may arise from industrial point sources such as accidental spillage during production, wastewater from pesticide production plants, leakage from old stockpiles, storage and transport of pesticides, or as leachate from former dumping sites and municipal waste. The disposal of unwanted or obsolete pesticide stocks has also resulted in many long-term contaminated sites with very high levels of pesticides. In contrast, sources of pollution arising from agricultural use are considered to be diffuse as the compounds are distributed over large areas and at rather low concentrations [3].

Pesticides, like other organic pollutants, are more easily bioavailable in freshly spiked soils, as compared to aged or long-term contaminated soils [4-6]. The stability and toxicity of many pesticides make them hazardous when incorporated into the food chain. There is also increasing concern about their transformation products because these can be present at higher levels in soil than the parent pesticide itself. In some cases these products are even more toxic and more mobile, representing a greater risk to the environment than parent molecules.

A persistent organic pollutant does not undergo biodegradation in certain environments, whereas a recalcitrant compound resists biodegradation in most environments studied so far [7]. While partial biodegradation can be achieved by only one or a few biochemical reactions, total biodegradation involves more extensive metabolism leading to mineralization. Slow degradation rates often limit the practical use of microorganisms in remediating contaminated sites. Therefore, pesticides removal using plants and their interactions with rhizospheric microorganisms appears to be a promising approach, based on three basic principles (Figure 2):

- the accessibility or bioavailability of the contaminants to the biological system (plant roots, microorganisms);
- the biochemistry of plants or microbes, which transform the contaminant to a less toxic product; and
- the possibility of optimising the biological activity to efficiently remove the organic pollutant [8-12].

As examples of the potential of plants and rhizospheric microorganisms to remove pesticides from contaminated environment, we shall consider here only two pesticides with different properties: atrazine (ATR, 2-chloro-4-(ethyl-amino)-6-isopropylamino *s* triazine) and lindane (1, 2, 3, 4, 5, 6 hexachlorocyclohexane), amenable to different phytoremediation strategies, phytodegradation and phytostimulation, respectively.

2. Atrazine

Triazines are a large family of herbicides, widely used over the last decades, especially atrazine, simazine, propazine, prometryn and a few others (Figure 3). Even if the exact and present figures are not freely available, the annual use of atrazine (ATR) alone has been estimated to be around 40,000 tons in the USA [13], 5,000 tons in China [14] and more than 2,000 tons in Europe [15]. Triazines are pre-emergence, selective systemic herbicides used mainly for the control of annual grasses and broad-leaf weeds in a variety of cultivated crops, such as maize, sorghum, fruit orchards, sugar cane or cotton. Once taken up by the roots, triazines are evenly distributed throughout the plant via the xylem, and act in leaves by inhibiting photosynthesis.

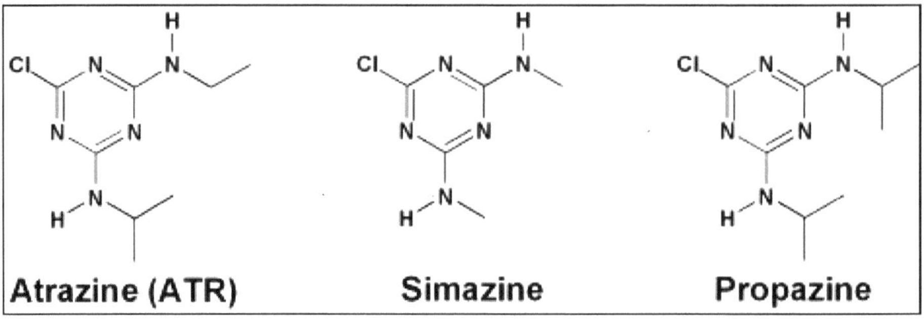

Figure 3. Chemical structure of some triazines (log Kow between 2 and 3).

To be effective, pre-emergence herbicides once applied to the soil, must have an agronomical remanence of several weeks or months, since they need time to exert their phytotoxicity and to kill any weeds germinating after their application. Herbicides are generally retained in the soil by adsorption in the superficial soil horizons, but washing can occur during the first rain or irrigation event following application. Additional losses occur through leaching to deeper soil horizons, and washing of soil particles from unvegetated surfaces. The potential danger of groundwater contamination has been recognized and it appears that atrazine and simazine application should be avoided in sandy soils, and should only be used in non-irrigated crops.

Leaching and agricultural runoff are therefore the primary mechanisms by which herbicides such as atrazine, and some of their degradation products, reach surface water and groundwater. Once atrazine reaches the surface water system, it is transported without substantial loss, and ends in lakes, reservoirs and alluvial aquifers, where it shows minimal loss by volatilization, sorption or transformation. Atrazine, other triazines and their metabolites are thus frequently detected in groundwater, lakes, streams and rivers of many countries [16].

Due to its persistence, atrazine is the second most common ground and surface water contaminant in the USA, and its metabolite deethylatrazine has also been found in groundwater and soils, sometimes at concentrations greater than that of the parent compound [17]. Moreover, atrazine is often used in combination with many other herbicides, such as alachlor, metolachlor, cyanazine, simazine, amitrole + simazine, or diuron + simazine. In most pesticide-contaminated agrochemical facilities, atrazine is thus found in combination with other widely used agricultural chemicals. Therefore, remediation strategies must often cope with a multiple-contaminant situation.

2.1. BIODEGRADATION, BIOREMEDIATION

In alkaline soils, a biological transformation of atrazine takes place, whereas in acidic soils both chemical and microbial transformations occur. By the action of bacteria, atrazine is mainly degraded to deethylatrazine (DEA, 6-Chloro-N^4-isopropyl-1,3,5-triazine-2,4-diamine) and deisopropylatrazine (DIA, 6-Chloro-N^2-ethyl-1,3,5-triazine-2,4-diamine) and to a lower extent to didealkylatrazine (DDA). In contrast, a chemical dechlorination into hydroxyatrazine occurs in anaerobic environments and acidic soils. Dealkylates are thus the main metabolites of atrazine found in soils, and since they still have some herbicidal activity and are mobile compounds, they can also contaminate aquifers [17, 18].

Predominance of degradation by dealkylation has been shown by Shapir and Mandelbaum [19]: significant atrazine disappearance is observed in subsurface soil due to the activity of indigenous microorganisms in the upper part of soil, but only 1% mineralization occurs. Panshin *et al.* [20] observed that after atrazine application, DEA is the dominant degradation product detected the following year. It was shown that cultivated common soil microorganisms such as *Pseudomonas*, *Nocardia*, and *Rhodococcus* sp. are able to degrade atrazine, predominantly by dealkylation of the side chains on the triazine ring.

The limiting factor of atrazine degradation seems to be the absence of efficient atrazine-mineralizing microorganisms in soils, whereas atrazine-dealkylating bacteria are active. Although a few bacterial strains (such as the *Pseudomonas* sp. strain ADP) able to cleave the triazine ring and mineralize atrazine completely have been isolated and characterized, and their genetic potential estimated, the majority of bacteria in atrazine-contaminated fields cannot further metabolize the dealkylated products [21-25]. These bacteria initiate atrazine degradation with hydrolytic dechlorination, catalysed by ATZA (atrazine chlorohydrolase), followed by two amidohydrolytic reactions catalysed by ATZB and ATZC, which altogether transform atrazine to cyanuric acid, which can be mineralized to CO_2 and NH_3 by three other hydrolases, ATZD (cyanuric acid amidohydrolase), ATZE (biuret hydrolase) and ATZF (allophanate hydrolase). However, if *atzA*, *B* and *C* genes are widely distributed in different bacterial species, the other genes are less common. Other known genes involved in triazine degradation include *trzD* and *trzN*; *trzN* encodes the triazine chlorohydrolase, whereas *trzD* encodes the cyanuric acid amidohydrolase that catalyses the ring cleavage of cyanuric acid.

Although atrazine has been applied in the field for decades, its persistence in the environment and the absence of large scale demonstrations of microbial mineralization indicate the difficulty of rapid microbial breakdown in the field and the rarity of atrazine mineralizers. Many studies investigating the efficacy of bioremediation have been carried out on a bench scale and under ideal laboratory conditions. However, environmental conditions such as soil pH, temperature, nutrient availability and contaminant bioavailability vary from site to site and greatly affect the bioremediation process and growth of atrazine-mineralizers. Therefore, the removal of atrazine in contaminated soils by bioremediation is at present only a dream, and there is still a long way to go before a bioaugmentation approach can be successfully applied. Other biological tools should thus be explored, and phytoremediation appears to be the most promising [26-29].

2.2. PHYTOREMEDIATION

Phytoremediation or the use of vegetation at waste sites or contaminated soils can overcome limitations of microbial cleanup such as low bacterial population or inadequate microbial activity. Improving the quality of surface water and reducing nonpoint source pollution can be achieved using wetland vegetation or vegetative filter strips to reduce herbicide runoff.

Plants can remediate environments contaminated with organic compounds directly via root uptake, detoxification by phytotransformation and conjugation with glutathione or sugars, and subsequent storage of nonphytotoxic metabolites in plant tissues; and/or indirectly by the release of exudates or enzymes which can enhance degradation by rhizosphere microorganisms [30, 31].

To be useful for the removal of herbicides from contaminated soil or preventing their runoff, a plant must first be able to grow in the presence of the target compounds without being harmed. The plant must not only be resistant to the pollutant, but also be able to remove it from the environment and to transform it into non-toxic metabolites or end-products. Differences in the ability of various plant species to accumulate and metabolize a particular pollutant have been shown, indicating that natural biodiversity should be better explored and exploited in order to choose the most appropriate plant species or variety in the development of any phytoremediation process [32]. Plant taxonomy and phytochemistry can help to use biochemical specificities of plants producing natural chemicals, whose structures are similar to xenobiotic compounds. Publications on plant metabolism of herbicides in species useful for phytoremediation are scarce compared to publications about plant metabolism for agronomic purposes. In the case of atrazine however, many results obtained in agronomy studies are useful when choosing the most appropriate families or genera for phytoremediation.

The use of vegetative filter strips is a low cost and practical option for improving the quality of runoff water from intensively farmed agricultural production areas [33]. Hybrid-poplar buffer strips were first initiated and planted in rows along a portion of a stream at the end of the 1980's [34]. Poplar is able to take up atrazine with transpiration stream, showing that such buffer strips are also effective in removing atrazine from agricultural percolation and runoff water. The only extensive study of plant metabolism of atrazine for a phytoremediation purpose is precisely in poplar (*Populus deltoides x nigra)*, able to transform the herbicide mainly into dealkylates and to a lesser extend, into polar ammeline [26, 27].

Grasses and semiaquatic plants can also remove nutrients and chemicals; reduce transport of contaminants like atrazine from runoff water by reducing flow which promotes deposition of sediment-adsorbed herbicides; and thus allow time for plant uptake and metabolism or infiltration of pollutants into soils and subsequent degradation before entering water systems. The use of common cattails (*Typha latifolia*) to remove simazine from contaminated water has also been sucessfully tested [31], whereas the atrazine mineralization potential of wetlands has been shown [35]; and a constructed wetland is able to treat efficiently atrazine present in nursery irrigation runoff [36].

Anderson and Coats [37] have evaluated the degradation of atrazine in rhizosphere soil of 15 plant species used for vegetative filter strips. Enhanced mineralization was found in rhizopshere soil collected from kochia (*Kochia acoparia*), common lambsquarters *(Chenopodium album*), foxtail barley (*Hordeum jubatum*), witchgrass (*Panicum capillare*), catnip (*Nepeta cataria*) and musk thistle (*Carduus nutans*). On the other hand, the efficiency of a natural filter of bluegrass (*Poa annua*) and fescue (*Festuca sp.*) strips located immediately down slope of a standard erosion plot of 9%

slope has been investigated [28]. Trapping efficiency of atrazine by a 4.5 m wide strip was 93%, in the same magnitude as dissolved phosphorus, nitrate, ammonium and sediments. This study emphasises the relevance of grass filters as buffer strips, but the mechanism underlying atrazine disappearance was not studied.

Decontamination of water polluted with 6 ppm atrazine by several marsh plants, common club-rush *(Schoeplectus lacustris)*, bulrush *(Typha latifolia)*, yellow iris (*Iris pseudacorus)* and common reed *(Phragmites australis)* was observed and the disappearance of atrazine was suggested to be due to the action of rhizosphere microorganisms [38]. The action of plants themselves was not explored, but was not excluded. Fernandez *et al.* [39] have also evaluated semiaquatic herbaceous perennial plants for their use in herbicide phytoremediation, such as canna *(Canna generalis)*, pickerel *(Pontaderia cordata*), and iris *(Iris x Charjoys Jan)*, and concluded that these taxa were not optimal for phytoremedation, since the plants exposed to herbicides showed significantly reduced biomass.

Phytoremediation can also prevent leaching of contaminants to groundwater from unplanted fields, after crop harvest [16, 31, 35]. On the other hand, higher rates of atrazine and simazine removal have been found in soil planted with *Pennisetum clandestinum* than in unplanted soil [29]. This could be due to plant uptake, degradation by enzymes secreted by plant roots, or increased microbial activity in the rhizosphere.

2.3. VETIVER AS A CANDIDATE FOR ATRAZINE PHYTOREMEDIATION

Vetiver (*Chrysopogon zizanioides* Nash) is a perennial tropical grass, native to India. Vetiver is by nature a hydrophyte, but often thrives under xerophytic conditions: it grows particularly well on river-banks and in rich marshy soil. It can withstand periods of flood, as well as extreme drought, survives temperatures of between –9°C and 45°C, is fire resistant, and is able to grow in any type of soil regardless of fertility, salinity, or pH. Vetiver is a tall (up to 3 meters high), fast growing perennial grass with stiff and tough stems which form a dense hedge with compact rhizome clumps (crown) when planted closely in rows [40].

The distribution of vetiver is pantropical, and some boundary strips are found in vetiver's native region of India. It was introduced recently in Southern regions of Europe, such as Italy, Portugal and Spain. Non-seeding vetiver plants are used in many countries for soil erosion control and many other applications: vetiver grass was first introduced for soil conservation and land stabilization in Fiji in the early 1950s. Recognizing the potential in combating land degradation, the World Bank has promoted the vetiver grass system since the mid 1980s, which is now used worldwide as a low-cost, low-tcchnology and effective means of soil and water conservation and land stabilization in developing countries. The US Board of Science and Technology for International Development has reported successful vetiver applications for stabilization of slopes, terraces and channel banks in numerous tropical and subtropical countries. Vetiver plantation for soil erosion control is mainly performed linearly, along fields, terraces, canals, streams, or rivers, where the erosive force of water is at its greatest, lakeshores, artificial embankments, and little canals for irrigation or water drainage. It can even be planted across the river itself to slow down the flow of water [41].

As a result of the available literature it was deemed relevant to study vetiver uptake potential to intercept and remove not only atrazine, but also dealkylates from soil. The uptake of DEA and DIA was also tested because of their toxicity; relatively high bioavailability due to low sorption on soil matrix (log Kow 1.38 for DEA, and 1.7 for DIA); and their frequent occurence in groundwater, suggesting possible penetration and translocation in plants [42].

The ability of vetiver to take up ATR, DEA and DIA was thus investigated in a model system: 8-month old vetiver plants cultivated in hydroponics and under sterile and moderate transpiring conditions [43]. The disappearance of atrazine and dealkylates from the hydroponic system was dependent on transpiring flux of the plant, showing that the influence of microorganisms was negligible. This relationship was not linear because the concentration of the tested compounds was not constant: water was refilled to the initial level before each sampling. When the herbicide concentration was decreasing in the medium, a progressively lower quantity of herbicide was absorbed. After 20 days, uptake of DEA and DIA per liter of transpired water was not significantly different from that of ATR (Table 1).

Table 1. Herbicide taken up by Vetiver plant per volume of transpired water. Values were calculated as cumulated quantity of herbicide taken up by the plant per total transpired water after 20 days exposure to 10 uM atrazine or dealkylates

Compound	Herbicide uptake per transpired water (mmol l^{-1})
ATR	5.20 ± 0.40
DEA	4.90 ± 0.05
DIA	4.80 ± 0.88

Vetiver could thus take up DEA and DIA in the same range as atrazine. This observation indicates that DEA and DIA uptake appeared to be largely a passive process closely associated with the movement of water, as reported by Wilson et al. [44] in *Canna hybrida*, by Raveton *et al.* [45] in *Zea mays* and *Acer platanus* protoplasts, and in *P. deltoides x nigra* by Burken and Schnoor [26, 27].

Since the uptake of DEA and DIA was passive and the transpiration stream constantly lowered their concentration in the roots, vetiver caused a global loss of dealkylates from the medium. The potential of vetiver for control of dealkylates produced by microorganisms in soil and atrazine runoff is believed to be high, since the plant was observed to absorb and tolerate atrazine, the first requirement of phytoremediation. The deep and dense root system of vetiver will physically retard the runoff of water loaded with atrazine and its primary metabolites, and retain soil and sediments on which atrazine and metabolites are adsorbed, allowing plant uptake and phytotransformation to occur, resulting in the removal of atrazine from the environment.

Since its leaf surface area is small compared to phreatophytes, and the highest uptake of atrazine and dealkylates is dependent on the volume of water transpired, vetiver is not thought to be very efficient for phytoremediation of highly contaminated soil or water. However, due to its highly dense root system, vetiver should be an ideal system against non-point pollution by atrazine and dealkylates. As vetiver is a huge grass, it is expected that this plant could remove pesticides in the same way as the smaller grasses, festuca and poa, in temperate climates. Vetiver is not a crop plant, but has already been shown to have a wide range of applications; an understanding of the effect of vetiver on atrazine and dealkylates should open a new application window. In the near future, vetiver could play an important ecological role for water protection, especially in developing tropical countries. To our knowledge, most of the plants studied until now for phytoremediation of atrazine are adapted to temperate climates of developed countries.

The resistance mechanism of vetiver to atrazine was investigated to further the assessment of its potential for phytoremediation of atrazine-contaminated environments. Plants known to metabolise atrazine rely on hydroxylation mediated by benzoxazinones, conjugation catalyzed by glutathione-S-transferases and dealkylation probably mediated by cytochromes P450. All three possibilities were thus explored in mature vetiver grown in hydroponics. Whereas the role of benzoxazinones in the chemical hydroxylation of atrazine is only marginal [46], conjugation to glutathione was found to play a major role in the detoxification of atrazine by vetiver [47]. Dealkylates are further conjugated with glutathione in sorghum. Vetiver, close to sorghum, could also detoxify DEA and DIA by conjugation. This would be highly beneficial for the environment.

The use of a hydroponic system is the first step towards a comprehensive knowledge of the fate of pesticides in plants, but it is also a useful tool for the assessment of phytotreatments of industrial wastewater, agricultural runoff, surface and groundwater contaminated with pesticides. On the other hand, over-concentration of atrazine was observed in oil from roots grown in soil, suggesting that during plant ageing, partition might play a non-negligible role in retaining atrazine from agricultural runoff. Studies with other pesticides are required to see if vetiver as a tool against pesticide runoff could be extended to include other contaminants.

3. Lindane

The organochlorine 1, 2, 3, 4, 5, 6 hexachlorocyclohexane (HCH) is an efficient insecticide, available in two formulations: technical-grade HCH (a mixture of different isomers, mainly α, β, δ, and γ-HCH) and lindane (almost pure γ-HCH). The eight possible isomers differ in the axial or equatorial orientations of the different chlorine atoms (Figure 4). Typically, the technical mixture consists of 60-70% α-HCH, 5-12% β-HCH, 10-15% γ-HCH, 6-10% δ-HCH and smaller amounts of other isomers. Lindane, the only isomer with insecticidal properties, is isolated from technical-grade HCH by crystallization. It is a rather hydrophobic compound, with a log Kow of 3.72.

HCH is toxic and considered as a potential carcinogen, but because of its low-cost production and its effective pesticide properties, it is ubiquitously used in tropical countries to reduce vector-transmitted diseases, to protect livestock and to increase agricultural yields. Global use of lindane and technical HCH are estimated to be as high as 6 and 11 millions of tons, respectively. Its simple application, efficacy, and economic return may explain the popularity of this broad-spectrum insecticide, and why it was produced in large quantities worldwide until the discovery of its toxicity.

Being a persistent organic pollutant, technical-grade HCH can stay as long as 15 years after the last application in the field; HCH is nowadays found all over the world in air, water and soil samples [1, 48-50]. In addition to agricultural soils, contaminated sites are found where isomers were disposed of in areas surrounding manufacturing centres. Remediation strategies are thus urgently needed to remove HCH isomers from environmental compartments so that they do not end up in food samples through growth of crop plants.

3.1. BIODEGRADATION, BIOREMEDIATION

The chemical structure and polarity of pesticides affect the solubility, sorption and volatility properties, and thus influence their transport, persistence and biodegradability. The ring structure of HCH prevents rotation around the C-C bonds, making vicinal elimination of chlorines dependent on the presence of axial chlorines oriented opposite to each other (anti-parallel positions). For example, β-HCH has no such chlorine pairs, since all chlorine atoms are equatorially oriented, which stabilize the molecule, and it is thus the most persistent isomer (Figure 4). In contrast, γ-HCH has 3 axial chlorines and 2 chlorine pairs, and α-HCH 4 axial chlorines and 1 chlorine pair, giving these isomers available sites for enzymatic attack (dehydrohalogenase). The physical properties and persistence in the environment of the different HCH isomers thus differ because of the different, axial or equatorial, chlorine orientations. Nevertheless, the four major isomers, having log Kow between 3.7 and 4.1, are considered as toxic and recalcitrant [50-52]. Even trickier, isomerisation of HCH can occur under both biotic and abiotic conditions. Bioisomerisation has been observed for bacterial cultures, as well as for bench-scale studies of sediment and soil slurries. This phenomenon should thus be taken into account in any bioremediation process, especially if the formation of the more stable β-HCH is significant.

HCH seems to be biodegradable under both oxic and anoxic conditions, but mineralization occurs only under oxic conditions. Biodegradation of HCH has been widely studied at laboratory scale, but information on full scale *in situ* bioremediation of industrial sites contaminated by HCH isomers, including lindane, is still very scarce [53]. Furthermore, the effect of HCH concentration on the biodegradation process is not yet known, especially in soils, but it has been recently reported that bacterial growth in liquid culture is decreased above 1 mM lindane and totally inhibited at 2.4 mM [54].

Lindane-degrading microorganisms have been isolated from different contaminated soils: among others, several *Clostridium* sp., *Pseudomonas* and especially *Sphingomonas paucimobilis* B90A (or *Sphingobium indicum*) and UT26 (or *Sphingobium japonicum*) [55]. Some of these strains are able to grow on γ-HCH as the only carbon source [52]. The aerobic degradation of lindane by *S. paucimobilis* is the best described and involves several novel enzymes encoded by *linA*, *linB*, *linC*, *linD*, *linE*, *linF*, and *linX* genes, leading to a possible mineralization. The LinA enzyme, a γ-HCH dehydrochlorinase, catalyses the dehydrochlorination of HCH to pentachloro-cyclohexene, then to 1,3,4,6-tetrachloro-1,4-cyclohexadiene. This metabolite is converted to 2,4,5-trichlorocyclohexenol, then to 2,5-dichloro-2,5-cyclohexadiene-1,4-diol (2,5-DDOL) by the LinB protein, a halidohydrolase. LinC, as well as LinX, catalyse the oxidation

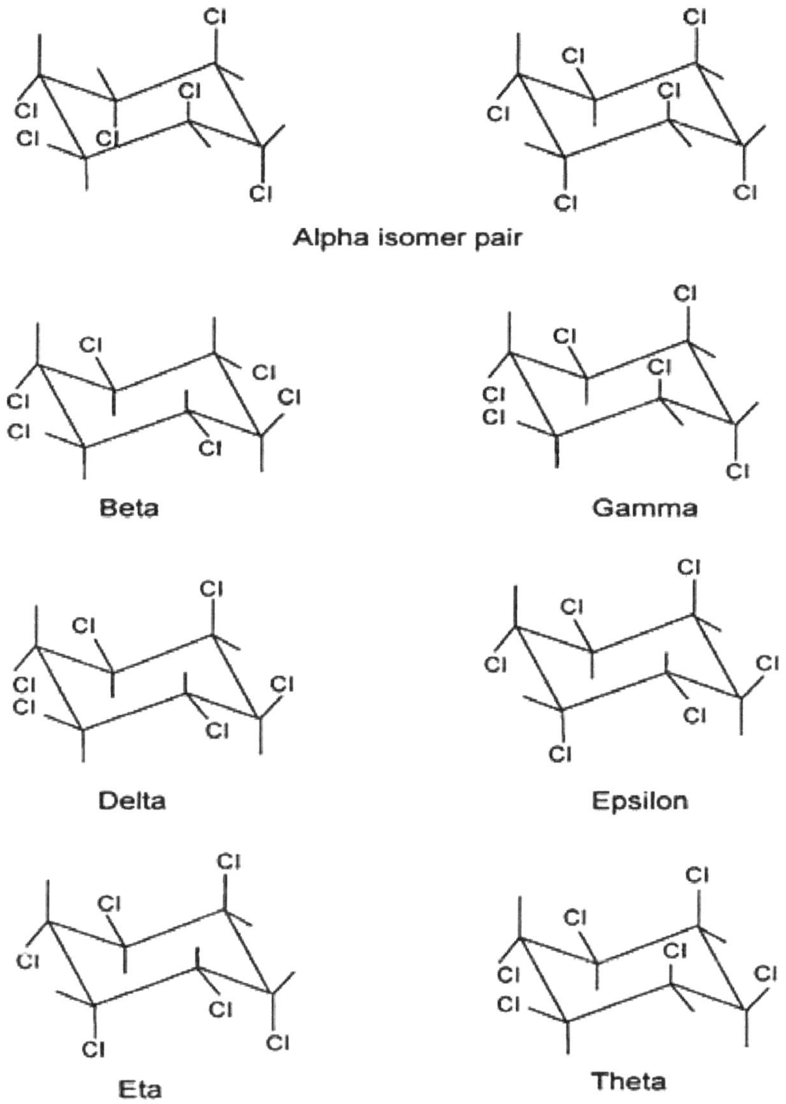

Figure 4. Chemical structures of HCH isomers, showing the equatorial and axial positions of the chlorine atoms: Alpha = aaaaee; Beta = eeeeee; Gamma = aaaeee; Delta = aeeeee; Epsilon = aeeaee; Eta = aaeaee; Theta = aeaeee.

of 2,5-DDOL to 2,5-dichlorohydroquinone (2,5-DCHQ) by a NAD^+-dependent dehydrogenase reaction. The *linD* gene encodes a reductive dehalogenase, catalysing the rapid conversion of 2,5-DCHQ to chlorohydroquinone, then very slowly to hydroquinone. The *linE* gene encodes a dioxygenase able to cleave the aromatic ring of both chlorohydroquinone and hydroquinone, forming 2-chloromaleylacetate or maleylacetate (MA), respectively. MA is then reduced to β-ketoadipate by an MA reductase, the product of gene *linF* [56]. *LinA*, *linB* and *linC* are genes expressed constitutively, existing separately from one another on the genome, and belong to the so-called "upstream pathway". *LinD*, *linE* and *linF* are inducible genes forming the "downstream pathway"; *linD* and *linE* are organised within an operon with a transcriptional regulator called LinR, and are induced at 7 mg l^{-1} γ-HCH, but not at 0.7 mg l^{-1} [52, 57, 58].

S. paucimobilis is not the only aerobic HCH-degrading bacterium. *Escherichia coli* is also able to degrade lindane but only up to 10% level. Another bacterium, *Rhodanobacter lindaniclasticus,* degrades lindane with higher removal capacity [52]. Recently, two bacteria able to remove from 45.5% to 90% of lindane within 2 to 8 weeks of incubation were identified as *Pandorea* sp. [59]. These bacteria can use lindane as a sole source of carbon and energy. Some cyanobacteria, e.g. *Anabaena sp.* strain PCC7120, *Nostoc ellipsosporum* also possess lindane-degradation capacity [60]. It should be mentioned however that mineralization of lindane by such microorganisms is rarely encountered, with the exception of *S. paucimobilis* strains.

Several anaerobic bacteria, such as *Clostridium sphenoides*, *C. rectum*, *Citrobacter freundii* and *Desulfovibrio* sp. have been shown to degrade lindane by reductive dechlorination, but no mineralization seems to occur [52].

Finally, lignin-degrading fungi like *Phanerochaete chrysosporium*, *Trametes hirsutus*, *Cyathus bulleri* and several *Pleurotus* species have been shown to degrade lindane by extracellular lignin-degrading non-specific peroxidases [52].

Until now however, bioaugmentation with indigenous microorganisms failed to enhance degradation of either isomer [52]. Survival and activity of the inoculum are not always guaranteed, but the cultivation of appropriate plant species should increase the success of such a strategy. Bioremediation of HCH-contaminated soil is probably not realistic for crop fields where application rates were low, since HCH is expected to be removed by natural attenuation. In contrast, bioremediation should be appropriate and useful on industrial post-production or waste dumping sites, and storage sites.

3.2. PHYTOREMEDIATION

The uptake of pesticides by plants depends on the physicochemical properties of the compound, mode of application, soil type, climatic factors and plant species. Once in the roots, the chemical may be translocated to shoot via xylem. The permeation from plant roots to xylem is optimal for moderately hydrophobic molecules, with a log Kow between 0.5 and 3.5. More hydrophobic chemicals tend to bind with lipid membranes or oil possibly present in plant roots. Translocation of non-ionic pesticides thus varies greatly between plant species and depends on the properties of the chemical. Therefore, the uptake and translocation of hydrophobic compounds (log Kow > 4) is limited, and consequently so is their phytodegradation. On the other hand, the transformation of pesticides by rhizosphere microorganisms could result in metabolites more efficiently absorbed and translocated by plants. Thus any factor enhancing rhizospheric microbial activity should also increase the overall efficiency of pesticide phytoremediation.

With regards to HCH isomers, phytoextraction should not be favoured, since they are hydrophobic chemicals, theoretically bound to soil particles and also to plant roots, thus preventing plant uptake [61-64]. However, it has been recently shown that the predicted partition of lindane with lipids using the log Kow is notably lower than the measured sorption in roots and shoots of ryegrass and wheat seedlings, due to underestimation of the plant lipid contents and to the fact that octanol is less effective than plant lipids as a partition medium [65].

Nevertheless, the persistence of all isomers was found to be lower in cropped plots of maize (*Zea mays*), wheat (*Triticum* sp.) or pigeon pea (*Cajananus cajan*) than in uncropped plots, therefore sustaining the idea of remediation of polluted sites with plants [66]. Revision of available literature concerning contamination of the aerial part of plants reveals that the α, β and γ isomers have been detected in many plants, including *Lactuca sativa* (lettuce), *Sesamum indicum* (sesame), *Hydrilla verticillata* (hydra), *Lagernia siceraira* (bottle gourd), *Memordica charantia* (bitter gourd), *Luffa cylindrical* (sponge gourd), *Citrullus varifistulosus* (tinda punjab), *Spinacia oleracea* (spinach) and *Brassica campestris* (rape). These species were not selected for testing in hydroponics however, since the detected residues were due to a direct contact with lindane, and not as a result of translocation from roots to shoots. Barriada-Pereira *et al.* [67] also attributed the presence of lindane in the shoots of *Rubus ulmifolius* and *Paranthropophytia* to atmospheric deposition and not translocation. However, chilli (*Capsicum annuum*) and coriander (*Coriander sativum*) cultivated in lindane-contaminated soil were reported to contain γ-HCH residues in their aerial parts [43]. This fact was considered as a clue for the selection of plants able to take up lindane from a hydroponic experimental system. Within 9 days, lindane concentration in the medium decreased by 70% with chilli and 86% with coriander (Figure 5). Adsorption on roots of chilli and coriander accounted for 29% and 40% of γ-HCH disappearance, respectively. The 23% and 30% remaining loss was called "plant effect", which could include the increasing of pH from 5.0 to 6.8, leading to enhanced hydrolysis, and a possible uptake of γ-HCH in unknown quantities with the transpiration flux [43, 68]. However, the translocation of lindane from roots to shoots should be low, due to lindane hydrophobicity, unless special molecules are produced by plants, able to increase the apparent aqueous solubility of hydrophobic pollutants, such as those shown by Campanella and Paul [69] for dioxin absorption by zucchini (*Cucurbita pepo* L.).

In further experiments, vetiver plants were grown in Hoagland solution with ^{14}C lindane (2 mg/l). At the end of 30 days, it was observed that 12% of lindane disappeared from the solution. Since lindane was found to accumulate in vetiver roots where essential oil is produced, an attempt was made to use a plant, which has oil in the shoot as well. Lemon grass (*Cymbopogan citrates*) was thus grown in Hoagland solution for 19 days in the presence of ^{14}C lindane, but again more ^{14}C residues were found in roots (11.4%) than in shoots (4.2%). All the plants tested for uptake of ^{14}C lindane, *Iris* sp., *Sesuvium portulacastrum*, *Zea mays* and *Cymbopogan* sp. showed that ^{14}C residues were retained in roots and there was no significant translocation to shoots.

Another study has shown that the concentration of lindane in ryegrass, cultivated in hydroponics, slowly increases with uptake time to reach a plateau after a few days, indicating that plant metabolism and formation of bound residues are minimal in such a system [70]. Other authors [71, 72] aimed to evaluate the bioaccumulation of HCH isomers in plants growing on areas surrounding a production centre and to investigate

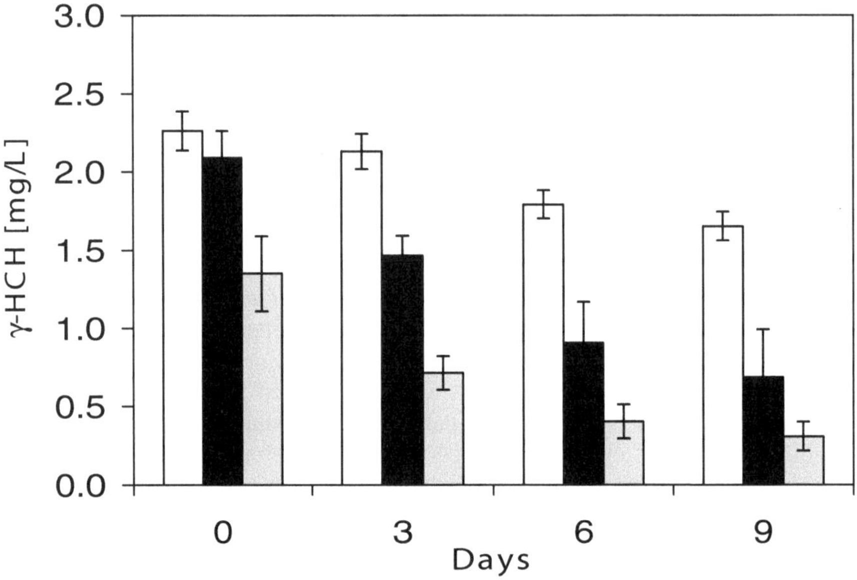

Figure 5. Time course experiment of lindane disappearance in solution with chilli and coriander. Controls (white columns), chilli (black columns), coriander (grey columns). Data points represent the means of 5 replicates determinations. Bar = SD.

the effect of the rhizosphere on these compounds. Five plant species were harvested and separated into roots, stems and leaves: *Chenopodium vulgare*, *Solanum nigrum*, *Cytisus striatus*, *Vicia sativa* and *Avena sativa*. Concentrations of total HCH in plant organs ranged between 1.7 and 62.5 mg kg^{-1}, depending on plant species and organs. Leaves systematically contained the highest amount of HCH, probably due to the volatilization of HCH isomers from the soil surface and subsequent sorption by leaves. In one species, C. striatus, the metabolites pentachlorocyclohexene, cyclopentiltrichloroethene and 6 trichloro, 4-en-hexanoic derivates were also detected in the all plant tissues analysed [71]. Data obtained from the bulk and rhizosphere soils from *C. striatus* and *A. sativa* suggest that both plant species tend to reduce the levels of all HCH isomers in the rhizosphere. This could be due to the enhanced biodegradation in the rhizosphere, root exudation of enzymes able to dechlorinate HCH isomers and/or sequestration by partitioning into the lipophilic plant tissues or uptake by the roots.

In the case of HCH, phytoremediation has thus shown its own limitations, even in the favourable case of hydroponic systems, where the pollutant is made highly bioavailable. Since rhizosphere microbial activity is known to aid the release of bound pesticide residues in soil, which can in turn enhance uptake and transformation, by plants, a combination of bioremediation and phytoremediation, or the phytostimulation of rhizosphere microorganisms, is likely to be more successful.

Phytostimulation of bacteria present or added in soil seems the most promising approach to remove lindane from contaminated sites. It has long been known that plants release a vast range of organic materials through roots into the rhizosphere. These exudates contain water soluble, insoluble and volatile compounds including sugars, amino acids, organic acids, flavonones, phenolic compounds and even enzymes [12, 73-77]. The root exudates can enhance the acquisition of nutrients by plants; stimulate microbial growth in the rhizosphere; and change pH, water flux and availability of oxygen. Microorganisms able to use phenolic compounds as a carbon source often have enzymes that can co-metabolise pollutants with similar structures. Plants have also a great capacity to release secondary metabolites having a surfactant activity, which is favourable for phytoremediation purposes. Thus the degradation of several chlorinated pesticides has been reported to be higher in a vegetated soil than a non-vegetated soil. For example, the biodegradation of HCH isomers is enhanced in rhizosphere soils of *Kochia* sp., as compared to bulk soil, even if the mechanism by which this occurs is not yet known [78].

A dynamic synergy does exist between plant roots and soil microorganisms. The microbial activity in the immediate vicinity of the root (rhizosphere) seems to offer a favourable environment for co-metabolism of soil-bound and recalcitrant chemicals. The microbial transformations of organic compounds are not always driven by energy needs, but also by the necessity to reduce toxicity for which microbes may have to suffer an energy deficit. Thus, the processes may be helped and driven by the abundant energy provided by root exudates. Certain soil microorganisms can also produce biosurfactant compounds that may facilitate the removal and degradation of organic chemicals by increasing their availability to plants. Plants can thus take advantage of an increased bioavailability of nutrients and degradation of phytotoxic soil contaminants.

A successful example of exploring rhizosphere microorganisms for the decontamination of pesticide-contaminated soils has been highlighted by the recent findings of a coordinated Indo-Swiss project [77]. The project was aimed at investigating possibilities to remediate agricultural soils contaminated with lindane, using a plant-rhizosphere system. Mineralisation of ^{14}C lindane by rhizospheric soils of plants growing in lindane-contaminated fields indicated that microorganisms capable of degrading the insecticide were present. Enrichment culture techniques resulted in the isolation of bacteria growing in the presence of lindane. *Klebsiella* sp., *Pseudomonas* sp., and *Pseudoarthrobacter* sp. degraded up to 50% of lindane with the formation of 2,3,4-trichlorobenzene. Root exudates of some plants stimulated the growth of lindane degrading bacterium *Pseudomonas* sp., indicating the need for an approach, which includes both plants and interacting microorganisms for an efficient degradation of pesticides. To treat lindane-contaminated soils, it thus appears that phytostimulation would be the most appropriate technique.

Soil composition influences sorption, soil pH, bulk density and water retention, all of which affecting aeration, nutrient availability and thus bioavailability and biodegradability of contaminants [61]. A high density of indigenous *S. paucimobilis* was found in the plant debris fraction of soil and it was postulated that plant organic matter integrated into the soil aggregates served as a microhabitat rich in growth substrates. This can be the basis for the application of plant-derived organic amendments to soil as a phytostimulation strategy, like rice straw or other cellulosic material, because of their efficiency, availability in large quantities and low cost [52]. For example, the Daramend® technology for bioremediation of HCH-contaminated soils is based on the application of solid plant-derived organic matter providing nutrients and a non-toxic habitat for indigenous microorganisms; it creates a concentration gradient that facilitates diffusion of organic contaminants from pockets of higher to lower concentrations on the amendment surface, where they are more bioavailable [79]. Since the texture, nutrient requirements and microbial populations of each soil are specific, the usefulness and composition of amendments should be assessed on a case-by-case basis.

Detailed strategies for optimising treatments on sites contaminated with HCHs remain to be established and could involve enhanced natural attenuation and optimisation of environmental conditions, to stimulate growth and biodegradation by indigenous microorganisms. Supplemental nutrients and/or organic amendments could be added to enrich the soil and stimulate the bacteria degrading HCH. Bioaugmentation with phytostimulation or a vegetative cover should also be tested, either by increasing the population of microorganisms able to degrade HCH isomers, or by increasing the bioavailability of the insecticide.

4. Conclusions and perspectives

Both examples developed here show that plants and soil microorganisms have certain limitations with respect to their individual abilities to remove and degrade organic pollutants like pesticides, and other molecules containing chlorines and/or aromatic ring structures. However, plants and bacteria have very specific and complementary metabolic pathways, and their combined appropriate use can breakdown many man-made chemicals. Therefore, a synergy between rhizospheric microorganisms leading to increased availability of hydrophobic compounds and plants leading to their removal and/or degradation, may overcome many of the limitations, thus providing a sound basis for enhancing biological remediation of contaminated environments.

Phytostimulation or rhizoremediation is of particular importance because it refers to an important contribution that microorganisms in the root-zone (rhizosphere) make to the overall breakdown and removal of organic pollutants by plants. Plant-microbial interactions in the rhizosphere are thus of utmost importance for the degradation of recalcitrant chemicals in the environment [12, 30, 80].

However, further research into the mechanisms by which plants can stimulate biodegradation and the complexity of the soil-plant-microbe system due to its interwoven nature is thus required to better explore and exploit their huge potential. Such studies must be done not only at laboratory scale, but also under real conditions, as demonstration projects, to optimise the phytoremediation process and convince regulators and the general public of the technique's feasibility [81, 82]. To increase its acceptance as a remediation concept, phytoremediation must also become an economically interesting approach and biomass disposal or use after the treatment is thus an important issue to consider. For example, the biomass of fibres, oil or fragrance producing plants like vetiver, could be used to recover these added-value products, if however their level of contamination is nil or low enough. Alternatively, contaminated biomass could be used for renewable energy generation, either by direct combustion, gasification or pyrolysis, or indirectly via biogas or biofuel production [83, 84].

References

[1] Courdouan, A; Marcacci, S; Gupta, S and Schwitzguébel, JP (2004) Lindane and technical HCH residues in Indian soils and sediments – A critical appraisal. Journal of Soils and Sediments 4: 192-196

[2] Pimental, D and Levitan, L (1986) Pesticides: amounts applied and amounts reaching pests. Bioscience 36: 86-91

[3] Chaudhry, Q; Schröder, P; Werck-Reichhart, D; Grajek, W and Marecik, R (2002) Prospects and limitations of phytoremediation for the removal of persistent pesticides in the environment. Environmental Science and Pollution Research 9: 4-17

[4] Alexander, M (2000) Aging, bioavailability and overestimation of risk from environmental pollutants. Environmental Science and Technology 34: 4259-4265

[5] Anhalt, JC; Arthur, EL; Anderson, TA and Coats JR (2000) Degradation of atrazine, metolachlor and pendimethalin in pesticides-contaminated soils: Effects of aged residues on soil respiration and plant survival. Journal of Environmental Science and Health B 35: 417-438

[6] Hatzinger, PB and Alexander, M (1995) Effect of aging chemicals in soil on their biodegradability and extractability. Environmental Science and Technology 29: 537-545

[7] Dua, M; Singh, A; Sethunathan, N and Johri, AK (2002) Biotechnology and bioremediation: successes and limitations. Applied Microbiology and Biotechnology 59: 143-152

[8] Atterby, H; Smith, N; Chaudhry, Q and Stead, D (2002) Exploiting microbes and plants to clean up pesticide contaminated environments. Pesticide Outlook 13: 9-13

[9] Davis, LC; Castro-Diaz, S; Zhang, QZ and Erickson, LE (2002) Benefits of vegetation for soils with organic contaminants. Critical Reviews in Plant Sciences 21: 457-491

[10] Belden, JB; Clark, BW; Phillips, TA; Henderson, KL; Arthur, EL and Coats, JR (2004) Detoxification of pesticides residues in soil using phytoremediation. Pesticide Decontamination and Detoxification ACS Symposium Series 863: 155-167

[11] Karthikeyan, R; Davis, LC; Erickson, LE; Al-Khatib, K; Kulakow, PA; Barnes, PL: Hutchinson, SL and Nurzhanova, AA (2004) Potential for plant-based remediation of pesticides-contaminated soil and water using nontarget plants such as trees, shrubs and grasses. Critical Reviews in Plant Sciences 23: 91-101

[12] Kuiper, I; Lagendijk, EL; Bloemberg, GV and Lugtenberg, BJJ (2004) Rhizoremediation: a beneficial plant-microbe interaction. Molecular Plant-Microbe Interactions 17: 6-15

[13] Radosevich, M; Traina, SJ; Hao, YL and Tuovinen, OH (1994) Degradation and mineralization of atrazine by a soil bacterial isolate. Applied and Environmental Microbiology 61: 297-302

[14] Jin, R and Ke, J (2002) Impact of atrazine disposal on the water resources of the Yang river in Zhangjiakou area in China. Bulletin of Environmental Contamination and Toxicology 68: 893-900

[15] Vighi, M and Funari, E (1995) Pesticide Risk in Groundwater. Lewis Publishers, Boca Raton, FL, USA, ISBN 0-87371-439-3, 275 p

[16] Coleman, JOD; Frova, C; Schröder, P and Tissut, M (2002) Exploiting plant metabolism for the phytoremediation of persistent herbicides. Environmental Science and Pollution Research 9: 18-28

[17] Mersie, W and Seybold, C (1996) Adsorption and desorption of atrazine, deethylatrazine, deisopropylatrazine and hydroxyatrazine on levy wetland soil. Journal of Agricultural and Food Chemistry 44: 1925-1929

[18] Qiao, X; Ma, L and Hummel, HE (1996) Persistence of atrazine and occurrence of its primary metabolites in three soils. Journal of Agricultural and Food Chemistry 44: 2846-2848

[19] Shapir, N and Mandelbaum, RT (1997) Atrazine degradation in subsurface soil by indigenous and introduced microorganisms. Journal of Agricultural and Food Chemistry 45: 4481-4486

[20] Panshin, SY; Carter, DS and Bayless, RE (2000) Analysis of atrazine and four degradation products in the pore water of the vadose zone, Central Indiana. Environmental Science and Technology 34: 2131-2137

[21] Singh, BK; Kuhad, RC; Singh, A; Lal, R and Tripathi, KK (1999) Biochemical and molecular basis of pesticide degradation by microorganisms. Critical Reviews in Biotechnology 19: 197-225
[22] Piutti, S; Hallet, S; Rousseaux, S; Philippot, L; Soulas, G and Martin-Laurent, F (2002) Accelerated mineralisation of atrazine in maize rhizosphere soil. Biology and Fertility of Soils 36: 434-441
[23] Rhine, ED; Fuhrmann, JJ and Radosevich, M (2003) Microbial community responses to atrazine exposure and nutrient availability: linking degradation capacity to community structure. Microbial Ecology 46: 145-160
[24] Martin-Laurent, F; Cornet, L; Ranjard, L; Lopez-Gutiérrez, JC; Philippot, L; Schwartz, C; Chaussod, R; Catroux, G and Soulas, G (2004) Estimation of atrazine-degrading genetic potential and activity in three French agricultural soils. FEMS Microbiology Ecology 48: 425-435
[25] Smith, D; Alvey, S and Crowley, DE (2005) Cooperative catabolic pathways within an atrazine-degrading enrichment culture isolated from soil. FEMS Microbiology Ecology 53: 265-273
[26] Burken, JG and Schnoor, JL (1996) Phytoremediation: plant uptake of atrazine and role of root exudates. Journal of Environmental Engineering 122: 958-963
[27] Burken, JG and Schnoor, JL (1997) Uptake and metabolism of atrazine by poplar trees. Environmental Science and Technology 31: 1399-1406
[28] Barfield, B; Blevins, R; Fogle, A; Madison, C; Inamdar, S; Carey, D and Evangelou, V (1998) Water quality impacts of natural filter strips. American Society of Agricultural Engineering 41: 371-381
[29] Singh, N; Megharaj, M; Kookana, RS; Naidu, R and Sethunathan, N (2004) Atrazine and simazine degradation in *Pennisetum* rhizosphere. Chemosphere 56: 257-263
[30] Van Eerd, LL; Hoagland, RE and Hall, JC (2003) Pesticide metabolism in plants and microorganisms. Weed Science 51: 472-495
[31] Wilson, PC; Whitwell, T and Klaine, SJ (2000) Metalaxyl and simazine toxicity to and uptake by *Typha latifolia*. Archives of Environmental Contamination and Toxicology 39: 282-288
[32] Marcacci, S and Schwitzguébel, JP (2005) Using plant phylogeny to predict detoxification of triazine herbicides, in Willey, N Ed., Phytoremediation: Methods and Reviews. Humana Press, NJ, USA, Chapter 20, (in press)
[33] Krutz, LJ; Senseman, A; Zablotowicz, RM and Matocha, MA (2005) Reducing herbicide runoff from agricultural fields with vegetative filter strips: a review. Weed Science 53: 353-367
[34] Nair, DR; Burken, JG; Licht, LA and Schnoor, JL (1993) Mineralization and uptake of triazine pesticides in soil-plant systems. Journal of Environmental Engineering 119: 842-854
[35] Anderson, KL; Wheeler, KA; Robinson, JB and Tuovinen, OH (2002) Atrazine mineralization potential in two wetlands. Water Research 36: 4785-4794
[36] Runes, HB; Jenkins, JJ; Moore, JA; Bottomley, PJ and Wilson, BD (2003) Treatment of atrazine in nursery irrigation runoff by a constructed wetland. Water Research 37: 539-550
[37] Anderson, TA and Coats, JR (1995) Screening rhizosphere soil samples for the ability to mineralize elevated concentrations of atrazine and metolachlor. Journal of Environmental Science and Health B 30: 473-484
[38] McKinlay, R and Kasperek, K (1998) Observations on decontamination of herbicide-polluted water by marsh plant systems. Water Research 33: 505-511
[39] Fernandez, TR; Whitwell, T; Riley, MB and Bernard, CR (1999) Evaluating semi-aquatic herbaceous perennials for use in herbicide phytoremediation. Journal of American Society of Horticultural Science 124: 539
[40] Bertea, CM and Camusso, W (2002) Vetiveria: anatomy, biochemistry and physiology, in Maffei, M Ed., Vetiveria. Taylor and Francis, London and New York, ISBN, 0-415-27586-5, pp. 19-43
[41] Truong, P (2002) Vetiver grass technology, in Maffei, M Ed., Vetiveria, Taylor and Francis, London and New York, ISBN, 0-415-27586-5, pp. 114-132.
[42] Briggs, GC; Bromilow, RH and Evans, AA (1982) Relationships between lipophilicity and root uptake and translocation of non-ionized chemicals by barley. Pesticide Science 13: 495-504
[43] Marcacci, S (2004) A phytoremediation approach to remove pesticides (atrazine and lindane) from contaminated environment, PhD thesis Nr 2950, EPFL, Lausanne, Switzerland.
[44] Wilson, PC; Whitwell, T and Klaine, SJ (1999) Phytotoxicity, uptake and distribution of 14C-simazine in *Canna hybrida* "Yellow King Humbert". Environmental Toxicology and Chemistry 18: 1462-1468
[45] Raveton, M; Ravanel, P; Serre, AM; Nurit, F and Tissut, M (1997) Kinetics of uptake and metabolism of atrazine in model plant system. Pesticide Science 49: 157-163

[46] Marcacci, S; Raveton, M; Ravanel, P and Schwitzguébel, JP (2005) The possible role of hydroxylation in the detoxification of atrazine in mature vetiver (*Chrysopogon zizanioides* Nash) grown in hydroponics. Zeitschrift für Naturforschung 60c: 427-434
[47] Marcacci, S; Raveton, M; Ravanel, P and Schwitzguébel, JP (2006) Conjugation of atrazine in vetiver (*Chrysopogon zizanioides* Nash) grown in hydroponics. Environmental and Experimental Botany (in press)
[48] Walker, K; Vallero, DA and Lewis, RG (1999) Factors influencing the distribution of lindane and other hexachlorocyclohexanes in the environment. Environmental Science and Technology 33: 4373-4378
[49] Wania, F; Mackay, D; Li, YF; Bidleman, TF and Strand, A (1999) Global chemical fate of α-hexachlorocyclohexane. 1. Evaluation of a global distribution model. Environmental Toxicology and Chemistry 18: 1390-1399
[50] Willett, KL; Ulrich, EM and Hites, RA (1998) Differential toxicity and environmental fates of hexachlorocyclohexane isomers. Environmental Science and Technology 32: 2197-2207
[51] Deo, PG; Karanth, NG and Karanth, NGK (1994) Biodegradation of hexachlorocyclohexane isomers in soil and food environment. Critical Reviews in Microbiology 20: 57-78
[52] Phillips, TM; Seech, AG; Lee, H and Trevors, JT (2005) Biodegradation of hexachlorocyclohexane (HCH) by microorganisms. Biodegradation 16: 363-392
[53] Van Liere, H; Staps, S; Pijls, C., Zwiep, G., Lassche, R. and Langenhoff, A. (2003) Full scale case: successful in situ bioremediation of a HCH contaminated industrial site in Central Europe (The Netherlands). Proceedings of the 7th International HCH and Pesticides Forum, Kyiv, Ukraine, 5-7 June 2003, ISBN 966-8187-31-8, pp. 128-132
[54] Pesce, SF and Wunderlin, DA (2004) Biodegradation of lindane by a native bacterial consortium isolated from contaminated river sediment. International Biodeterioration and Biodegradation 54: 255-260
[55] Pal, R; Bala, S; Dadwahl, M; Kumar, M; Dhingra, G; Prekash, O; Prabagaran, SR; Shivaji, S; Cullum, J; Holliger, C and Lal, R (2005) The hexachlorocyclohexane-degrading bacterial strains *Sphingomonas paucimobilis* B90A, UT26 and Sp. having similar *lin* genes are three distinct species, *Sphingobium indicum sp. nov.*; *Sphingobium japonicum sp. nov.*; and *Sphingobium francense sp. nov.* and reclassification of *Sphingomonas chungburkensis* as *Sphingobium chungbukense comb. nov.* International Journal of Systematic and Evolutionary Microbiology (in press)
[56] Endo, R; Kamakura, M; Miyauchi, K; Fukuda, M; Ohtsubo, Y; Tsuda, M and Nagata, Y (2005) Identification and characterization of genes involved in the downstream degradation pathway of γ-hexachlorocyclohexane in *Sphingomonas paucimobilis* UT26. Journal of Bacteriology 187: 847-853
[57] Miyauchi, K; Lee, HS; Fukuda, M; Takagi, M and Nagata, Y (2002) Cloning and characterization of *linR*, involved in regulation of the downstream pathway for γ-hexachlorocyclohexane degradation in *Sphingomonas paucimobilis* UT26. Applied and Environmental Microbiology 68: 1803-1807
[58] Suar, M; van der Meer, JR; Lawlor, K; Holliger, C and Lal, R (2004) Dynamics of multiple *lin* gene expression in *Sphingomonas paucimobilis* B90A in response to different hexachlorocyclohexane isomers. Applied and Environmental Microbiology 70: 6650-6656
[59] Okeke, BC; Siddique, T; Arbestain, MC and Frankenberger, WT (2002) Biodegradation of γ-hexachlorocyclohexane (Lindane) and α-hexachlorocyclohexane in water and soil slurry by a *Pandoraea* species. Journal of Agricultural and Food Chemistry 50: 2548-2555
[60] Kuritz, T (1999) Cyanobacteria as agents for the control of pollution by pesticides and chlorinated organic compounds. Journal of Applied Microbiology 85: 1865-1925
[61] Agnihotri, NP and Barooah, AK (1994) Bound residues of pesticides in soil and plant – A review. Journal of Scientific and Industrial Research 53: 850-861
[62] Bromilow, RH and Chamberlain, K (1995) Principles governing uptake and transport of chemicals, in Trapp, S and McFarlane, JC Eds., Plant Contamination – Modeling and Simulation of Organic Chemical Processes. Lewis Publishers, Boca Raton, FL, USA, ISBN 1-56670-078-7, pp. 37-68
[63] Sicbaldi, F; Sacchi, GA; Trevisan, M and Del Re, AAM (1997) Root uptake and xylem translocation of pesticides from different chemical classes. Pesticide Science 50: 111-119
[64] Burken, JG (2003) Uptake and metabolism of organic compounds: green-liver model, in McCutcheon, SC and Schnoor, JL Eds., Phytoremediation: Transformation and Control of Contaminants. Wiley Interscience, Hoboken, NJ, USA, ISBN 0-471-39435-1, pp. 59-84
[65] Li, H; Sheng, G; Chiou, CT and Xu, O (2005) Relation of organic contaminant equilibrium sorption and kinetic uptake in plants. Environmental Science and Technology 39: 4864-4870

[66] Singh, G; Kathpal, T; Spencer, W and Dhankar, J (1991) Dissipation of some organochlorine insecticides in cropped and uncropped soil. Environmental Pollution 70: 219-239
[67] Barriada-Pereira, M; Concha-Grana, E; González-Castro, MJ; Muniategui-Lorenzo, S; López-Mahía, P; Prada-Rodríguez, D and Fernández-Fernández, E (2003) Microwave-assisted extraction versus Soxhlet extraction in the analysis of 21 organochlorine pesticides in plants. Journal of Chromatography A 1008: 115-122
[68] Marcacci, S; Paratte, S and Schwitzguébel, JP (2002) Phytoextraction of lindane by chilli and coriander in hydroponic system, in T Macek, M Mackova and K Demnerova, Eds, Proceedings of the 12th International Biodeterioration and Biodegradation Symposium, Prague, Czech Republic, CSBMB Prague, JPM Tisk, p. 193, ISBN 80-86313-08-5
[69] Campanella, B and Paul, R (2000) Presence, in the rhizosphere and leaf extracts of zucchini (*Cucurbita pepo* L.) and melon (*Cucumis melo* L.), of molecules capable of increasing the apparent aqueous solubility of hydrophobic pollutants. International Journal of Phytoremediation 2: 145-158
[70] Li, H; Sheng, G; Sheng, W and Xu, O (2002) Uptake of trifluralin and lindane from water by ryegrass. Chemosphere 48: 335-341
[71] Barriada-Pereira, M; González-Castro, MJ; Muniategui-Lorenzo, S; López-Mahía, P; Prada-Rodríguez, D and Fernández-Fernández, E (2005) Organochlorine pesticides accumulation and degradation products in vegetation samples of a contaminated area in Galicia (NW Spain). Chemosphere 58: 1571-1578
[72] Monterroso, MC; Camps Arbestain, M; Calvelo Pereira, R; Gomez Garrido, B; Lorenzo, SM; Lopez-Mahia, P; Prada, D and Macias, F (2002) Environmental fate and behavior of HCH isomers in a soil-plant system in a contaminated site. Organohalogen Compounds 59: 307-310
[73] Fletcher, JS and Hedge, RS (1995) Release of phenols by perennial plant roots and their potential importance in bioremediation. Chemosphere 31: 3009-3016
[74] Yoshitomi, KJ and Shann, JR (2001) Corn (*Zea mays* L.): root exudates and their impact on ^{14}C-pyrene mineralization. Soil Biology and Biochemistry 33: 1769-1776
[75] Singer, AC; Crowley, DE and Thompson, IP (2003) Secondary plant metabolites in phytoremediation and biotransformation. Trends in Biotechnology 21: 123-130
[76] Valant-Vetschera, KM; Roitman, JN and Wollenweber, E (2003) Chemodiversity of exudates flavonoids in some members of the Lamiaceae. Biochemical Systematics and Ecology 31: 1279-1289
[77] Chaudhry, Q; Blom-Zandstra, M; Gupta, S and Joner, EJ (2005) Utilising the synergy between plants and rhizosphere microorganisms to enhance breakdown of organic pollutants in the environment. Environmental Science and Pollution Research 12: 34-48
[78] Singh, N (2003) Enhanced degradation of hexachlorocyclohexane isomers in rhizosphere soil of *Kochia* sp. Bulletin of Environmental Contamination and Toxicology 70: 775-782
[79] Phillips, TM; Seech, AG; Trevors, JT and Piazza, M (2000) Bioremediation of soils containing hexachlorocyclohexane, in Wickramanayake, GB; Gavaskar, AR; Gibbs, JT and Means JL Eds., Case Studies in the Remediation of Chlorinated and Recalcitrant Compounds. Battelle Press, Columbus, OH, USA, pp. 285-292
[80] Schwitzguébel, JP (2001) Hype of hope: the potential of phytoremediation as an emerging green technology. Remediation 11(4): 63-78
[81] Van der Lelie, D; Schwitzguébel, JP; Glass, DJ; Vangronsveld, J and Baker, A (2001) Assessing phytoremediation's progress in the United States and Europe. Environmental Science and Technology 35: 446A-452A
[82] Schwitzguébel, JP; Van der Lelie, D; Baker, A; Glass, DJ and Vangronsveld, J (2002) Phytoremediation: European and American trends, success, obstacles and needs. Journal of Soils and Sediments 2: 91-99
[83] Singhal, V and Rai, JPN (2003) Biogas production from water hyacinth and channel grass used for phytoremediation of industrial effluents. Bioresource Technology 86: 221-225
[84] Schwitzguébel, JP (2004) Potential of phytoremediation, an emerging green technology: European trends and outlook. Proceedings of the Indian National Science Academy B70: 131-152

PHYTOREMEDIATION OF VOLATILE ORGANIC COMPOUNDS

JOEL G. BURKEN AND XINGMAO MA
Department of Civil, Architectural & Environmental Engineering; Butler Carlton Civil Engineering Hall; University of Missouri-Rolla; Rolla, Missouri; 65409-0030 USA
Phone 573-341-6547; FAX 573-341-4729; E-mail: burken@umr.edu

1. Introduction

Many volatile organic compounds pose a unique type of threat to our environment, existing over vast areas in multiple media: water, soil and vapour. Phytoremediation applications are also unique as they are both highly visible and publicly acceptable and are also both long-term and low-energy to apply. These attribute pairings are quite unusual, being mutually exclusive for all other remedial approaches. The other attribute that is unmatched by any other technology is the ability to concurrently remediate and ecologically restore the environment. The listed phytoremediation traits do not yet include: rapid, highly effective, or well understood. Research and engineering findings covered herein address these concerns and reveal how and why phytoremediation can offer an unmatched possibilities to treat volatile organic compounds in particular.

Numerous contaminants are classified as 'volatile organic compounds' that are commonly referred to as VOCs. The definition of a VOC is generally a compound that has a vapour pressure of greater than 0.1 mm of mercury. As this definition is broad and only considers one chemical property, the term VOC in this work specifically refers to research performed on chlorinated solvents and a number of petroleum hydrocarbons. The extensive use as solvents and fuels lead to widespread releases through spills and leaks in the handling and transport due to the immense volumes. Additionally, in previous decades these compounds were disposed via methods that are quite improper in light of today's knowledge, but at the time simple dumping on unused land or even dumping out of the back door of a dry cleaner or automechanic shop were common disposal practices and accepted. Release to the environment occurred in almost all places where these compounds were used. The inherent physico-chemical properties that made these compounds so desirable are the same properties that make these compounds a bane to current efforts to address the earlier releases. The chlorinated solvents are particularly problematic because these compounds with low reactivity, relatively low water solubility, dense liquid phases, and high volatility now pose large dissolved groundwater plumes with overlying vapour contamination of the vadose (unsaturated) zones, and pools of pure compounds that are difficult to access. Large enduring plumes of these compounds are difficult to address, even though decades of effort were spent to remediate contaminated environments.

M. Mackova et al. (ed.), Phytoremediation Rhizoremediation, 199–216.

Current remedial approaches for VOC-contaminated groundwater and unsaturated zones include: air sparging, vacuum-enhanced recovery, aqueous pump and treat, in situ chemical reactions (reduction or oxidation), *in-situ* bioremediation (microbial-based), permeable reactive barriers, and almost all conceivable combinations of these and even a few other technologies. Many of these technologies are certainly effective at removing and/or destroying these compounds, but become cost prohibitive and energy intensive when considering the scope of many VOC plumes or do not address the long-term and multimedia problem that the VOCs pose. In addition, many of the existing sites have a relatively low risk associated, as the most atrocious sites have been addressed through existing technologies once sites were identified. The resulting legacy is many large, relatively low risk sites that include multiple contaminated media, but still a treatment is desired. Phytoremediation can fit this niche, offering a long-term, relatively low cost treatment option.

Increased understanding of phytoremediation applied to volatile organic compounds (VOCs) has been developed, showing volatilisation to the atmosphere as a dominant fate, occurring via diffusion from the trunk/stem/roots and leaves. Other findings include uptake from the vadose zone profile. These findings generate a new view of VOC phytoremediation. This new view includes fate considerations, new sampling and monitoring potential, and evidence that all three media (water, soil, vapour) can be addressed via phytoremediation. This treatment is unique in that it offers a highly visible and yet publicly acceptable treatment option and, as shown in this body of work, can address the multiple-media nature of VOC contaminants. Phytoremediation for VOCs can not be considered as a quick option, with clearly identified endpoints. Time and careful consideration of the impacts expected are requirements of phytoremediation.

2. Background

The recent advancements in the field of VOC phytoremediation has come not in developing the concept. VOC phytoremediation is not conceptually new, as full-scale applications and pilot-scale testing have been ongoing for roughly a decade. One of the first large scale applications was planted in 1996 at the J-Field Site at Aberdeen Proving Ground (APG). The site, planted by Applied Natural Systems®, is also the field site discussed later in this chapter. A pilot-scale test system initiated at the same time also revealed tremendous potential, as the control-dosed system showed greater than 90% sustained reduction in mass [1]. Recent discovery in VOC phytoremediation has occurred in understanding the fate and unravelling the specific plant-contaminant interactions that are unique to volatile compounds, many of these interactions were not considered and certainly not fully understood at the time the application concept was developed.

Phytoremediation for organic compounds focuses on two main process: uptake and rhizodegradation. For many organic compounds, the rhizodegradation is a dominant fate, and overall the *in-situ* destruction is a desired fate. The specific research findings and discussion in this chapter focus on the uptake and relate briefly to rhizosphere degradation, but detailed specifics rhizodegradation are covered in great detail in separate sections of the compilation.

When considering vegetative uptake in remediation, there are a number of subsequent fates, which will be discussed herein; but the uptake and entry into plants is a universal step to all plant-mediated removal or metabolic degradation. Plant uptake of organics was one of the initial research topics attracting interested in the phytoremediation of organics.

Table 1. The Mass Balance of TCE in Phytoremediation Experiments. Quantitative values presented from published laboratory studies.

Note	Rhizodegradation	Uptake & Transpiration	Metabolism	Residues*	Others
1	23%	56%#	ND°	ND	20%
2	N/A	N/A	>1.5% (CO_2)	3.7%	88.1%
3	N/A	N/A	>1.5% (CO_2)	3.9%	94.4%
4	1.06%	0.083%	ND	0.55%	91.6%
5	0.84%	0.084%	ND	0.37%	92.9%
6	1.06%	0.027%	ND	0.237%	93.0%
7	1.00%	ND	ND	0.17%	95.4%
8	0.71%	ND	ND	0.143%	96.57%
9	N/A	63.1±7.4%	ND	0.6±0.4%	9.1±2.1%
10	N/A	44.7±10.0%	ND	0.7±0.4%	4.4±1.6%
11	N/A	21.4±5.3%	N/A	3.4±0.9%	10.8±3.6%

1. Laboratory study with alfalfa [10].
2-3. Data from batch axenic cell culture experiment [6].
4-8. Data from whole plant study with hybrid poplars. [7].
9-10. Data from tomato study with two different doses, 0.58 μC_i and 0.15μC_i [11].
11. Data from whole plant study with hybrid poplars. [3]
*Residues include both extractable and non-extractable ^{14}C.
#The percentage was calculated from evapotranspiration (evaporation + transpiration).
N/A = Not Available, ND = Not Detectable.

From the initial research and applications there were inherent disconnects in the research approach for VOCs in lab studies conducted and in the field scale evaluation. This disconnect was partly due to the development paths for phytoremediation concepts. Initial phytoremediation studies were targeting non-volatile compounds, and much of the research was adapted and developed from existing agrochemical research on the fate and activity of herbicides and insecticides, inherently non-volatile compounds. The transport of these compounds followed the transpiration stream from the subsurface and the resulting fate was largely in the leaf tissues, where the compounds would reside either intact or as a metabolite that was stored in the leaf tissues. From the earliest information on plant uptake and volatilisation of VOCs, notation regarding the specific transport pathway referred to the leaves as the potential source of the VOCs emanating from the plant tissues [2].

Initial research on phytoremediation of specific VOCs such as trichloroethylene (TCE) targeted uptake and two specific fates: persistence/degradation in plant tissues and volatilisation from the leaves [3-7]. These studies all lead to valuable findings including; overall mass balances showing that the compounds were indeed translocated, TCE levels in plant tissues, metabolic products and pathways, presence of bound residues in tissues, and volatilisation to the atmosphere. The relative weighting of each fate varied considerably in these independent studies, Table 1, which is not surprising as each effort focused upon different aspects of experiment design in complex, whole-plant systems. To further complicate the dissimilarity, sampling of the field-scale systems that were already in place revealed an even greater disconnect, showing little agreement with the laboratory studies [1, 8]. In full-scale and pilot-scale systems TCE was not detected in high quantities in evapotranspiration from leaves nor was TCE or metabolites found at levels to explain removal rates anticipated. Metabolism of TCE and other compounds does occur in plants and the reactions have been outlined. Trichloroacetic acid (TCAA) has been identified repeatedly as a primary metabolite. Metabolites have been shown to reach levels in tissues up to 30 mg per kg [1, 8, 9]. Subsequent studies revealed VOC transport mechanisms and pathways unique to volatile compounds that had not been considered or evaluated previously. These transport mechanisms and pathways are discussed in detail herein.

3. VOC: laboratory studies

3.1. UPTAKE AND DIFFUSION FROM PLANT TISSUES

Contaminant presence in plant tissues when dosed in laboratory studies or when growing over VOC-contaminated groundwater clearly indicated the uptake occurs for a variety of organic compounds and specifically VOCs [3, 4]. In this line of research a number of compounds were identified in plant tissues and the volatilisation of benzene, TCE, nitrobenzene, toluene, ethylbenzene and *m*-xylene were shown. Later research revealed that the concentrations in plant tissues were linearly related to the aqueous exposure concentration in laboratory experiments, Figure 1. Vroblesky *et al.* [12] had used this direct concept in field investigation and discovered the presence of a decreasing concentration gradient with elevation. Volatile compounds were proposed to escape to the atmosphere directly from the transpiration stream in tree trunks [12]. This discovery helped to increase attention on vapour phase mechanisms, including diffusion. Earlier efforts have been presumptuous in that volatile compounds emanated from leaves, along with the transpired water [2-4, 6, 13]. Once the VOC-specific diffusion mechanisms were considered and targeted in research, a more thorough understanding developed.

Detailed research to elucidate VOC transport sought to isolate and quantify the vapour phase mechanisms, which potentially lead to the observed gradient. An initial step was to directly isolate the diffusion and volatilisation from the stem of poplar cuttings growing in contaminated media. The experimental arrangement was developed to directly capture and measure the flux of chlorinated VOCs in laboratory studies. The experimental apparatus employed to directly measure the efflux from tissues was termed a “diffusion trap” and was installed on hybrid poplar cuttings. The cuttings were planted in 1 litre glass reactors, filled with soil to promote natural rooting condition. The reactors were watered with a TCE-spiked feed solution. Prior to dosing, the upper portions of the cuttings were each fitted with 2 of the diffusion traps, which consisted of a 2.5 cm quartz glass tube around the cutting and sealed with two Teflon lined septa to seal the trap to the cutting. Gas flow was controlled through the trap via two hypodermic needles in the septa, and the purged gas phase was passed through an activated carbon trap to accumulate any TCE diffused through the stem. The diffusion traps were placed at two different heights to evaluate the height specific diffusion expected if the gradient observed was indeed due to diffusion along the transpiration stream. A detailed schematic of the experimental arrangement has been previously published [15]. The reactors were dosed in pairs at a variety of concentrations to further evaluate the impact of concentration. After 28 days of dosing the feeding of TCE was curtailed and clean water was dosed to the plants to follow the diffusive flux with time. Plants were grown under controlled conditions in a walk-in fume hood, to eliminate any potential leaks from the soil compartments, which were sealed, but minimal loss of the highly volatile TCE was expected and placement in the hood would ensure that any leaked TCE was rapidly swept away.

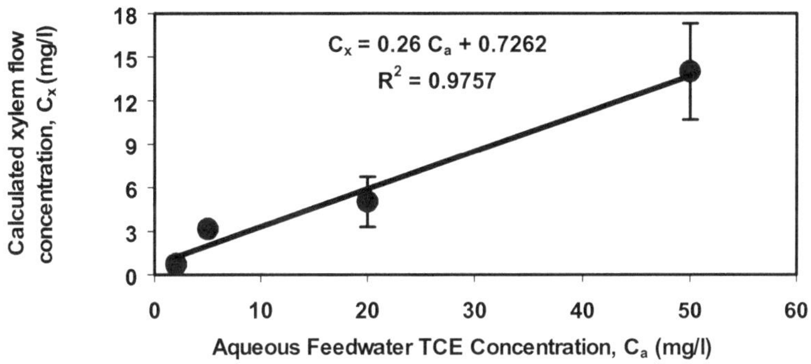

Figure 1. Concentration in xylem flow calculated from headspace sampling plotted versus the aqueous concentration of TCE in controlled laboratory study. Figure adapted from Ma and Burken [14].

Findings from the uptake and diffusion experiments confirmed that TCE did directly volatilise from the cutting, and that the diffusive flux was related to height. Results shown in Figure 2 reveal a slow increase in diffusive flux over the first 7 to 10 days. After this period the diffusion from the cuttings remained constant while the TCE feed was constant. After the TCE was removed from the feedwater, the flux quickly decreased. In most cases the TCE diffusion was non-detectable after 10 days of feeding with clean water. The rapid flushing of the TCE from the system indicates that there is not a high transient storage in the plant tissues and that the transport and residence of the TCE is minimal in this system. Longer residence would be expected in a full-scale tree, as will be discussed in later sections on modelling and field scale sampling.

The uptake and diffusion experiment also showed two other relationships. Diffusive flux was clearly related to height, with the upper trap accumulating less TCE than the lower trap in all sampling events for all plants. This trend is exhibited clearly in figure 2, and in all reactor sets [15]. The gradient with height supports that the diffusive loss is a major fate of TCE translocated from the subsurface, in this experiment and in field settings. Another clear relationship was observed as the linear correlation of diffusion to exposure concentration. Linear relationship was noted between the feed concentration and the mass diffused. In associated hydroponic experiments however, decreased transpiration due to apparent toxicity at higher exposure concentrations indicated that transpiration rates were also linked, which agrees with accepted theory regarding uptake and chemical translocation. The apparent toxicity in the hydroponic study occurred at concentrations of 0.4 mM, which is in the range of observed impacts but below the zero-growth concentration of 0.9 mM noted by Dietz and Schnoor [16], so decreased transpiration is not unexpected.

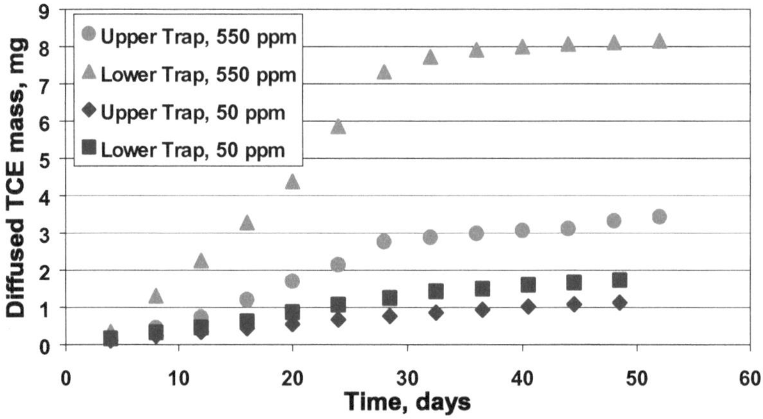

Figure 2. TCE volatilisation from hybrid poplar stems in a controlled laboratory. Diffusion and volatilisation reached consistent levels within 8 days of dose initiation and levels off within 10 days following the removal of TCE from the feedwater on day 28. Details of experimental arrangement appear in Ma and Burken [15]

Related research was also performed for methyl-*t*-butyl ether (MTBE), presented in detail elsewhere [17]. MTBE was observed to be taken up and to diffuse from the transpiration stream of the xylem tissue. Diffusion was also related to height of the diffusion trap on the stem and the dose concentration, Figure 3. The transpiration rate was also a clear impact on the diffusive flux, as the diffusion normalized with the rate of water use was shown to be a linear relationship, Figure 3b.

Observing MTBE diffusion and volatilisation processes confirms that the vapour pressure of the VOCs in question is of significance. Earlier research evaluated vapour pressure and the Henry's law constant as the primary properties relating to volatilisation extent in phytoremediation experiments [4]. MTBE volatilisation suggests that the vapour pressure correlates better with diffusion and volatilisation from the transpiration stream than does the Henry's constant. In previous work the degree of volatilisation was

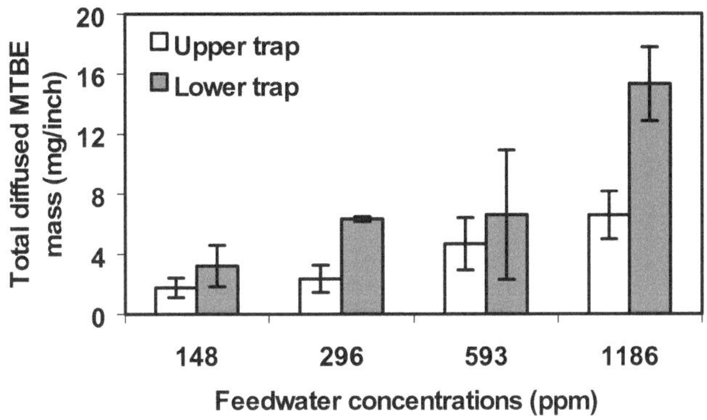

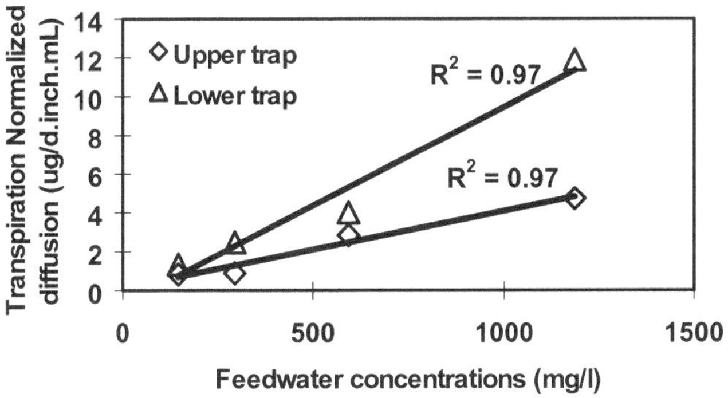

Figure 3a. Total MTBE collection from the vacuum diffusion traps over the 7-day experiment. The error bars represent the maximum and minimum values of two replicates; 3b. MTBE accumulation in plants normalized for the volume of water transpired. Figure adapted from Ma and Burken, 2004 [17].

noted to be related to the vapour pressure, although the impact of a compound's Henry's constant, relating activity in air and in water, was not discounted. When considering MTBE's vapour pressure of 0.33 atm relative to the empirical relationship previously proposed, volatilized of MTBE from plant tissues is expected, but it should also be noted that the V_p for MTBE is higher than the compounds previously evaluated. Conversely, the low Henry's law constant for MTBE, $K_H = 0.02$ dimensionless at 25° C, would indicate very limited volatilization could occur if the relationship of volatilization to Henry's law constant was valid. In the more recent MTBE research, volatilization was dominant when compared to the concentration in plant tissues [17]. Therefore, when considering the potential diffusion and volatilization of organic compounds from plant tissues, compounds with a higher vapour pressure pose the highest potential for volatilization.

3.2. CHLORINATED VOC DEGRADATION IN PLANT TISSUES

Plant tissues are also capable of degrading chlorinated solvents. Degradation products of TCE detected in plant tissues include trichloroethanol, trichloroacetic acid, dichloroacetic acid and trichloroethanol glycoside [1, 5, 7, 8, 18-21]. These oxidative metabolites have been identified in controlled cell culture experiments, whole plant laboratory experiments and full-scale field settings. When evaluating oxidative metabolite levels in plants, caution and background sampling should take place as TCAA is also produced in the atmosphere and is a common atmospheric pollutant, and has been detected in plant tissues world-wide at concentrations up to 300 μg per kg [22, 23].

Degradation pathways for TCE in particular have been evaluated; however no established pathway is presented to date. Common metabolites listed above strongly suggest an oxidative pathway. Mammalian and bacterial metabolic pathways have also shown the metabolites as intermediates. Shang *et al.* [9] outlined these pathways in detail, including a number of the enzymes involved. The plant metabolism model posed for anthropogenic compounds could also explain the metabolite signature that is common to many experiments and field sites. The metabolic process termed the "Green liver model" has been supported in numerous studies focusing on contaminant transport. The main metabolites observed and the production of bound residues does coincide with the three phase metabolic processes, Table 2.

In this process xenobiotics are initially transformed to intermediates that can be readily conjugated. The resulting conjugates undergo two potential fates. Some conjugates are covalently bound into the cell wall via lignification and some are targeted by transporter enzymes in the vacuole membrane, resulting in removal from the cell cytosol and any interference with metabolic activities. While not directly identified as the reactions taking place, the comparison of the commonly identified metabolites of TCE and the established reactions clearly shows that plant metabolism of TCE can be explained by the green-liver phases.

4. Diffusion model considerations

Whole-plant and tissue-specific models have been generated and experimentally verified. A fundamental model was developed, and while the trends in model output

Table 2. Green-liver model phases with comparison of established reactions to identified metabolites in TCE phytoremediation studies.

Green Liver Phase	Established Reactions (Enzymes)	Metabolites noted for TCE
I: Activation or transformation	Oxidation or hydroxylation (Cytochrome P-450) or reduction (dehalogenase)	Trichloroacetic acid, trichloroethanol, dichloroacetic acid
II: Conjugation	Glucoside (*O*- or *N*-glucosyltransferase) or Glutathione (glutathione *S*-transferase)	Trichloroethanol glycoside
III: Sequestration or compartmentalisation	Glucoside- or Glutathione-conjugate transporter enzymes or lignification	Vacuole-bound xenobiotic-conjugates and Bound residues

would result in an exponential decrease in VOC concentrations with elevation along the tree trunk, an explicit solution for the model was not attained. A simplified model was based on assumptions that the radial diffusion could be estimate as planar diffusion with some geometric simplifications [24]. Model input values included the concentration in the xylem flow at ground level, dimensions of the tree stem, the transpiration rate for the cutting and the height of the diffusion trap used in analysis and data collection. The lone fitting parameter in the model was D, termed effective diffusivity. The resulting model generated a plant tissue profile that very closely mimicked the concentration profiles for TCE measured in the tree tissues for hybrid poplars grown in the laboratory and dosed with aqueous TCE at various concentrations. The model also generated a profile for diffusive flux from the tree. The profile closely followed the trend of the data but under-predicted the flux, by approximately 10-40% [24]. The underestimation is hypothesized to result three primary causes. One cause is that both dispersion and diffusion result in VOC transport in the xylem flow As the two processes can not be isolated in live plant experiments, the term "effective diffusivity" was adopted, as noted above. Another potential cause has to do with the continuous nature of the model and the discrete nature of the diffusion traps, which covered a 2.5 cm length of the stem. The measured value was plotted at the median height of the trap. A third potential cause is the tissue differentiation of xylem – vascular cambium-phloem. The tissue boundaries that exist in this region and possible gas phase in the outer periderm can pose a considerable diffusive resistance, causing lower than predicted diffusive efflux from this simpler model, which considers only one diffusivity value for all tissues. The deviation in the data and model is likely a combination of the posed causes. To evaluate the diffusivity independently, separate experiments are being conducted on individual tissues rather than actively transpiring, whole plants. Diffusivity values for common contaminants have been calculated in a number of studies, ranging from tests with dried sections of xylem tissue to small scale, active plant systems [24-27]. Reported diffusivities range widely in these studies, due to experimental arrangements and approaches, so care should be taken in interpreting these results. Trapp *et al.* [28] and Mackay and Gschwend [26] present predictive relationships. These relationships should be thought of as "order of magnitude" estimates from chemical and tissue properties, measured values could stray greatly from these predictions. The complex hetrogeneous nature of plant tissues likely prohibits one or two variant models from being truely representative.

Overall these modelling approaches applied to whole plant tests reveal that chemical concentrations in the stems and flux to the atmosphere for trees taking up VOCs can be modelled considering the processes of translocation in and diffusion through the xylem tissues. The processes of metabolism, phloem transport and sorption were also incorporated into the model development discussed above [24]. In solving the model for this arrangement however, these processes were considered to be in equilibrium (sorption), at magnitudes that were not significant (metabolism and phloem transport). These processes are discussed in greater detail later in this chapter.

5. Vapour phase uptake

Uptake of VOCs into plant tissues was clearly displayed in the work outlined above, however comprehensive understanding of how VOCs enter plant tissues in the subsurface has had considerable recent discovery and is still not completely understood. Early work on the uptake of organics targeted herbicide uptake and activity [29], and the transpiration stream concentration factor (TSCF) concept was developed. The TSCF is calculated as the aqueous transpiration stream concentration divided by the bulk aqueous concentration in the root zone. The TSCF therefore range from 0 for a compound that is not taken up and translocated to 1.0 for a compound that transports without impediment. The TSCF values for compounds were compared to many chemical properties, with a bell-shaped curve as a function of log K_{ow} being the most commonly accepted, and inherent relation to K_{ow} is expected as uptake includes movement through and past organic phases as part of an aqueous stream. The evaluation of TSCF values and relationships for VOCs led to widely different values. TSCF values range from 0.02 to 0.75 for TCE in laboratory studies [3, 6, 7, 9, 19, 30]. The wide range of values determined for TCE uptake casts doubt on the methods utilized. Recent research findings suggest that vapour phase transport is a major transport process that was not fully considered and could explain some of the variation noted.

Vapour phase uptake was identified during tree coring investigation at the New Haven Missouri Riverfront site. Results showed a disconnect between the tree coring analysis for tetrachloroethylene (PCE) and the underlying aqueous concentrations [31]. Further investigation of the contaminant distribution at the site revealed that the tree cores collected at the site were in better agreement with the vadose zone PCE concentrations. Uptake and transport of contaminants from the vadose zone was hypothesized to occur and be a dominant process at the field site. Laboratory studies to investigate the process in controlled setting, clearly showed that PCE is taken up from the vadose zone [32]. In this study identical PCE mass was introduced either to the saturated zone or the vadose zone of soil filled reactors planted with hybrid poplar cuttings. Diffusion from the cuttings was monitored and the tissue concentrations were determined at the termination of the experiment. The study also revealed that increasing gas exchange rate in the vadose zone decreased the translocation and volatilization of PCE. Recent research has shown the processes also apply to a wide variety of VOCs; including benzene, toluene, xylene, naphthalene TCE and MTBE [33]. These results suggest that VOCs can not only be taken up from the vadose zone, but also released to the vadose zone from roots following uptake from a deeper soil profile. This concept, termed hydraulic lift, is well studied for water as plants can transport water from the saturated zone and release to the upper vadose zone when transpiration rates decrease in dark periods. The same concept can be considered for VOCs. A potential transport can be explained considering hydraulic transport and the chemical activity of the compounds. VOCs transport into root tissues via two pathways, an advective transport with the water flow into roots systems and a diffusive transport as chemical activity in the root reaches equilibrium between the surrounding contaminated soil. Compounds are then transported in the roots, primarily via advection, through a relatively clean soil profile and the VOCs diffuse from the roots to approach chemical activity equilibrium.

The proposed vapour phase transport explains data observed in the laboratory studies conducted and the disparity observed in the early uptake studies to determine the TSCF values. Diffusion of vapour phase compounds was not considered in the initial studies conducted regarding TSCF values for volatile compounds. These studies varied greatly in how the root zone was treated, some being sealed air tight and some being actively aerated. The definition of TSCF inherently excludes any vapour phase transport, either to or from root tissues. Therefore, the "TSCF-concept" of should not be applied for volatile compounds.

6. Field scale: contaminant fate and hydraulic control

Full-scale phytoremediation of VOCs has been active at contaminated sites since 1995 and 1996. In the decade since then, findings at the field sites have varied. In part the results are sketchy as some processes that are important were not previously considered and data collection is limited and largely unpublished. The full-scale phytoremediation site at the APG J-field site in Maryland is among the best characterized and analysed. Sampling at the site provided some of the first clear delineation of the vapour phase mechanisms taking place, summarised in Table 2. Sampling of the trees on-site have shown VOC efflux from the stems and leaves of trees, a decreasing trend in VOC concentrations in the transpiration stream with height, and a gradient in the radial direction in the trunks of tree. The gradient of contaminants in the tissue going from the centre of the trunks to the outer bark, as shown in Figure 4, reveals a clear driving force for diffusion, and allows estimation of diffusivities in modelling efforts. Using tedlar bags, VOCs were detected in leaf bags in the canopy, at low levels [8], and diffusion traps placed on trees verified both TCE and TeCA volatilisation from the trunk via three different analytic methods; solid phase microextraction (SPME), 2-liter summa canister, and gas phase purge with collection onto activated carbon or Tenax [34]. The flow-through purge and trap captured over 100 μg/hour while sampling only a 0.3 m section of tree trunk [35]. A rapid decrease of VOC concentrations in root tissues was also shown in the subsurface when root tissues were sampled for TCE and TeCA. Overall the contaminant fate at the APG site has shown VOCs are largely volatilized to the atmosphere from plant tissues, but are also subject to degradation in plant tissues and potentially redistribution into the vadose zone. Hydraulic control of the plume at the APG J-Field site has also been noted, as discussed below.

While APG is a well-studied site, numerous other sites also have measurements and estimates for contaminant removal rates and contaminant fate. Volatilisation and phytodegradation were also shown at Hill Air Force base. The main VOC removal mechanisms was noted to be phytovolatilisation with rate estimates ranging from 2 to 50 grams per tree per year, while metabolism in the plants was also a noted fate with metabolites measured in some but not all trees sampled [19]. In other full-scale research, performed at Carswell Airfield, uptake and volatilisation were again noted, but not quantified. Degradation in the root zone was cited as a primary fate and is discussed below in the rhizoremediation section [36].

Even at these well-studied sites, quantifiable estimates of remediation rates, contaminant removal rates and periods for site remediation remain unclear. The the complexity of water transport between the soil zones (vadose and saturated) and into plant roots from these zones makes quantification quite difficult water transport alone, much less the compounding complexity of VOCs being present in three phases. In field testing, the importance of rainfall or other water sources has also been noted, as infiltration of rainwater is used preferentially to groundwater, [19, 37]. The source of transpired water should be carefully considered in field-scale sampling of any potential phytoremediation mechanism.

At numerous sites, phytoremediation impacts on groundwater levels and control are undeniable. In pilot scale work, Newman and colleagues revealed the hydraulic impacts of hybrid poplars when planted cells the pilot scale system were dewatered at the peak of the growing season, making sampling of groundwater impossible [1]. This work clearly shows the potential of phytoremediation by presenting a comprehensive water budget for the pilot scale system. Use and management of contaminated plumes is the goal of many remediation efforts. At one chlorinated solvents site, Phytokinetics® designed a phytoremediation application to work in tandem with a pump and treat system. Trees planted at the site removed an estimated 6.5 x 10^6 litres over the 2004 growing season, allowing the pumps to run at a much lower capacity [38]. VOC fate was not analysed.

Plume capture was most clearly shown at the APG J-Field site as work by Schneider *et al.* [39] clearly revealed a complete hydraulic gradient reversal through the progression of the growing season. Hydraulic gradients were not only inward horizontally from the plume boundaries, but also upward from strata below the contaminated zone. The hydraulic profile data was clearly response to the phyto-remediation design on the site as a diurnal fluctuation was observed, with the lowest hydraulic potential occurring in late afternoon (average at 17:30) and the greatest potential occurring at daybreak (average at 04:00).

When the hydraulic findings from the APG site are paired with the chemical analysis performed on site, the results undeniably support VOC phytoremediation with respect to control of large plumes, Table 3. Methods to evaluate transpiration from phytome-diation applications and the hydraulic impacts are now becoming established [39, 40] and further comprehension and proof of hydraulic control will follow.

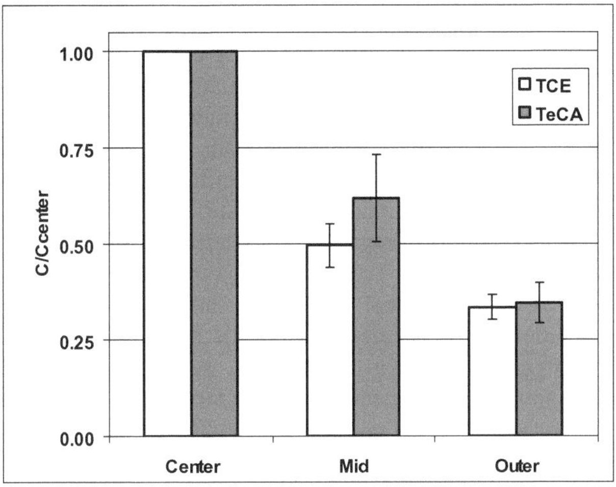

Figure 4. Gradient of 1,1,2,2-tetrachloroethane (TeCA) and trichloroethylene (TCE) in xylem tissue at the Aberdeen Proving Ground J-Field site, supporting the mechanistic understanding of the volatilization measured from trees on site.

Table 3. Delineation of processes and hydraulic impacts at the Aberdeen Proving Ground J-Field site, Edgewood Maryland.

Process	Evidence	Citation
Uptake and translocation	VOCs and metabolites in plant tissues	[8, 15, 34]
Plant Metabolism	Metabolites (TCAA) in plant tissues	[8]
Diffusion to Atmosphere	Decreasing VOC profile with height, measured VOCs in diffusion traps	[15, 34]
Subsurface Vapour Exchange	Decreasing root VOC concentrations	[15, 34]
Hydraulic plume control	Reversing hydraulic gradient, diurnal fluctuations in water levels.	[39, 41]

7. VOC rhizodegradation

While major processes are discussed in detail in this work with specific data presented from the authors' research, there are other aspects that should be considered in approaching research or implementation. Degradation of contaminants prior to contacting the plants is a desirable fate, limiting transfer into above ground tissues and to the atmosphere. The rhizosphere effect is covered in detail elsewhere in this volume (Leigh MB). If rhizodegradation is complete, all plant – VOC interactions noted above are insignificant. In research at Carswell Air Force Base, reductive dechlorination of chlorinated solvents was noted to occur, preceding plant mediated removal [42]. Under the phytoremediation plot, ground water oxidation reduction potential was sharply reduced, background electron acceptor pools were depleted and products of reductive dechlorination (dichloroethylene, chloride ions) were clearly detected. The potential for plant systems to lead to reducing conditions in the subsurface is established, as plants can contribute a tremendous flux of readily degradable organic carbon. While the reductive dechlorination is desirable, the predominant trend is for aerobic conditions to be enhanced in the rhizosphere through soil dewatering and diurnal fluctuations of the water table drawing air and oxygen into the subsurface. This trend towards aerobic conditions is a general statement regarding terrestrial, not wetland, applications of phytoremediation.

In a number of cases the aerobic conditions can lead to enhanced hydrocarbon degradation. Degradation of hydrocarbon fuel spills has been noted [43, 44] and can decrease the role of processes that are directly plant mediated, like uptake and translocation. While numerous applications of phytoremediation for petroleum are in place and VOC fractions are likely present, little published work exists on the VOCs in particular. Increased benzene degradation in planted reactors was clearly noted in a laboratory study, although the artefacts associated with the laboratory study likely increase the extent of degradation enhancement [45]. To evaluate the impacts of terrestrial phytoremediation on hydrocarbon VOCs, more research is needed.

8. Engineering improvements

While understanding the plant-VOC interactions for existing systems is important, applying this understanding for engineering improvements to phytoremediation and generating new approaches hold the greatest potential. Engineering can specifically target the plant or microbial populations involved through genetic engineering or can target the engineering the plant introduction to specifically target the contaminated media, improving the plant-contaminant interactions.

8.1. ADVANCING PLANT-CONTAMINANT INTERACTIONS

The upward transport of contaminated groundwater noted at APG is a substantial finding as phytoremediation is commonly cited to be limited to shallow contamination. Accessing deep groundwater contamination is the goal of engineered phytoremediation systems. At the Portsmouth Gas Diffusion Plant, Piketon, Ohio, a phytoremediation system was designed to target contaminated strata roughly 10 meters below ground surface. The targeted TCE plume was primarily in the sandy-gravel Gallia aquifer, overlain with a clay aquitard. Augured holes were placed to connect the contaminated aquifer and trenches at the ground surface, and the holes were backfilled with sand to allow a low resistance transport path between the contaminated zone and the trenches, which were planted with hybrid poplars. Hydraulic impact on the contaminated aquifer was recorded with pressure transducers. The system was shown to significantly impact the deep, contaminated Gallia aquifer, revealing a daily piezometric drop of up to 0.4m, and the daily piezometric drop was correlated to the solar radiation recorded on the site.

Over the course of one growing season, over 2.1 x 10^6 litres were extracted as recorded with sap flow measurements [46]. Results from this site show that once the specific mechanisms are known, the mechanisms and site specific conditions can be exploited through engineering to increase the effectiveness and potential for phytoremediation to treat VOC plumes. Other improvements could be as simple as decreasing infiltration in contaminated areas, to promote evapotranspiration of contaminated groundwater rather than clean, infiltrating rainfall.

8.2. MOLECULAR ENGINEERING ADVANCEMENTS.

Concurrently as the understanding of VOC phytoremediation have advanced, molecular engineering approaches to enhance processes are occurring and innovative applications to exploit specific plant systems have been designed. One line of molecular engineering has been to take advantage of the plant-microbe symbiosis. By selecting root-colonizing bacteria as recipients for enhanced degradation genes, the improved degradation of TCE has been observed in sealed reactors containing plants inoculated with recombinant strains as compared to plants alone or plants with the wild type microbes [47-50]. In these studies, toluene-*o*-monooxygenase (TOM) was incorporated into the chromosomal DNA of the wild type bacteria to provide an aerobic degradation pathway for degradation. The plant-bacterial symbiosis was also observed to depend on the host plant selection, as wheat and poplar colonizers were not able to thrive in the root zone of other plant species [48]. While degradation potential was clear, the survival rate of the inoculants and the residence time of the VOC in the root zone were a concern for the systems tested.

Another recombinant bacterial approach tested recently targeted endophytic bacteria, organisms colonizing the interior plant cells, rather than root-colonizers. In work by Barac and colleagues endophytic bacteria were isolated from different tissues of hybrid poplars and engineered with a TOM gene [51]. In the inoculated whole plant studies, toluene degradation was proven to be enhanced and the presence of the recombinant organisms was show using a fluorescent tag. The engineering of endophytic bacteria has a clear advantage in terms of survival as competition in the plant interior is greatly reduced. However the compounds must be able to pass into the plant tissues where the bacteria colonize. For VOC applications, the xylem tissues are ideal, as compounds have the longest residence times in the xylem.

Another molecular engineering approach is to target the plant metabolic pathways. In doing so, the competition of the microbial populations is avoided, and survival and thriving of the plant is much easier to evaluate and promote. Work on TCE degradation includes GM plants to express a mammalian cytochrome P450 gene with high activity for TCE [52]. Enhanced TCE degradation and maintenance of the gene was shown to occur. Although not targeting VOC in the particular tests, rice has been successfully modified to express a human P450 CYP2B6 [53]. The degradation potential for GM plants is literally bounded only by the known, identified degradation genes and the ability to integrate these genes into plants. However the compounds of interest, in the same aspect of the endophytic approach, must pass into the tissues where the genes are expressed and have a residence time that allows for degradation.

9. Summary

Expansive plumes associated with many volatile contaminants make phytoremediation an attractive approach. The current state of understanding shows that certain compounds can be removed from the subsurface through plant uptake, which includes uptake from both the saturated zone and the unsaturated vadose zone. Vadose zone uptake may be useful in efforts to address vapour transport and vapour intrusion concerns. In targeting dissolved contaminants, understanding hydraulic aspects of phytoremediation reveal that engineering can target groundwater and can clearly increase phytoremediation impacts on groundwater plumes, including accessing deeper groundwater profiles. Following uptake, volatilization to the atmosphere is a major fate and occurs from the transpiration stream directly as well as from leaf tissues. Phytodegradation within the leaf tissues is significant for the contaminants reaching the leaves, with trichloroacetic acid being a most common metabolite identified in TCE applications.

The complexity of site hydrogeologic conditions, contaminant distribution between media (vapor, aqueous or sorbed) and in different zones (vadose or saturated), along with variable rooting patterns of plants makes quantification of remediation at a larger site very difficult. Improved quantifiable estimates are being developed using recent VOC fate measurements and advancements in hydraulic understanding of phytoremediation applications.

References

[1] Newman, LA; Wang, XP; Muiznieks, IA; Ekuan, G; Ruszaj, M; Cortellucci, R; Domroes, D; Karscig, G; Newman, T; Crampton, RS; Hashmonay, RA and Yost, MG (1999) Remediation of trichloroethylene in an artificial aquifer with trees: A controlled field study. Environ Sci Technol 33(13): 2257-2265

[2] McFarlane, JC; Pfleeger, T and Fletcher, J (1987) Transpiration Effect on the Uptake and Distribution of Bromacil, Nitrobenzene, and Phenol in Soybean Plants. J Environ Qual 16(4): 372-376

[3] Burken, JG and Schnoor, JL (1998) Predictive Relationships for Uptake of Organic Contaminants by Hybrid Poplar Trees. Environ Sci Technol 32(21): 3379-3385

[4] Burken, JG and Schnoor, JL (1999) Distribution and volatilization of organic contaminants following uptake by hybrid poplar trees. Int J Phytoremediation 1(2): 139-152

[5] Gordon, M; Choe, N; Duffy, J; Ekuan, G; Heilman, P; Muiznieks, I; Ruszaj, M; Shurtleff, BB; Strand, S; Wilmoth, J and Newman, L (1998) Phytoremediation of Trichloroethylene with Hybrid Poplars. Environ Health Perspect 106(Aug): 1001-1004

[6] Newman, LA; Strand, SE; Choe, N; Duffy, J; Ekuan, G; Ruszaj, M; Shurtleff, BB; Wilmoth, J; Heilman, PE and Gordon, MP (1997) Uptake and Biotransformation of Trichloroethylene by Hybrid Poplars. Environ Sci Technol 31(4): 1062-1067

[7] Orchard, BJ; Doucette, WJ; Chard, JK and Bugbee, B (2000) Uptake of trichloroethylene by hybrid poplar trees grown hydroponically in flow-through plant growth chambers. Environ Toxicol Chem 19(4): 895-903

[8] Compton, HR; Haroski, DM; Hirsch, SR and Wrobel, JG (1998) Pilot-Scale Use of Trees to Address VOC Contamination, in GB Wickramanayake and RE Hinchee, Eds., International Conference on Remediation of Chlorinated and Recalcitrant Compounds, Battelle Press: Monterey, CA. p. 245-250 ISBN 1-57477-059-4

[9] Shang, TQ; Newman, LA and Gordon, MP (2003) Fate of Trichloroethylene in Terrestrial Plants, in SC McCutcheon and JL Schnoor, Eds., Phytoremediation: Transformation and Control of Contaminants, John Wiley and Sons Inc.: Hoboken, New Jersey. p. 529-561. ISBN 0-471-39435-1

[10] Narayanan, M; Davis, LC and Erickson, LE (1995) Fate of Volatile Chlorinated Organic Compounds in a Laboratory Chamber with Alfalfa Plants. Environ Sci Technol 29(9): 2437-2444

[11] Schnabel, WE; Dietz, AC; Burken, JG; Schnoor, JL and Alvarez, PJ (1997) Uptake and Transformation of Trichloroethylene by Edible Garden Plants. Water Res 31(4): 816-824
[12] Vroblesky, DA; Neitch, CT and Morris, JT (1999) Chlorinated Ethenes from Groundwater in Tree Trunks. Environ Sci Technol 33(1): 510-515
[13] McFarlane, JC; Pfleeger, T and Fletcher, JS (1990) Effect, Uptake and Disposition of Nitrobenzene in Several Terrestrial Plants. Environ Toxicol Chem 9: 513-520
[14] Ma, X and Burken, JG (2002) VOCs Fate and Partitioning in Vegetation: Use of Tree Cores in Groundwater Analysis. Environ Sci Technol 36(21): 4663-4668
[15] Ma, X and Burken, JG (2003) Diffusion of TCE to the Atmosphere in Phytoremediation Applications. Environ Sci Technol 37(11): 2534-2539
[16] Dietz, AC and Schnoor, JL (2001) Phytotoxicity of Chlorinated Aliphatics to Hybrid Poplar. Environmental Toxicology and Chemistry 20(2): 389-393
[17] Ma, X; Richter, A; Burken, JG and Albers, S (2004) Phytoremediation of MTBE with Hybrid Poplar Trees. Int J Phytoremediation 6(2): 157-167
[18] Doucette, W; Bugbee, B; Hayhurst, S; Plaehn, W; Downey, D; Taffinder, S and Edwards, R (1998) Phytoremediation of Dissolved-Phase Trichloroethylene Using Mature Vegatation, in GB Wickramanayake and RE Hinchee, Eds., The First International Conference on Remediation of Chlorinated and Recalcitrant Compounds, Battelle Press: Columbus Ohio. p. 251-256. ISBN 1-57477-059-4
[19] Doucette, WJ; Bugbee, BG; Smith, SC; Pajak, CJ and Ginn, JS (2003) Uptake, Metabolism, and Phytovolatilization of Trichloroethylene by Indigenous Vegetation: Impact of Precipitation, in SC McCutcheon and JL Schnoor, Eds., Phytoremediation: Transformation and Control of Contaminants, John Wiley and Sons Inc.: Hoboken, New Jersey. p. 561-588. ISBN 0-471-39435-1
[20] Nzengung, VA and Jeffers, P (2001) Sequestration, Phytoreduction, and Phytooxidation of Halogenated Organic Chemicals by Aquatic and Terrestrial Plants. Int J Phytoremediation 3(1): 13-40
[21] Shang, TQ and Gordon, MP (2002) Transformation of [C-14] Trichloroethylene by Poplar Suspension Cells. Chemosphere 47(9): 957-962
[22] Forczek, ST; Uhlířová, H; Gryndler, M; Albrechtová, J; Fuksová, K; Vágner, M; Schröder, P and Matucha, M (2004) Trichloroacetic Acid In Norway Spruce/Soil-System. II. Degradation in the Plant. Chemosphere 56: 327-333
[23] Chlor, E (2000) Trichloroacetic Acid in the Environment. http://www.eurochlor.org/chlorine/publications/publications.htm: Brussels, Belgium. p. 46
[24] Ma, X and Burken, JG (2004) Modeling of TCE Diffusion to the Atmosphere and Distribution in Plant Stems. Environ Sci Technol 38(17): 4580-4586
[25] Trapp, S; Miglioranza, K and Mosebek, H (2001) Sorption of Lipophilic Organic Compounds to Wood and Implications for Their Environmental Fate. Environ Sci Technol 35: 1561-1566
[26] Mackay, AA and Gschwend, PM (2000) Sorption of Monoaromatic Hydrocarbons to Wood. Environ Sci Technol 34(5): 839-845
[27] Davis, LC; Zhang, Q and Erickson, LE (2002) Benefits of Vegetation for Soils with Organic Contaminants. Critical Reviews in Plant Sciences 21(5): 457-491
[28] Trapp, S and Karlson, U (2001) Aspects of Phytoremediation of Organic Pollutants. J Soil and Sediments 1: 37-43.
[29] Shone, MGT and Wood, AV (1972) Factors Affecting Absorbtion and Translocation of Simazine by Barley. J Exp Bot 23(74): 141-51
[30] Davis, LC; Vanderhoff, S; Dana, J; Selk, K; Smith, K; Golpen, B and Erickson, LE (1998) Movement of Chlorinated Solvents and Other Volatile Organics Through Plants Monitored by Fourier Transform Infrared (FT-IR) Spectrometry. J Hazardous Waste 1-4: 1-19
[31] Schumacher, JG; Struckhoff, GC and Burken, JG (2004) Assessment of Subsurface Chlorinated Solvent Contamination Using Tree Cores at the Front Street Site and Former Dry Cleaning Facility at the Riverfront Superfund Site, New Haven Missouri, 1999-2003. US Geological Survey. p. 35
[32] Struckhoff, G; Burken, JG and Schumacher, J (2005) Phytoremediation of Vadose Zone VOCs. Environ Sci Technol 39(6): 1563-1568
[33] Breite, SR (2005) Uptake and Transport of Vapor Phase VOCs in Phytoremediation, in Civil Architectural and Environmental Engineering. University of Missouri Rolla: Rolla, MO
[34] Burken, JG and Ma, X (2002) Chlorinated Solvents Phytoremediation: Uptake and Diffusion, in AR Gavakar and ASC Chen, Eds., Remediation of Chlorinated and Recalcitrant Compounds, Battelle Press: Columbus, OH, p 2B-24 ISBN 1-57477-132-9

[35] Burken, J (2005) Phytoremediation of VOC's: Understanding Fate and Innovative Applications, in Third International Phytotechnologies Conference. 2005. Atlanta, GA, USA: http://www.cluin.org/phytoconf/
[36] Eberts, SM; Harvey, GJ; Jones, SA and Beckman, SW (2003) Multiple Process Assessment for a Chlorinated-Solvent Plume, in SC McCutcheon and JL Schnoor, Eds., Phytoremediation: Transformation and Control of Contaminants, John Wiley and Sons Inc.: Hoboken, New Jersey, p. 589-633 ISBN 0-471-39435-1
[37] Vroblesky, DA; Clinton, BD; Vose, JM; Casey, CC; Harvey, GJ and Bradley, PM (2004) Groundwater Chlorinated Ethenes, In Tree Trunks: Case Studies, Influence of Recharge and Potential Degradation Mechanism. Ground Water Monit Remediation 24(3): 124-139
[38] Ferro, AM; Zollinger, N and Thompson, B (2005) Groundwater Phytoremediation System Performance at the SRSNE Superfund Site, in Third International Phytotechnologies Conference. 2005. Atlanta, GA, USA: http://www.cluin.org/phytoconf/
[39] Schneider, WH; Hirsh, SR; Compton, HR; Burgess, AE and Wrobel, JG (2002) Analysis of Hydrologic Data to Evaluate Phytoremediation System Performance, in AR Gavakar and ASC Chen, Eds., Remediation of Chlorinated and Recalcitrant Compounds, Battelle Press: Columbus, OH.
[40] Vose, JM; Harvey, GJ; Elliot, KJ and Clinton, BD (2003) Measuring and Modeling Tree and Stand Level Transpiration, in SC McCutcheon and JL Schnoor, Eds., Phytoremediation: Transformation and Control of Contaminants, John Wiley and Sons Inc.: Hoboken, New Jersey. p. 263-282. 0-471-39435-1
[41] Schneider, WH; Wrobel, JG; Hirsh, SR; Compton, HR and Haroski, DM (2000) The influence of an integrated remedial system on groundwater hydrology, in GEA Wickramanayake, Ed., Battelle International Conference on Remediation of Chlorinated and Recalcitrant Compounds, Battelle Press: Columbus, OH, p. 477-484 ISBN 1-57477-094-2
[42] Godsy, E; Warren, E and Paganelli, V (2003) The role of microbial reductive dechlorination of TCE at a phytoremediation site. Int J Phytoremediation 5(1): 73-87
[43] Carman, E; Crossman, T and Gatliff, E (1998) Phytoremediation of No. 2 Fuel-Oil Contaminated Soil. J Soil Contam 7(4): 455-466
[44] Wiltse, C; Rooney, W; Schwab, A and Banks, M (1998) Greenhouse Evaluation of Agronomic and Crude Oil-Phytoremediation Potential among Alfalfa Genotypes. J Environ Qual 27:169-173
[45] Burken, JG; Ross, C; Harrison, LM; Marsh, A; Zetterstrom, L and Gibbons, JS (2001) Benzene Toxicity and Removal, in Laboratory Phytoremediation Studies, ASCE Practice Periodical of Hazardous, Toxic, and Radioactive Waste Management 5(3): 161-171
[46] Sokol, J; Rieske, D; Ritchey, J; Childers, S and Galanti, M (2005) Measuring Effectiveness of Phytoremediation For a TCE-Contaminated Groundwater Plume Using Sap Flow Instrumentation, in 2005 Third International Phytotechnologies Conference. 2005. Atlanta, GA, USA: http://www.cluin.org/phytoconf/
[47] Wood, TK; Shim, H; Ryoo, D; Burken, JG and Gibbons, JS (200) Root colonizing genetically engineered bacteria for trichloroethylene phytoremediation, in Battelle International Conference on Remediation of Chlorinated and Recalcitrant Compounds. 2000. Monterey, CA
[48] Shim, H; Chauhan, S; Ryoo, D; Bowers, K; Thomas, S; Canada, K; Burken, J and Wood, T (2000) Rhizoshpere Competitiveness of Trichloroethylene-Degrading, Poplar-Colonizing Recombinant Bacteria. Appl Environ Microbiol. 66(11): 4673-4678
[49] Yee, DC; Maynard, JA and Wood, TK (1998) Rhizoremediation of Trichloroethylene by a Recombinant, Root-Colonizing *Pseudomonas fluorescens* Strain Expressing Toluene *ortho*-Monooxygenase Constitutively. Appl Environ Microbiol 64(1): 112-118
[50] Gilbertson, AW; Burken, JG; Fitch, MW and Wood, TK (2002) Fluorescent, Root-Colonizing Recombinant Bacteria to Enhance the Rhizosphere Degradation of TCE, in AR Gavakar and ASC Chen, Eds., Remediation of Chlorinated and Recalcitrant Compounds, Battelle Press: Columbus, OH. 1-57477-132-9
[51] Barac, T; Taghavi, S; Borremans, B; Provoost, A; Oeyen, L; Colpaert, J; Vangronsveld, J and van der Lelie, D (2004) Engineered endophytic bacteria improve phytoremediation of water-soluble, volatile, organic pollutants. Nature Biotechnology 22(5): 583-588
[52] Doty, SL; Shang, TQ; Wilson, AM; Tangen, J; Westergreen, AD; Newman, LA; Strand, SE and Gordon, MP (2000) Enhanced metabolism of halogenated hydrocarbons in transgenic plants containing mammalian cytochrome P450 2E1. P Natl Acad Sci USA 97(12): 6287-6291
[53] Hirose, S; Kawahigashi, H; Ozawa, K; Shiota, N; Inui, H; Ohkawa, H and Ohkawa, Y (2005) Transgenic Rice Containing Human CYP2B6 Detoxifies Various Classes of Herbicides. Agr Food Chem 53: 3461-3467

IN VITRO PROPAGATION OF WETLAND MONOCOTS FOR PHYTOREMEDIATION

MIHÁLY CZAKÓ[1], XIANZHONG FENG[2], YUKE HE[2], SHARADA GOLLAPUDI[1], AND LÁSZLÓ MÁRTON[1]
[1]Department of Biological Sciences, University of South Carolina, 700 Sumter St, Columbia, SC 29208, USA. Fax: 1-803-777-4002; E-mail: marton@biol.sci.edu [2]National Laboratory of Plant Molecular Genetics, Shanghai Institute of Plant Physiology, Chinese Academy of Sciences, 300 Fenglin Road, Shanghai 200032, People's Republic of China

1. Introduction

Wetlands are home to very important natural remediators of pollutants. Plants in wetlands are often exposed to pollutants because of their very location in an ecosystem that often occupies the boundaries of land and water. Monocot plants that can form monocultures are important components of wetland plant communities. Genetic improvement of wetland monocots by introduction of key transgenes can significantly improve their remediation abilities. Tissue culture is prerequisite for genetic manipulation, and advances are reported here in culture and propagation *in vitro* of wetland monocots of various ecological tolerances regarding water depth, flow, seasonal fluctuations and salinity. The monocots represent numerous genera in various families such as Cyperaceae, Juncaceae, Poaceae, and Typhaceae. The reported species are in various stages of *in vitro* propagation including embryogenic tissue, multishoot culture, micropropagation, and genetic engineering.

Phytoremediation is the use of plants to remediate pollutants in contaminated soil, water and air, encompassing a number of methods for the degradation (phyto- and rhizodegradation), removal (phytoextraction, rhizofiltration and phytovolatilisation), or immobilization (hydraulic control and phytostabilization) of contaminants [1].

Phytoremediation of natural ecosystems like wetlands is a not straightforward because the ecological balance and species diversity is supposed to be restored to original or at least retained at the pre-remediation state. Remediation of diverse wetlands is not likely to be achievable by a single all-purpose wetland species. Instead, several species from each habitat concerned should be considered.

Grass like plants are important as ornamentals, sources of energy and biomass, raw materials, animal and potentially to carry out industrial processes such as phytoremediation

M. Mackova et al. (eds.), *Phytoremediation Rhizoremediation*, 217–225.

[2]. Environmentally important wetland monocots have been neglected by plant breeders; however, their remediation potential could certainly be expanded by somaclonal breeding [3-7] and introduction of key genes to increase the remediation ability [8].

Therefore, a method is needed for propagation that is independent of fertility and seasons, and is more rapid and more efficient than conventional methods and is sustainable at a high rate of multiplication. *In vitro* culture, if properly optimized, can satisfy all these requirements and it also facilitates genetic manipulation, e.g. for enhanced remediation ability. Tissue culture and transgenic technology needs to be developed for multiple species.

In vitro propagation of the great majority of monocot (mostly Poaceae) species that have been reported, was based on callus initiated on high concentrations of a strong auxin, such as 2,4-dichlorophenoxyacetic acid (2,4-D) [9]. Induction of the regenerable callus cultures from monocot species is conventionally attempted from germinating seeds, immature embryos or immature inflorescences. These explants have been used successfully in a number of wetland monocots (Table 1).

Our laboratory has a program for developing cell culture, propagation and gene transfer technologies for numerous species from various wetlands, including salt marshes, brackish water, rivers, riverbanks, lakes, ponds, and bogs. The monocots represent multiple genera in several families including Cyperaceae, Juncaceae, Poaceae, and Typhaceae. Here we report progress on *in vitro* culture methods for 41 wetland monocot species.

1.2. MATERIALS AND METHODS

Immature inflorescences of Poaceae and Typhaceae were harvested while unemerged and still covered by leaf sheaths. For Cyperaceae and Juncaceae, inflorescences were gathered while the flowers were small early in their development. The explants were surface sterilized by dilute bleach solution then cultured on primary explant medium DM-8 in the dark at 26-28°C. DM-8 medium [10] contained (in mg l^{-1}, unless indicated otherwise): Murashige and Skoog (MS) salts [11], Sigma Fine Chemicals) 4,300; Miller's salt solution [6% (w/v) KH_2PO_4], 3 ml; *myo*-inositol, 100; Vitamix [12], 2 ml; sucrose, 30,000; supplemented with the plant growth regulators adenine hemisulfate, 400 μM; picloram, 0.12; indole-3-butyric acid, 1; 2,4-dichlorophenoxyacetic acid, 0.5; isopentenyladenine, 0.5; *trans*-zeatin, 0.5; thidiazuron (TDZ) 3.0; and solidified with Phytagel (Sigma Fine Chemicals), 2000. The secondary medium (0201) contained fewer plant growth regulators at reduced levels: 0.2 2,4-D and 0.1 μM TDZ [8]. The tertiary, shoot induction/multiplication medium contained no vitamins or *myo*-inositol and only 0.1 μM TDZ as plant growth regulator [8]. The pH of all tissue culture media was adjusted to 5.8 before sterilization in a pressure cooker (109°C, 35 kPa, 25 min). Callus was subcultured every four weeks for maintenance. Regenerated and rooted plants were separated, potted in the greenhouse, and initially kept under plastic wrap cover to help acclimatization.

Table 1. In vitro culture of wetland monocots reported previously.

Species	Family	References	Species	Family	References
Cyperaceae			Lemnaceae		
Carex acutiformis Ehrh.		[8]	*Lemna spp.*		[13]
Carex divisa Huds.		[8]	Poaceae		
Carex gracilis Curt.		[8]	*Arundo donax* L.		[8,14,15]
Carex hyalinolepis Steud.		[8]	*Distichlis spicata* Greene		[6,16]
Carex lurida Wahlenberg		[17]	*Erianthus giganteus* P. Beauv.		[8]
Carex melanostachya Willd.		[8]	*Erianthus strictus* Baldwin		[8]
Carex nigra Reichard		[8]	*Miscanthus sinensis* Andersson		[8,14,18-21]
Carex pendula Huds.		[8]	*Oryza sativa* L.		[22]
Carex riparia Curtis		[8]	*Panicum dichotomum* L.		[8]
Carex spicata Huds.		[8]	*Paspalum urvillei* Steud.		[8]
Carex vulpine L.		[8]	*Phragmites australis* Steud.		[8,23-27]
Cladium jamaicense Crantz		[8]	*Saccharum officinarum* L.		[28,29]
Cyperus giganteus Vahl.		[8]	*Setaria gigantea* Makino		[8]
Cyperus haspan L.		[8]	*Sorghum halepense* Pers.		[8]
Cyperus iria L.		[8]	*Spartina alterniflora* Loisel		[7,8,30,31]
Cyperus pseudovegetus Steud.		[8]	*Spartina argentinensis* Parodi		[32-36]
Cyperus retrorsus Chapm.		[8]	*Spartina cynosuroides* Roth.		[8,37]
Eleocharis palustris Roem. & Schult.		[8]	*Spartina patens* Muhl.		[8,38-40]
Eleocharis vivipara Link		[8]	*Spartina pectinata* Link		[8,41]
Lepironia articulate Domin.		[8]	*Spartina spartinae* Hitchc.		[8]
Schoenoplectus californicus Soják		[8]	*Sporobolus virginicus* Kunth		[5,42]
Schoenoplectus tabernaemontani Palla		[8,43]	*Vetiveria zizanioides* Nash		[8,44]
Scirpus americanus Pers.		[8]	Typhaceae		
Scirpus polyphyllus Vahl.		[17]	*Typha angustifolia L.*		[8,17]
Scirpus robustus Pursh.		[45]	*Typha dominguensis* Pers.		[8]
Juncaceae			*Typha x glauca* Godr.		[8]
Juncus acuminatus Michaux		[8,46]	*Typha latifolia* L.		[2,8,17]
Juncus articulatus L.		[8]			
Juncus compressus Jacq.		[8]			
Juncus dichotomus Elliott		[8]			
Juncus effusus L.		[8,47]			
Juncus roemerianus Scheele		[8]			
Juncus tenuis Willd.		[8]			

2. *In vitro* culture and micropropagation

Developing cell culture and micropropagation protocols for a diverse group of wetland monocot species was facilitated by the availability of a universal cell culture initiation medium, which had successfully been used for a great number of species [8, 10, 15]. The method comprised using immature inflorescences as explants and cultivating the tissue on a primary medium to produce totipotent embryogenic tissue culture, cultivating the embryogenic tissue on a secondary medium to produce regenerating cultures, then sustaining shoot multiplication and production of complete plantlets having roots and shoots of tertiary medium before acclimating the plantlets in soil.

Establishment of primary callus from the 41 species listed in Table 2 table was straightforward. It was preferred that the explant be obtained from an immature inflorescence. The tips of field-grown or greenhouse-grown pre-flowering shoots of Poaceae and Typhaceae with leaf sheaths completely enclosing a developing but yet unemerged immature inflorescence, whose surface has been sterilized, were stripped of the leaves and the inflorescences were cut into cross-sectional pieces, which were then cultivated on a solid-type primary medium containing plant hormones. Primary callus and, in certain species, shoot formation, but not elongation, occur on the primary medium, and so the method is therefore suitable for sustained maintenance and propagation of the totipotent tissue culture.

3. Phytoremediation

Wetland plants in general are considered to be useful for phytoremediation [2]. *Spartina alterniflora* in particular has received more attention [7] and was subject to biochemical studies. *S. alterniflora*'s natural capacities for pollutant remediation have been evaluated before. The chemical similarity of selenium to sulfur allows Se to enter the DMSP pathway and this is the basis of *S. alterniflora's* potential use for abatement of selenium pollution by phytovolatilisation [48, 49]. *S. alterniflora* also has tolerance to certain man-made organic pollutants. It is uniquely adapted to the salt marsh which is impacted both by halocompounds elaborated by marine organisms and also man-made halogenated-organics [50]. *Spartina*'s natural biodegradation activities towards 2,4,6-trichlorophenol (TCP) [8, 51] and trichloroethene (TCE) [52], respectively, have been assessed in hydroponic experiments. With the introduction of organomercurial lyase and mercuric reductase genes, *Spartina alterniflora* has become a potentially multipurpose plant for phytoremediation [8], not only for salt marshes, but for non-saline constructed wetlands as well as it grows best under freshwater conditions when competition is absent. *Spartina alterniflora* has successfully been introduced into temperate and subtropical regions of all continents [53]. Therefore, the potential areas of Spartina's application include nearly all wetlands globally.

4. Gene transfer

Spartina alterniflora was an attractive target species because it is a globally widespread, large, perennial grass, which forms monocultural stands in salt marshes [54]. Transgenic *S. alterniflora* lines have been generated [8]. The bacterial organomercurial lyase (*merB*) [55] and mercuric reductase (*merA*) [56] genes were introduced by *Agrobacterium*-mediated gene transfer. Co-introduction and co-expression of two genes from separate *Agrobacterium* strains was possible and the transgenic lines were resistant to both organomercurials and ionic mercury salts. Although the overall frequency of transformation was low, the co-transformation is of significance because essentially the complete mercury detoxification pathway has been incorporated into an ecologically important species in one step. Highly poisonous organic mercury compounds form naturally via biological processes in inorganic mercury contaminated sites. Therefore a transgenic plant that can mineralise organomercurials to much less toxic inorganic mercury to be removed by volatilisation after reduction is very advantageous [48].

The generation of other transgenic wetland monocots from the species listed in the Tables 1 and 2 will be certainly facilitated by the availability of embryogenic cell cultures and plant regeneration systems.

Table 2. Wetland monocots newly introduced into in vitro culture and progress in previously reported species.

Species	**Family**	**Primary callus**	**Embryogeni c callus**	**Multishoot culture**	**Acclimatised plants**
	Cyperaceae				
Baumea articulata S. T. Blake		+			
Bolboschoenus maritimus Palla		+			
Bolboschoenus robustus Soják		+			
Carex divisa Huds.		+	+	+	+
Carex elata All.		+	+		
Carex gracilis Curt.		+	+	+	
Carex hirta L.		+			
Carex lacustris Willd.		+			
Carex leporina L.		+			
Carex lurida Wahlenberg		+			

Carex melanostachya Willd.	+	+	+	+
Carex nigra Reichard	+	+	+	+
Carex riparia Curtis	+	+	+	+
Carex spicata Huds.	+	+	+	+
Carex spissa L. H. Bailey	+			
Carex stenophylla Wahlenberg	+			
Carex vesicaria L.	+	+		
Carex vulpine L.	+	+	+	+
Cyperus longus L.	+	+	+	
Cyperus pseudovegetus Steud.	+	+	+	
Eleocharis palustris Roem. & Schult.	+	+	+	
Lepironia articulate Domin.	+	+	+	+
Kyllinga brevifolia Rottb.	+	+		
Schoenoplectus acutus Muhl.	+			
Schoenoplectus lacustris Palla	+			
Scirpus atrovirens Willd.	+	+		
Scirpus cyperinus Kunth.	+	+	+	
Scirpus polyphyllus Vahl.	+	+		
Scirpus silvaticus L.	+	+		
Juncaceae				
Juncus acuminatus Michaux	+	+		
Juncus glaucus Wahlenberg	+			
Juncus gymnocarpus Coville	+			
Juncus inflexus L.	+			
Poaceae				
Miscanthus sinensis Anderss. subsp. Condensatus T. Koyama "Cosmopolitan"	+	+	+	+
Miscanthus sacchariflorus Hack.	+			
Phragmites communis Trin. var. *variegata* Hitchc. ex L.H. Bailey	+			
Vetiveria zizanioides Nash	+	+		
Zizaniopsis miliacea Doell & Ascherson	+			
Typhaceae				
Typha minima Funck ex Hoppe	+			

Acknowledgements

This research was supported by grants from the South Carolina Sea Grant Consortium.

References

[1] Cunningham, SD; Shann, JR; Crowley, DE and Anderson, TA (1997) Phytoremediation of contaminated water and soil, in Kruger, EL Anderson, TA and Coats, JR, Ser. Eds., Vol. 664 from ACS Symposium Series, Phytoremediation of Soil and Water Contaminants, pp. 2-17, Oxford University Press, Cambridge, MA, ISBN 0-8412-3503-1

[2] Rogers, SD; Beech, J and Sarma, KS (1998) Shoot regeneration and plant acclimatization of the wetland monocot cattail (*Typha latifolia*). Plant Cell Rep 18: 71-75

[3] Larkin, PJ and Scowcroft, WR (1981) Somaclonal variation - a novel source of variability from cell culture for plant improvement. Theor Appl Genet 60: 197-214

[4] McClintock, B (1984) The significances of responses of the genome to challenge. Science 226: 792-801

[5] Seliskar, DM (1998) Natural and tissue culture-generated variation in the salt marsh grass *Sporobolus virginicus*: Potential selections for marsh creation and restoration. Hortscience 33: 622-625

[6] Seliskar, DM and Gallagher, JL (2000) Exploiting wild population diversity and somaclonal variation in the salt marsh grass *Distichlis spicata* (Poaceae) for marsh creation and restoration. Am J Bot 87: 141-146

[7] Wang, JB; Seliskar, DM and Gallagher, JL (2003) Tissue culture and plant regeneration of *Spartina alterniflora*: Implications for wetland restoration. Wetlands 23: 386-393

[8] Czakó, M; Feng, X; He, Y; Liang, D and Márton, L (2005) Genetic modification of wetland grasses for phytoremediation. Z Naturforsch C 60c: 285-291

[9] Conger, BV and Kuklin, AI (1995) *In vitro* culture and plant regeneration in gramineous crops, in Terzi, M; Cella, R and Falavigna, A, Eds, Current Issues in Plant Molecular and Cellular Biology: Proceedings of the VIIIth International Congress on Plant Tissue and Cell Culture, Florence, I, pp. 59-68, Kluwer Academic Publishers BV, Dordrecht, The Netherlands, ISBN 0-7923-332-25

[10] Czakó, M and Márton, L (2001) A heartwood pigment in Dalbergia cell cultures. Phytochemistry 57: 1013-1022

[11] Murashige, T and Skoog, F (1962) A revised medium for rapid growth and bio-assays with tobacco tissue cultures. Physiol Plant 15: 473-497

[12] Márton, L and Browse, J (1991) Facile Transformation of Arabidopsis. Plant Cell Rep 10: 235-239

[13] Márton, L and Czakó, M (2004) Sustained totipotent culture of selected monocot genera. U.S. Patent No. 6,821,782 B2.

[14] Meagher, RB (2000) Phytoremediation of toxic elemental and organic pollutants. Curr Opin Plant Biol 3: 153-162

[15] Ansede, JH; Pellechia, PJ and Yoch, DC (1999) Selenium biotransformation by the salt marsh cordgrass *Spartina alterniflora*: Evidence for dimethylselenopropionate formation. Environ Sci Technol 33: 2064-2069

[16] Sanger, DM, Holland, AE and Scott, GI (1999) Tidal creek and salt marsh sediments in South Carolina coastal estuaries: I. Distribution of trace metals. Arch Environ Contam Tox 37: 445-457

[17] Márton, L; Chen, YP and Czakó, M (2000) Method for decomposing toxic organic pollutants U.S. Patent No 6,087,547.

[18] Macek, T; Francova, K; Kochankova, L; Lovecka, P; Ryslava, E; Rezek, J; Sura, M; Triska, J; Demnerova, K and Mackova, M (2004) Phytoremediation: Biological cleaning of a polluted environment. Reviews on Environmental Health 19: 63-82

[19] Chung, CH (1989) Ecological engineering of coastlines with saltmarsh plantations, in Mitsch, WJ and Jorgensen, SE, Eds, Ecological Engineering: an Introduction to Ecotechnology, pp. 255-289, John Wiley and Sons, New York, ISBN 0-4716-255-90

[20] Mobberley, D (1956) Taxonomy and distribution of the genus Spartina. Iowa State Coll J Sci 30: 471-574

[21] Bizily, SP; Rugh, CL; Summers, AO and Meagher, RB (1999) Phytoremediation of methylmercury pollution: merB expression in *Arabidopsis thaliana* confers resistance to organomercurials. Proc Natl Acad Sci USA 96: 6808-6813

[22] Rugh, CL; Wilde, HD; Stack, NM; Thompson, DM; Summers, AO and Meagher, RB (1996) Mercuric ion reduction and resistance in transgenic *Arabidopsis thaliana* plants expressing a modified bacterial *merA* gene. Proc Natl Acad Sci USA 93: 3182-3187

[23] Stomp, A-M and Rajbhandari, N (2000) Genetically Engineered Duckweed. U.S. Patent No. 6,040,498.

[24] Tóth, S and Mix-Wagner, G (1998) Embryogenic callus induction of different explants of *Miscanthus sinensis, Miscanthus* x *giganteus*, and *Arundo donax* genotypes. Sust Agric Food, Energy, and Ind 249-253

[25] Straub, PF; Decker, DM and Gallagher, JL (1989) Tissue culture and regeneration of *Distichlis spicata* (Gramineae). Am J Bot 76: 1448-1451

[26] Rogers, SMD (2003) Tissue culture and wetland establishment of the freshwater monocots Carex, Juncus, Scirpus, and Typha. *In Vitro* Cell Dev Biol Plant 39: 1-5

[27] Lewandowski, I (1997) Micropropagation of *Miscanthus x giganteus*, in Bajaj, YPS, Ser. Ed., Vol. 39, Biotechnology in Agriculture and Forestry, High-Tech and Micropropagation V, pp. 239-255, Springer, New York, ISBN 3-540-61606-3

[28] Gawel, NJ; Robacker, CD and Corley, WL (1987) Propagation of *Miscanthus sinensis* through tissue culture. Hortscience 22: 1137-1137

[29] Gawel, NJ; Robacker, CD and Corley, WL (1990) *In vitro* propagation of *Miscanthus sinensis*. Hortscience 25: 1291-1293

[30] Holme, IB and Petersen, KK (1996) Callus induction and plant regeneration from different explant types of *Miscanthus* x *ogiformis* Honda "Giganteus". Plant Cell Tissue Organ Cult 45: 43-52

[31] Heaton, ACP; Rugh, CL; Kim, T; Wang, NJ and Meagher, RB (2003) Toward detoxifying mercury-polluted aquatic sediments with rice genetically engineered for mercury resistance. Environ Toxicol Chem 22: 2940-2947

[32] Lauzer, D; Dallaire, S and Vincent, G (2000) *In vitro* propagation of reed grass by somatic embryogenesis. Plant Cell Tissue Organ Cult 60: 229-234

[33] Mathe, C; Hamvas, MM; Grigorszky, I; Vasas, G; Molnar, E; Power, JB; Davey, MR and Borbely, G (2000) Plant regeneration from embryogenic cultures of *Phragmites australis* (Cav.) Trin. ex Steud. Plant Cell Tissue Organ Cult 63: 81-84

[34] Poonawala, IS; Jana, MM and Nadgauda, RS (1999) Factors influencing bud break and rooting and mass-scale micropropagation of three Phragmites species: *P. karka, P. communis* and *P. australis*. Plant Cell Rep 18: 696-700

[35] Sangwan, R and Gorenflot, R (1975) *In vitro* culture of Phragmites tissues: callus formation, organ differentiation and cell suspension culture. Z Pflanzenphysiol 75s: 256-269

[36] Straub, PF; Decker, DM and Gallagher, JL (1988) Tissue culture and long-term regeneration of *Phragmites australis* (Cav.) Trin. ex Steud. Plant Cell Tissue Organ Cult 15: 73-78

[37] Arencibia, AD; Carmona, ER; Tellez, P; Chan, MT; Yu, SM; Trujillo, LE and Oramas, P (1998) An efficient protocol for sugarcane (*Saccharum spp.* L.) transformation mediated by *Agrobacterium tumefaciens*. Transgenic Res 7: 213-222

[38] Chengalrayan, K; Gallo-Meagher, M and English, RG (2001) Novel selection agents for sugarcane transformation. Soil and Crop Science Society of Florida Proceedings 60: 81-87

[39] Utomo, HS; Wenefrida, I and Croughan, TP (2001) Smooth cordgrass synthetic seeds: Production, storage and potential use for coastal erosion controls. TurfGrass Trends 10: 12-15

[40] Croughan, TP; Cao, HX; Regan, RP; Meche, MM; Wang, XH; Xie, QJ; Trumps, DB and Materne, MD (1993) Biotechnological applications to coastal erosion control. Ann Res Rep LA State Univ Baton Rouge 85: 157-160

[41] Bueno, MS; Petenello, C; Feldman, SR and Ortiz, JP (2005) Obtención de variantes de *Spartina argentinensis* Parodi mediante el cultivo *in vitro* de inflorescencias inmaduras, in BairesBiotec 2005, Congreso internacional, VI Simposio Nacional de Biotecnología- REDBIO Argentina 2005. Encuentro Trinacional REDBIO Argentina-Chile-Uruguay. Buenos Aires

[42] Bueno, MS; Chincuini, DAE; Soumoulou, M; Feldman, SR and Ortiz, JP (2002) Obtención de callos a partir de hojas de *Spartina argentinensis* Parodi, in XXII Reunión anual de la Sociedad de Biología de Rosario. Diciembre 2002

[43] Bueno, MS; Chincuini, DAE; Soumoulou, M; Feldman, SR and Ortiz, JP (2003) Estudio del comportamiento *in vitro* de diferentes explantos de *Spartina argentinensis* Parodi, in II Congreso Nacional de Pastizales Naturales. VI Jornada Regional. IV Reunión de la Asociación Argentina de Prosopis

[44] Bueno, MS; Chincuini, DAE; Soumoulou, M; Feldman, SR and Ortiz, JP (2003) Citodiferenciación de callos de *Spartina argentinensis* Parodi, in Proc. of XXIII Reunión anual de la Sociedad de Biología de Rosario

[45] Bueno, MS; Feldman, SR and Ortiz, JP (2004) Utilización de cariopses como explantos para el cultivo de tejidos de *Spartina argentinensis* Parodi, in Proc. of Simposio Internacional de Biotecnología. II Simposio Argentino-Italiano de Bacterias Lácticas, Tucumán
[46] Li, XG and Gallagher, JL (1996) Tissue culture and plant regeneration of big cordgrass, *Spartina cynosuroides*: Implications for wetland restoration. Wetlands 16: 410-415
[47] Li, XG; Seliskar, DM; Moga, JA and Gallagher, JL (1995) Plant regeneration from callus cultures of salt-marsh hay, Spartina patens, and its cellular-based salt tolerance. Aquat Bot 51: 103-113
[48] Rao, JD; Seliskar, DM and Gallagher, JL (1995) The effect of phytohormones, NaCl, and carbon source on plant shoot regeneration from callus cultures of *Spartina patens*. Plant Physiol 108: 48-48
[49] Rao, JD; Seliskar, DM and Gallagher, JL (1996) *In vitro* shoot regeneration from callus cultures of *Spartina patens*, a halophytic C4 grass. Plant Physiol 111: 486-486
[50] Hogan, CJ (1988) Preprophase bands in a suspension culture of the monocot *Spartina pectinata*. Exp Cell Res 175: 216-222
[51] Straub, PF; Decker, DM and Gallagher, JL (1992) Characterization of tissue-culture initiation and plant regeneration in *Sporobolus virginicus* (Gramineae). Am J Bot 79: 1119-1125
[52] Gollapudi, S (2005) Development of a tissue culture system in great bulrush for biotechnological applications. M.Sc. Thesis, The University of South Carolina, Columbia, SC, USA
[53] Maffei, M (2002) Vetiveria: The Genus Vetiveria, Taylor & Francis, New York, ISBN 0-415-27586-5
[54] Wang, JB; Seliskar, DM and Gallagher, JL (2004) Plant regeneration via somatic embryogenesis in the brackish wetland monocot *Scirpus robustus*. Aquat Bot 79: 163-174
[55] Sarma, KS and Rogers, SMD (1998) Plant regeneration and multiplication of the emergent wetland monocot *Juncus accuminatus*. Plant Cell Rep 17: 656-660
[56] Sarma, KS and Rogers, SMD (2000) Plant regeneration from seedling explants of *Juncus effusus*. Aquat Bot 68: 239-247

MODIFYING A PLANT'S RESPONSE TO STRESS BY DECREASING ETHYLENE PRODUCTION

BERNARD R. GLICK
Department of Biology, University of Waterloo
200 University Avenue West, Waterloo, ON, Canada N2L 3G1
Fax: +1-519-746-0614, E-mail: glick@sciborg.uwaterloo.ca

1. Introduction

Plants that are relatively tolerant of various environmental contaminants are often used as components of phytoremediation strategies. Nevertheless, as a consequence of the stress that the environmental contaminant(s) imposes on plants, they are often unable to proliferate to any great extent and remain small in the presence of high levels of the contaminant. To remedy this situation, plant growth-promoting bacteria that facilitate the proliferation of plants under environmentally stressful conditions may be added to the system. These bacteria have been selected or engineered to lower the level of growth-inhibiting stress ethylene within the plant and also to directly promote plant growth, usually by either providing the plant hormone indoleacetic acid or facilitating the acquisition of iron from the soil. The net result of adding these bacteria to plants is a significant increase in both the number of seeds that germinate and the amount of biomass that the plants are able to attain in contaminated soil, making phytoremediation in the presence of plant growth-promoting bacteria a much faster and more efficient process. In addition, some of the benefits conferred upon plants by plant growth-promoting bacteria may be provided by genetically modifying plants to have lower ethylene levels.

1.1. PLANT GROWTH-PROMOTING BACTERIA

Beneficial free-living soil bacteria are often referred to as plant growth-promoting rhizobacteria, or PGPR, and are commonly found in association with the roots of plants [1, 2]. There is usually a high concentration of bacteria in the plant rhizosphere because of the presence of high levels of nutrients that are exuded from the roots of most plants, and these nutrients can be used to support bacterial growth and metabolism [3-5].

Plant growth-promoting bacteria can positively influence plant growth and development in two different ways: indirectly or directly [1, 2]. The indirect promotion of plant growth occurs when these bacteria decrease or prevent some of the deleterious effects of phytopathogenic organisms. For example, many biocontrol bacteria indirectly promote

M. Mackova et al. (eds.), Phytoremediation Rhizoremediation, 227–236.

plant growth by synthesizing pathogen-inhibiting antibiotics. Bacteria can directly promote plant growth by providing the plant with a compound synthesized by the bacterium or by facilitating the uptake of nutrients from the environment by the plant. Thus, plant growth-promoting bacteria may fix atmospheric nitrogen; synthesize siderophores which can solubilize and sequester iron from the soil; synthesize phytohormones including auxins and cytokinins which can enhance plant growth; solubilize phosphorus from the soil; and contain enzymes that can modulate plant growth and development [1, 2, 6-11]. A particular bacterium may affect plant growth and development using any one, or more, of these mechanisms and may utilize different mechanisms under different conditions. For example, bacterial siderophore synthesis is induced only in soils that do not contain sufficient levels of iron. Similarly, bacteria do not fix nitrogen when sufficient fixed nitrogen is available.

1.2. ACC DEAMINASE AND THE REDUCTION OF PLANT ETHYLENE

In higher plants, ethylene is produced from L-methionine via the intermediates, *S*-adenosyl-L-methionine (SAM) and 1-aminocyclopropane-1-carboxylic acid (ACC) [12]. The enzymes involved in this metabolic sequence are SAM synthetase, which catalyzes the conversion of methionine to SAM [13]; ACC synthase, which is responsible for the hydrolysis of SAM to ACC and 5-methylthioadenosine [14]; and ACC oxidase, which metabolizes ACC to ethylene, carbon dioxide and cyanide [15].

In 1978, an enzyme capable of degrading ACC was isolated from *Pseudomonas* sp. strain ACP [16]. Since then, ACC deaminase has been detected in several fungi and yeasts [17, 18] as well as in bacterial strains [1, 2, 19-26] This enzyme cleaves the plant ethylene precursor, ACC, to produce ammonia and α-ketobutyrate. It has been proposed [11] that microorganisms that contain the enzyme ACC deaminase can all promote plant growth since they act as a sink for ACC and thereby lower ethylene levels in a developing or stressed plant.

Ethylene is important for normal development in plants as well as for their response to stress [27]. Ethylene is important during the early phase of plant growth; it is required by many plant species for seed germination, and the rate of ethylene production increases during germination and seedling growth [28]. Ethylene also induces some plant defences including induced systemic resistance [29]. However, high levels of ethylene can lead to inhibition of root elongation and the onset of senescence.

Strains of ACC deaminase-containing plant growth-promoting bacteria can reduce the amount of ACC within plant tissues that is detectable by HPLC, and hence the ethylene levels in plants are also lowered [5, 30]. As a consequence of this activity, ACC deaminase-containing plant growth-promoting bacteria promote root elongation in a variety of (ethylene sensitive) plants [31]. In addition to lowering ethylene levels during plant development, ACC deaminase-containing plant growth-promoting bacteria decrease the levels of "stress ethylene" — the accelerated biosynthesis of ethylene associated with biological and environmental stresses and pathogen attack [32]. Thus, the deleterious effects of flooding, high salt or drought on tomato plants [26, 33, 34] were decreased and the shelf life of the petals of ethylene sensitive cut flowers was prolonged, following treatment with ACC deaminase-containing plant growth promoting bacteria [35]. Moreover, biocontrol strains of bacteria carrying ACC deaminase genes were better able to protect plants against various phytopathogens [36]. In addition, canola seedlings grown in the presence of high levels of nickel, produced much less ethylene when the seeds were inoculated with an ACC deaminase-containing nickel-resistant plant growth-promoting strain that also produced indoleacetic acid and high levels of siderophores [22]. In each of these situations, the "stress ethylene" produced and the damage caused by it, was reduced by the activity of ACC deaminase.

2. Plant growth-promoting bacteria and phytoremediation

While plants grown on metal contaminated soils might be able to withstand some of the inhibitory effects of high concentrations of metals within a plant, two features of most plants could result in a decrease in plant growth and viability. In the presence of plant inhibitory levels of metals, most plants (i) synthesize stress ethylene and (ii) become severely depleted in the amount of iron that they contain. Fortunately, ACC deaminase-containing plant growth-promoting bacteria may be used to relieve some of the toxicity of metals to plants. This can occur in two different ways (i) a decrease in the level of stress ethylene in plants growing in metal-contaminated soil and (ii) utilization by plants of complexes between bacterial siderophores and iron. Plant siderophores bind to iron with a much lower affinity than bacterial siderophores so that in metal-contaminated soils a plant is generally unable to accumulate a sufficient amount of iron (and often becomes chlorotic) unless bacterial siderophores are present.

In one study, in an effort to overcome the inhibition of plant growth by nickel, a bacterium was isolated from a nickel contaminated soil sample; the bacterium was (i) nickel-resistant, (ii) capable of synthesizing the auxin indoleacetic acid, (iii) able to grow at the cold temperatures (i.e., 5–10°C) that one expects to find in nickel contaminated soil environments in northern climes such as Canada, and (iv) an active producer of ACC deaminase [22]. In order to isolate plant growth-promoting bacteria, all of the nickel-resistant bacterial isolates from a nickel-contaminated rhizosphere soil sample were tested for the ability to grow on minimal medium with ACC as the sole source of nitrogen [37]. Nickel-resistant bacterial strains that were also able to grow on ACC were tested for the ability to produce siderophores and grow in cold temperatures. It was ascertained in laboratory tests, that the selected bacterium could promote plant growth (both roots and shoots) in the presence of high levels (1–6 mM) of nickel [22, 38].

Subsequently, a spontaneous siderophore overproducing mutant of this bacterium was selected. When the wild-type bacterium and the siderophore overproducing mutant were tested in the laboratory, both of them were observed to promote the growth of tomato, canola and Indian mustard plants in soil that contained otherwise inhibitory levels of nickel, lead or zinc. In addition, the siderophore overproducing mutant decreased the inhibitory effect of the added metal on plant growth significantly more than the wild-type bacterium. Metal contamination of soils is often associated with iron-deficiency of the plants grown in these soils [39]. The low iron content of plants that are grown in the presence of high levels of metals generally results in these plants becoming chlorotic, since iron deficiency inhibits both chloroplast development and chlorophyll biosynthesis [40]. Moreover, iron deficiency is a stress that causes the plant to synthesize stress ethylene. However, once they have bound iron, bacterial iron-siderophore complexes can be taken up by plants and thereby serve as an iron source for plants [41].

Thus, there is (at least) a dual role for bacteria that facilitate plant growth in metal-contaminated soils. On the one hand, the bacteria lower the level of stress ethylene in the plant thereby allowing it to develop longer roots and thus better establish itself during early stages of growth [11, 22]. On the other hand, the bacterium helps the plant to acquire sufficient iron for optimal plant growth, in the presence of levels of metals that might otherwise make the acquisition of iron difficult [42]. When the siderophore overproducing mutant was tested in the field with nickel-contaminated soil, it was observed that both the number of seeds that germinated, and the size that the plants were able to attain was increased by 50–100% by the presence of the bacterium.

In another study, the common reed, *Phragmites australis*, a plant that has often been suggested for use in the phytoremediation of wetlands, was grown from seed in the laboratory in copper-contaminated soil. It was observed that the addition of a copper-resistant strain of *Pseudomonas asplenii* that had been genetically transformed to express a bacterial ACC deaminase gene significantly stimulated seed germination in the presence of high levels of copper where the native form of this bacterium had no stimulatory effect on seed germination (M.L.E Reed, B. Warner and B.R. Glick, submitted for publication). This is consistent with the notion that one reason that plant germination is often inhibited by the presence of high levels of soil contaminants is that a high level of ethylene is produced in seeds as a response to the contaminant. In this case, lowering seed ethylene levels so that they are no longer inhibitory should promote seed germination in the presence of a range of contaminants. Moreover, in these experiments, both the native and the transformed strain of *Pseudomonas asplenii* had a small but reproducible stimulatory effect on *Phragmites australis* root and shoot growth. This indicates that, at least for *Phragmites australis*, growth inhibition by copper is not solely a consequence of stress ethylene synthesis but rather likely mainly reflects copper inhibition of plant metabolic processes.

In a separate but similar study, the growth of canola roots and shoots in copper-contaminated soil was stimulated (significantly) but to the same extent by both native and ACC deaminase-transformed *Pseudomonas asplenii* (M.L.E. Reed and B.R. Glick, submitted for publication). In this case, the promotion of plant growth by both native and transformed *Pseudomonas asplenii* was attributed to the production of indoleactic acid by the added bacteria since this strain does not produce siderophores (and therefore could not be involved in providing iron to the plant). Only the transformed strain has ACC deaminase activity so that, as with *Phragmites australis*, decreasing ethylene levels is not a factor in growth promotion, and bacterial indoleacetic acid has previously been shown to be capable of directly promoting plant growth [43].

Given the extreme toxicity of arsenate to most plants and bacteria, the development of a phytoremediation scheme for the detoxification of arsenate-contaminated soils is not a simple matter. For example, unlike what has been observed with nickel, lead, copper and zinc, arsenate-resistant plant growth-promoting bacteria do not significantly protect plants from arsenate inhibition.

Polycyclic aromatic hydrocarbons (PAHs) are a particularly recalcitrant group of contaminants and are known to be highly persistent in the environment. *In situ* microbial remediation (i.e., bioremediation) has been attempted, but it is difficult to generate sufficient biomass in natural soils to achieve an acceptable rate of movement of hydrophobic PAHs (which are often tightly bound to soil particles) to the microbes where they can be degraded. In addition, relatively few microorganisms can use high molecular weight PAHs as a sole carbon source. More recently, there have been some improvements in the strategies for bacterial remediation of contaminated soil, including inoculation with bacteria that were selected from PAH contaminated sites, or supplementing contaminated soils with nutrients [44]. However, there has only been limited success with these techniques. For bioremediation to be effective, the overall rate of PAH removal and degradation must be accelerated above current levels. One way to achieve this is to increase the amount of biomass in the contaminated soil. For this reason, the use of phytoremediation has received considerable attention [45-47].

Although using plants for remediation of persistent contaminants may have advantages over other methods, many limitations exist for the large-scale application of this technology. For example, many plant species are sensitive to contaminants including PAHs so that they grow slowly, and it is time consuming to establish sufficient biomass for meaningful soil remediation. In addition, in most contaminated soils, the number of microorganisms is depressed so that there are not enough bacteria either to facilitate contaminant degradation or to support plant growth. To remedy this situation, both degradative and plant growth-promoting bacteria may be added to the plant rhizosphere. Phytoremediation (where contaminant degradation is dependent solely on plants) is not significantly faster than bioremediation (where biodegradation of the organics is by microorganisms independent of plants) for removal of PAHs or TPHs (Total Petroleum Hydrocarbons) [48-50]. However, cultivating plants together with plant growth-promoting bacteria allows the plants to germinate in the presence of soil contaminants to a much greater extent than they would otherwise, and then to grow well under stressful conditions and accumulate a larger amount of biomass than plants grown in the absence of plant growth-promoting bacteria. In addition, the plant growth-promoting bacteria in these experiments significantly increase the amount of PAH or TPH that is removed from the soil. In heavily contaminated soils, plant growth-promoting bacteria increased seed germination and plant survival, increased the plant water content, helped plants to maintain their chlorophyll contents and chlorophyll a/b ratio, and promoted plant root and shoot growth. In the case of PAHs, this is most likely due to a combination of the direct promotion of plant growth by bacterial indoleacetic acid and a lowering of the concentration of stress ethylene by bacterial ACC deaminase (MLE Reed and BR Glick, submitted for publication). As a consequence of the treatment of plants with plant growth-promoting bacteria, the plants provide a greater sink for the contaminants since they are better able to survive and proliferate.

3. Phytoremediation with plants engineered to produce less ethylene

If ACC deaminase-containing plant growth-promoting bacteria, bound to plant roots, can act as a sink for some of the excess ACC produced as a consequence of environmental stress, then transgenic plants expressing a bacterial ACC deaminase gene should behave similarly and have a level of stress ethylene lower than non-transformed plants and consequently be less susceptible to the deleterious effects of the stress. In fact, in two separate studies transgenic plants expressing ACC deaminase were shown to proliferate to a much greater extent than the comparable non-transformed plants in the presence of metals [51, 52]. In one study, transgenic tomato plants expressing a bacterial ACC deaminase gene under the transcriptional control of two tandem 35*S* cauliflower mosaic virus promoters (constitutive expression), the *rolD* promoter from *Agrobacterium rhizogenes* (root specific expression) or the pathogenesis related PRB-1*b* promoter from tobacco, were compared to non-transgenic tomato plants in their ability to grow in the presence of cadmium, cobalt, copper, magnesium, nickel, lead or zinc and to accumulate these metals [51]. These transgenic tomato plants acquired a greater amount of metal within the plant tissues, and were less subject to the inhibitory effects of the metals on plant growth than were non-transformed plants. Moreover, plants in which the ACC deaminase gene was under the transcriptional control of the *rolD* promoter were more resistant to the various metals than were the other transgenic plants.

Of course, there is no expectation that transgenic tomato plants will ever become part of a phytoremediation strategy. Nevertheless, the results that were obtained with tomato plants were intriguing and served as a starting point for the development of other transgenic plants with lowered ethylene concentrations that could be used as a component of a phytoremediation scheme. Thus, both transgenic tobacco (because of its potentially large leaf biomass) and canola (because of its previously demonstrated ability to be a moderate accumulator of numerous metals) were transformed with bacterial ACC deaminase genes under the transcriptional control of either the 35*S* or *rolD* promoters (Li, Q, Shah, S, Saleh-Lakh, S and GLick, BR, submitted for publication; Stearns, JC, Shah, S, Dixon, DG, Greenberg BM and Glick, BR, submitted for publication). When they were tested, in laboratory and greenhouse experiments, the transgenic tobacco and canola plants responded similarly to the presence of nickel in the soil to the previously constructed transgenic tomatoes. In all instances, transgenic plants in which the exogenous ACC deaminase gene was controlled by the *rolD* promoter demonstrated the highest level of resistance to growth inhibition by nickel. Moreover, *rolD* canola plants were also resistant to growth inhibition by high levels of salt in the soil (Sergeeva, E, Shah, S and Glick, BR, submitted for publication). Reminiscent of the protection from salt stress that is afforded by a salt-resistant plant growth-promoting bacterium [34]. From these and other data, it appears that the behaviour of plants to a variety of stresses (metals, salt, flooding and pathogens), transformed with an exogenous ACC deaminase gene controlled by the *rolD* promoter, is similar to the way in which these plants respond when ACC deaminase-containing plant growth-promoting bacteria have colonized the plant roots. In both cases, root-associated ACC deaminase acts as a sink for ACC and thereby prevents the formation of growth inhibitory levels of stress ethylene. The major difference between these two scenarios is that, in addition to lowering ethylene levels, the bacteria can directly promote plant growth by providing the hormone indoleacetic acid or siderophores that help the plant to obtain a sufficient amount of iron. In fact, in laboratory and greenhouse experiments, ACC deaminase-containing plant growth-promoting bacteria generally are a greater stimulus to plant growth under a range of stressful and potentially inhibitory conditions than are ACC deaminase transgenes expressed exclusively in the roots. Unfortunately, as a consequence of a number of environmental factors (such as weather and the presence of predators in the soil) plant growth-promoting bacteria may not always be as persistent in field conditions as they are in the greenhouse. One way around this problem may be to select or engineer endophytic bacterial strains that promote plant growth by employing some of the above mentioned bacterial mechanisms [53-55]. Finally, it should also be noted that plant ethylene levels may be decreased through a variety of genetic manipulations (e.g., the use of antisense versions of ACC oxidase) other than ACC deaminase. [29].

In another study, the growth of canola plants expressing ACC deaminase under the control of two tandem 35*S* cauliflower mosaic virus promoters in the presence of arsenate was monitored [52]. About 70-80% of the transgenic plants germinated while a maximum of 25-30% of the non-transformed plants germinated. Although a small ethylene pulse is important in breaking seed dormancy in many plants, too much ethylene can inhibit plant seed germination [56]. In the presence of arsenate, ACC deaminase may enhance the process of germination by hydrolyzing any excess ACC that forms as a consequence of the stress, hence lowering the inhibitory level of ethylene in seeds. Transgenic canola also had much higher fresh and dry weights of roots and shoots, and higher leaf chlorophyll contents, than non-transformed canola grown in the presence of arsenate. Moreover, the addition of plant growth-promoting bacteria to the roots of transgenic canola plants grown in arsenate-contaminated soils helped the plants to grow to a slightly larger size. In this case, growth promotion is probably attributable to the bacterial indoleacetic acid. When biomass and rate of seed germination are considered in calculating arsenate accumulation, for each seed planted, transgenic canola expressing ACC deaminase takes up approximately eight times as much arsenate as non-transformed canola. This notwithstanding, considerable work remains to be done before a practical system for the phytoremediation of arsenic can be implemented.

4. Summary and conclusions

Microbial activities exerted in the rhizosphere can influence plant growth, development and metabolism at both the root and the shoot levels, and can reduce the effects of various stresses. More specifically, traits that directly contribute to the promotion of plant growth and stress reduction include the synthesis of indoleacetic acid, siderophores and the enzyme ACC deaminase. Several strains of plant growth-promoting bacteria with different properties are already commercially available and are being used to increase crop yields.

Given the current reluctance on the part of many consumers worldwide to embrace the use of foods derived from genetically modified plants, it may be advantageous to use either natural or genetically engineered plant growth-promoting bacteria as a means to promote growth or reduce disease through induction of resistance, rather than genetically modifying the plant itself to the same end. Moreover, given the large number of different plants, the various cultivars of those plants and the multiplicity of genes that would need to be engineered into plants, it is not feasible to genetically engineer all plants to be resistant to all pathogens and environmental stresses. Rather, it seems more logical to engineer plant growth-promoting bacteria to do this job; the first step in this direction could well be the introduction of appropriately regulated ACC deaminase genes. While ethylene signalling is required for the induction of systemic resistance elicited by rhizobacteria, a significant increase in the level of ethylene is not. Hence, lowering of ethylene levels by bacterial ACC deaminase is not incompatible with the induction of systemic resistance. Indeed, some bacterial strains possessing ACC deaminase also induce systemic resistance.

Acknowledgements

Work from the author's laboratory was supported by grants from the Natural Science and Engineering Research Council, CRESTech (a province of Ontario Centre of Excellence), Ontario Hydro, and Inco. The following individuals contributed to the work reviewed here: Genrich Burd, George Dixon, XiaoDong Huang, Sibdas Ghosh, Bruce Greenberg, Varvara Grichko, Jiping Li, Qiaosi Li, Wenbo Ma, Shimon Mayak, Barbara Moffatt, Lin Nie, Cheryl Patten, Donna Penrose, Lucy Reed, Saleema Saleh-Lakha, Elena Sergeeva, Saleh Shah, Jennifer Stearns, Tsipi Tirosh, Chunxia Wang and Barry Warner.

References

[1] Glick, BR (1995) The enhancement of plant growth by free-living bacteria. Can J Microbiol 41: 109-117.

[2] Glick, BR; Patten, CL; Holguin, G and Penrose, DM (1999) Biochemical and genetic mechanisms used by plant growth promoting bacteria. Imperial College Press, London, UK, ISBN 1-86094-152-4.

[3] Whipp, JM (1990) Carbon utilization, in Lynch JM, Ed. The rhizosphere. pp 59-97, John Wiley, Chichester, UK, ISBN 0471925489.

[4] Bayliss, C; Bent, E; Culham, DE; MacLellan, S; Clarke, AJ; Brown, GL and Wood, J (1997) Bacterial genetic loci implicated in the *Pseudomonas putida* GR12-2R3-canola mutualism: identification of an exudate-inducible sugar transporter. Can J Microbiol 43: 809-18.

[5] Penrose, DM and Glick, BR (2001) Levels of 1-aminocyclopropane-1-carboxylic acid (ACC) in exudates and extracts of canola seeds treated with plant growth-promoting bacteria. Can J Microbiol 47: 368-72.

[6] Brown, ME (1974) Seed and root bacterization. Annu Rev Phytopathol 12: 181-97.

[7] Davison, J (1988) Plant beneficial bacteria. Bio/technology 6: 282-286.

[8] Kloepper, JW; Lifshitz, R and Zablotowicz, RM (1989) Free-living bacterial inocula for enhancing crop productivity. Trends Biotechnol 7: 39-43.

[9] Lambert, B and Joos, H (1989) Fundamental aspects of rhizobacterial plant growth promotion research. Trends Biotechnol 7: 215-9.

[10] Patten, CL and Glick, BR (1996) Bacterial biosynthesis of indole-3-acetic acid. Can J Microbiol 42: 207-20.

[11] Glick, BR; Penrose, DM and Li, J (1998) A model for the lowering of plant ethylene concentrations by plant growth promoting bacteria. J Theor Biol 190: 63-8.

[12] Yang, SF and Hoffman, NE (1984) Ethylene biosynthesis and its regulation in higher plants. Annu Rev Plant Physiol 35: 155-89.
[13] Giovanelli, J; Mudd, SH and Datko, AH (1980) Sulphur amino acids in plants, in: Miflin, BJ, Ed, Amino acids and derivatives. The biochemistry of plants: a comprehensive treatise, Vol. 5, pp 435-505, Academic Press, New York, USA, ISBN 0-12-675416-0.
[14] Kende, H (1989) Enzymes of ethylene biosynthesis. Plant Physiol 91: 1-4.
[15] John, P (1991) How plant molecular biologists revealed a surprising relationship between two enzymes, which took an enzyme out of a membrane where it was not located, and put it into the soluble phase where it could be studied. Plant Mol Biol Rep 9: 192-4.
[16] Honma, M and Shimomura, T (1978) Metabolism of 1-aminocyclopropane-1-carboxylic acid. Agric Biol Chem 42: 1825-31.
[17] Honma, M (1993) Stereospecific reaction of 1-aminocyclopropane-1-carboxylate deaminase, in: Pech, JC; Latché, A and Balagué, C, Eds. Cellular and molecular aspects of the plant hormone ethylene. pp 111-6 Kluwer Academic Publishers, Dordrecht, The Netherlands, ISBN 0-7923-2169-3.
[18] Minami, R; Uchiyama, K; Murakami, T; Kawai, J; Mikami, K; Yamada, T; Yokoi, D; Ito, H; Matsui, H and Honma, M (1998) Properties, sequence, and synthesis in *Escherichia coli* of 1-aminocyclopropane-1-carboxylate deaminase from *Hansenula saturnus*. J Biochem 123: 1112-8.
[19] Klee, HJ and Kishore, GM (1992) Control of Fruit Ripening and Senescence in Plants. United States Patent Number: 5,702,933.
[20] Jacobson, CB; Pasternak, JJ and Glick, BR (1994) Partial purification and characterization of 1-aminocyclopropane-1-carboxylate deaminase from the plant growth promoting rhizobacterium *Pseudomonas putida* GR12-2. Can J Microbiol 40: 1019-25.
[21] Campbell, BG and Thomson, JA (1996) 1-Aminocyclopropane-1-carboxylate deaminase genes from Pseudomonas strains. FEMS Microbiol Lett 138: 207-10.
[22] Burd, GI; Dixon, DG and Glick, BR (1998) A plant growth promoting bacterium that decreases nickel toxicity in plant seedlings. Appl Environ Microbiol 64: 3663-8.
[23] Belimov, AA; Safronova, VI; Sergeyeva, TA; Egorova, TN; Matveyeva, VA; Tsyganov VE, Borisov, AY; Tikhonovich, IA; Kluge, C; Preisfeld, A; Dietz, KJ and Stepanok VV (2001) Characterization of plant growth promoting rhizobacteria isolated from polluted soils and containing 1-aminocyclopropane-1-carboxylate deaminase. Can J Microbiol 47: 642-52.
[24] Ghosh, S; Penterman, JN; Little, RD; Chavez, R and Glick, BR (2003) Three newly isolated plant growth-promoting bacilli facilitate the growth of canola seedlings. Plant Physiol Biochem 41: 277-81.
[25] Ma, W; Sebestianova, S; Sebestian, J; Burd, GI; Guinel, F and Glick, BR (2003) Prevalence of 1-aminocyclopropane-1-carboxylate deaminase in *Rhizobia* spp. Antoine van Leeuwenhoek 83: 285-91.
[26] Mayak, S; Tirosh, T and Glick, BR (2004) Plant growth-promoting bacteria that confer resistance to water stress in tomato and pepper. Plant Sci 166: 525-530.
[27] Deikman, J (1997) Molecular mechanisms of ethylene regulation of gene transcription. Physiol Plant 100: 561-6.
[28] Abeles, FB; Morgan, PW and Saltveit, Jr, ME (1992) Ethylene in plant biology. 2nd ed.: Academic Press New York, USA, ISBN 0-12-041451-1.
[29] Stearns, J and Glick, BR (2003) Transgenic plants with altered ethylene biosynthesis or perception. Biotechnol Adv 21: 193-210.
[30] Penrose, DM; Moffatt, BA and Glick BR (2001) Determination of 1-aminocyclopropane-1-carboxylic acid (ACC) to assess the effects of ACC deaminase-containing bacteria on roots of canola seedlings. Can J Microbiol 47: 77-80.
[31] Hall, JA; Peirson, D; Ghosh, S and Glick BR (1996) Root elongation in various agronomic crops by the plant growth promoting rhizobacterium *Pseudomonas putida* GR12-2. Isr J Plant Sci 44: 37-42.
[32] Morgan, PW and Drew, CD (1997) Ethylene and plant responses to stress. Physiol Plant 100: 620-30.
[33] Grichko, VP and Glick, BR (2001) Amelioration of flooding stress by ACC deaminase-containing plant growth-promoting bacteria. Plant Physiol Biochem 39: 11-7.
[34] Mayak, S; Tirosh, T and Glick, BR (2004) Plant growth-promoting bacteria that confer resistance in tomato to salt stress. Plant Physiol Biochem. 42: 565-572.
[35] Nayani, S; Mayak, S and Glick, BR (1998) The effect of plant growth promoting rhizobacteria on the senescence of flower petals. Ind J Exp Biol 36: 836-9.

[36] Wang, C; Knill, E; Glick, BR and Défago, G (2000) Effect of transferring 1-aminocyclopropane-1-carboxylic acid (ACC) deaminase genes into *Pseudomonas fluorescens* strain CHA0 and its *gacA* derivative CHA96 on their growth promoting and disease-suppressive capacities. Can J Microbiol 46: 898-907.
[37] Penrose, DM and Glick, BR (2003) Methods for isolating and characterizing ACC deaminase-containing plant growth-promoting rhizobacteria. Physiol Plant 118: 10-15.
[38] Ma, W; Zalec, K and Glick, BR (2001) Effects of the bioluminescence-labelling of the soil bacterium *Kluyvera ascorbata* SUD165/26. FEMS Microbiol Ecol 35: 137-44.
[39] Mishra, D and Kar, M (1974) Nickel in plant growth and metabolism. Bot Rev 40: 395-452.
[40] Imsande, J (1998) Iron, sulfur, and chlorophyll deficiencies: a need for an integrative approach in plant physiology. Physiol Plant 103: 139-44.
[41] Bar-Ness, E; Chen, Y; Hadar, Y; Marschner, H and Romheld V (1991) Siderophores of *Pseudomonas putida* as an iron source for dicot and monocot plants. Plant Soil 130: 231-41.
[42] Burd, GI; Dixon, DG and Glick, BR (2000) Plant growth-promoting bacteria that decrease heavy metal toxicity in plants. Can J Microbiol 46: 237-45.
[43] Patten, CL and Glick BR (2002) The role of bacterial indoleacetic acid in the development of the host plant root system. Appl Environ Microbiol 68: 3795-801.
[44] Suthersan, SS (2002) Natural and enhanced remediation systems. pp 239-267, CRC Press, Boca Raton, USA, ISBN 1566702828.
[45] Cunningham, SD and Berti, WR (1993) Remediation of contaminated soils with green plants: an overview. In Vitro Cell Dev Biol 29P: 207-12.
[46] Cunningham, SD; Berti, WR and Huang, JW (1995) Phytoremediation of contaminated soils. Trends Biotechnol 13: 393-7.
[47] Cunningham, SD and Ow, DW (1996) Promises and prospects of phytoremediation. Plant Physiol 110: 715-9.
[48] Huang, X-D; El-Alawi, Y; Penrose, DM; Glick, BR and Greenberg, BM (2004) Responses of plants to creosote during phytoremediation and their significance for remediation processes. Environ Pollut 130: 453-463.
[49] Huang, X-D; El-Alawi, Y; Penrose, DM; Glick, BR and Greenberg, BM (2004) Multi-process phytoremediation system for removal of polycyclic aromatic hydrocarbons from contaminated soils. Environ Pollut 130: 465-476.
[50] Huang, X-D; El-Alawai, Y; Gurska, J; Glick, BR and Greenberg, BM (2004) A multi-process phytoremediation system for decontamination of Persistent Total Petroleum Hydrocarbons (TPHs) from soils. Microchem J, in press
[51] Grichko, VP; Filby, B and Glick, BR (2000) Increased ability of transgenic plants expressing the bacterial enzyme ACC deaminase to accumulate Cd, Co, Cu, Ni, Pb, and Zn. J Biotechnol 81: 45-53.
[52] Nie, L; Shah, S; Burd, GI; Dixon, DG and Glick, BR (2002) Phytoremediation of arsenate contaminated soil by transgenic canola and the plant growth-promoting bacterium *Enterobacter cloacae* CAL2. Plant Physiol Biochem 40: 355-61.
[53] Glick, BR (2004) Teamwork in phytoremediation. Nature Biotechnol 22: 526-527.
[54] Barac, T; Taghavi, S; Borremans, B; Provoost, A; Oeyen, L; Colpaert, JV; Vangronsveld, J and van der Lelie, D (2004) Engineered endophytic bacteria improve phytoremediation of water-soluble, volatile, organic pollutants. Nature Biotechnol 22: 583-588.
[55] Sessitsch, A; Coenye, T; Sturz, AV; Vandamme, P; Ait Barka, E; Wang-Pruski, G; Faure, D; Reiter, B; Glick, BR and Nowak, J (2005) *Burkholderia phytofirmins* sp. Nov., a novel plant-associated bacterium with plant beneficial properties. Int J Syst Evol Microbiol, in press.
[56] Bewley, JD and Black, M (1985) Dormancy and the control of germination, in Seeds: physiology of development and germination, pp. 175-235, Plenum Press, New York, USA, ISBN 0-30-641687-5.

MYCORRHIZAL FUNGI AS HELPING AGENTS IN PHYTOREMEDIATION OF DEGRADED AND CONTAMINATED SOILS

MIROSLAV VOSÁTKA, JANA RYDLOVÁ, RADKA SUDOVÁ AND MARTIN VOHNÍK
Institute of Botany, Academy of Sciences of the Czech Republic, 252 43 Průhonice, Czech Republic, E-mail: vosatka@ibot.cas.cz

1. Introduction

Plant roots were defined from phytoremediation point of view as "exploratory, liquid-phase extractors that can find, alter and/or translocate elements and compounds against large chemical gradients" [1]. Since the roots of majority of higher plants live naturally in symbiosis with different types of mycorrhizal fungi [2], this association should be regarded as an organic component of the phytoremediation systems [3, 4, 5]. However, relatively few studies have focused on the effects of mycorrhiza on phytoremediation and *vice versa*, despite the widely acknowledged importance of mycorrhizal symbionts for plant growth and fitness particularly in harsh environments.

By various mechanisms mycorrhizal fungi are able to take either direct or indirect part in different processes of phytoremediation of contaminated soils including phytostabilisation, phytoextraction or phytodegradation. Indirectly, mycorrhiza can increase plant ability to withstand soil phytotoxicity due to improved nutrition, particularly in the soils with relatively immobile phosphorus, protect plants against root pathogens and drought stress and enhance soil aggregation and consequently increase retention of xenobiotics. These functions are of particular importance mainly in degraded and contaminated soils that are often poor in nutrients, with low water holding capacity and adverse physical conditions. Through altered root exudation, mycorrhiza may also affect composition and activity of microbial communities in the rhizosphere towards the microflora more effective in xenobiotics degradation or microflora, which stimulates plant growth (plant growth promoting rhizobacteria such as nitrogen-fixing bacteria). Furthermore, mycorrhizal fungi also directly help the plant to escape from the build-up of phytotoxic concentrations of certain pollutants by secreting specific detoxifying compounds (e.g. organic acids) or by binding the pollutants into fungal tissues associated with the roots and thus creating a physical barrier against their translocation to the shoots of the host plant. Important role in direct interactions of all types of mycorrhiza with soil contamination plays so called extraradical fungal mycelium (ERM) radiating from the colonised root cortex far into the surrounding soil. This mycelial network represents extensive interface between roots and pollutants

M. Mackova et al. (eds.), Phytoremediation Rhizoremediation, 237–257.

dispersed in soil. Until now, some experimental data were published on possible role of different types of mycorrhiza in phytoremediation of xenobiotics, however, majority of those represents results of microcosm experiments and is focused mainly on heavy metals. This chapter is not a detailed review of the literature which is rather extensive in particular for ectomycorrhizal symbiosis, but it focuses on basic features of the role of mycorrhizas in phytoremediation, shifts in occurrence of mycorrhizal fungi in degraded and contaminated soils, the effects of xenobiotics (heavy metals – HMs, polyaromatic hydrocarbons – PAHs and polychlorinated biphenyls – PCBs) on development of symbioses and the effects of symbioses on plant tolerance and fitness in contaminated environment.

The role of mycorrhiza in phytoremediation should be viewed from the perspective that there are different types of mycorrhizal symbiosis varying in host plant species and each mycorrhizal type can also exhibit different mechanisms of interaction with xenobiotics. Two basic types of mycorrhizal symbiosis are endomycorrhiza and ectomycorrhiza. The endomycorrhiza is further divided into several subtypes among them arbuscular mycorrhizal and ericoid mycorrhizal being the most important and having a relevant role in phytoremediation, other subtypes such as orchideoid or arbutoid mycorrhizas are rather marginal from this point of view.

2. Potential of mycorrhizal fungi in phytoremediation of xenobiotics

2.1. ARBUSCULAR MYCORRHIZAL (AM) FUNGI

The arbuscular mycorrhiza is the most ubiquitous as the arbuscular mycorrhizal fungi belonging to the phylum *Glomeromycota* [6] form symbiosis with more than 80% of vascular plant species [2]. This type of mycorrhiza occurs in majority of the terrestrial ecosystems, however, it is difficult to be observed by naked eye in nature as the AM fungi form only spores in the soil and not the fruiting bodies and symbiosis does not change morphology of the colonised roots.

The AM fungi have not been considered as an important component of HM phytoremediation protocols in early studies because many metal hyperaccumulators, often small and slowly growing, belong to plant families *Brassicaceae* or *Caryophyllaceae* that predominantly do not form mycorrhizal symbiosis [7]. However, a range of plant species intended for phytoremediation practices has been extended also to the plants capable of forming AM symbiosis and readily producing large biomass (e.g. maize, tobacco, hemp, sunflower or some hardwood trees), partly along with the progress in genetic engineering and possibility to introduce genes inducing hyperaccumulation of xenobiotics [8]. Regarding different phytoremediation strategies feasible for HM contaminated soils, AM fungi are of interest especially for phytoextraction and phytostabilisation, whereas their potential for rhizofiltration technique is negligible due to a limited ability of AM fungi to survive under long-term flooding [4].

2.2. ERICOID MYCORRHIZAL (ERM) FUNGI

Ericoid mycorrhiza is a distinctive subtype of endomycorrhiza formed mostly by ascomycetous fungi inhabiting fine roots of the members of *Ericaceae* [9]. Ericaceous species have worldwide distribution occurring and sometimes dominating over large areas in both Northern and Southern Hemisphere. Their habitats are characterized by challenging environmental conditions including acidic, nutrient poor substrates with high carbon/nitrogen ratio, low decomposition rate, poor drainage and not least by elevated level of toxic compounds including heavy metals [2, 9].

The key role of the ErM fungi for the establishment of ericaceous vegetation at HM polluted habitats is widely accepted [10, 11, 12], as ericoid endophytes show high resistance to metal toxicity as compared to other mycorrhizal fungi [13]. Since ericaceous species occurring at HM contaminated sites are relatively slowly growing dwarf shrubs, their role in phytoextraction is unlikely to reach significant levels on industrial scale. However, the ErM fungi are suggested to be the key factors enabling ericaceous dwarf shrubs to dominate at HM polluted sites [11]. The ericaceous plants are often the first colonisers of such habitats [12], which together with their ability to grow at nutrient poor substrates underline their phytostabilisation effect in natural succession on contaminated sites.

2.3. ECTOMYCORRHIZAL (ECM) FUNGI

Ectomycorrhiza occurs in majority of coniferous trees and also in numerous broadleaved trees mainly from the families *Fagaceae*, *Betulaceae* and *Ulmaceae*. The EcM associations are formed mainly by basidiomycetous or ascomycetous fungi, most of them produce fruiting bodies known as mushrooms. Unfortunately, the diversity of fruiting bodies aboveground is not necessarily connected with the diversity of fungi forming ectomycorrhizas underground. In contrast to other types of mycorrhiza, these associations change distinctly morphology of the root system by forming typically thickened ectomycorrhizal root tips, from which dense fungal mycelium spreads into surrounding soil exploiting its nutrient sources. Extraradical ectomycorrhizal mycelium often forms thick and dense hyphal cords called rhizomorphs serving for water and nutrient transport between distant parts of mycelium.

Plants growing on contaminated substrates benefit from the associations with the EcM fungi, which can support their growth by improving their nutrient and water uptake and also alleviate HM toxicity. Areas polluted by HM are often re-vegetated with trees and EcM fungi are regarded as positive factors influencing survival of tree species at such areas [14] thus significantly affecting site phytostabilisation.

3. Effect of xenobiotics on occurrence of mycorrhizal fungi and development of mycorrhizal symbioses

3.1. ARBUSCULAR MYCORRHIZAL FUNGI

There are relatively scarce data on the effects of soil contamination on field abundance and diversity of populations of AM fungi. It was found that contamination of soils with heavy metals decreased numbers of AM propagules and reduced mycorrhizal infectivity of soils as compared to uncontaminated sites [15, 16, 17]. Nevertheless, spores of the AM fungi and relatively high levels of root colonisation were reported even from highly contaminated mine spoils as well as polluted agricultural soils [18, 19, 20, 21, 22]. A great variability in HM tolerance of different AM fungi isolates have been shown [23, 24, 25], with higher HM tolerance often reported for indigenous isolates obtained from polluted soils as compared with non-indigenous ones [26, 27, 28]. Especially the isolates capable to tolerate extreme soil contamination represent a valuable material for potential inoculation of plants within phytoremediation programmes. More attention should be, however, paid to the stability of high HM tolerance of these isolates during sub-culturing process since significant shifts in tolerance to contamination have been recently reported under keeping in conditions without selection pressure. For example, a decline in Mn tolerance of a *Glomus* sp. isolate from a Mn contaminated substrate was shown after its 2-year maintenance in metal-free substrate, compared with the lineage grown constantly under HM stress in the original substrate [28]. A similar decrease in Al tolerance of *Glomus clarum* isolates was reported, with the tolerance decreasing with increasing time under conditions without metal exposure [25].

Negative effects of heavy metals were observed at any ontogenetic stage of AM development, from spore germination, germ tube growth, root penetration and colonisation to proliferation of the extraradical mycelium into surrounding soil and formation of new spores. Considerable sensitivity of spore germination to HMs was firstly described almost thirty years ago [29, 30] and this stage was repeatedly shown as more sensitive than subsequent germ tube growth [23, 31]. A majority of results on HM effect on AM development refers, however, to the level of root colonisation in HM contaminated substrate. Generally lower percentage of root colonisation in contaminated soil is a complex phenomenon caused by inhibition of spore germination, lower extension of hyphae, unsuccessful root penetration and disruption of the internal hyphae [11]. In comparison to intraradical phase of colonisation, only little attention has been paid to the effect of HMs on the development of extraradical mycelium. Both higher [17, 32] and lower [33] sensitivity of ERM to HM exposure has been reported in comparison to root colonisation.

As regards the interaction of AM fungi with organic pollutants, in one of the few studies [34] there was found that mycorrhizal colonisation of several plant species by indigenous AM population was not significantly affected by addition of a single PAH (anthracene) in concentrations up to 10 $g.kg^{-1}$. However, colonisation of clover and leek decreased when industrial PAH polluted soil was added into unpolluted soil, while maize and ryegrass colonisation was not affected. Spiking of soil with a mixture of three PAHs reduced colonisation of clover by a non-adapted AM fungus *Glomus mosseae*

BEG69 to a half of that in non-spiked soil [35]. Uptake of phosphorus was maintained in mycorrhizal clover when PAHs were added, but was reduced in non-inoculated clover and in inoculated clover plants that received surfactant to increase PAH availability. This may indicate higher PAH sensitivity of clover as compared to the AM fungus. No such effects were observed for ryegrass. By contrast to the previous results, the colonisation of alfalfa by *G. caledonium* was not significantly affected in soil artificially contaminated with a single PAH (benzopyrene) up to 10 $mg.kg^{-1}$ but significantly decreased at 100 $mg.kg^{-1}$ [36]. In two pot experiments with PCB contaminated soils (mixture of Delor 103 and Delor 106, total PCB concentrations 25 and 94 $g.kg^{-1}$, respectively), relatively high colonisation of tobacco (up to 98%) and alfalfa (up to 52%) roots by two non-adapted AM isolates was observed, depending on plant species, cultivar and AM fungal isolate [37].

3.2. ERICOID MYCORRHIZAL FUNGI

The data on the effects of HMs on occurrence of ericoid mycorrhizal fungi in the soil are very limited due to rather difficult quantification of ericoid mycorrhiza in the field samples – the ErM fungi generally do not form spores or fruiting bodies. However, occurrence of ericoid mycorrhizal fungi can be linked together with occurrence of their ericaceous hosts, which are usually highly mycorrhizal at both disturbed and undisturbed sites.

For the ErM fungi, there are only few reports on how heavy metals influence development of symbiosis and growth of the host plants. Screening the effects of lead on the growth of ericoid mycorrhizal mycobiont *Hymenoscyphus ericae* cultivated *in vitro* showed that this fungus was able to grow on media containing up to 400 $\mu g.ml^{-1}$ Pb [38]. The same fungus was able to grow over all iron concentrations tested (0-144 $\mu g.ml^{-1}$), exhibiting greater resistance than its host plants [39]. The growth of another ErM fungus *Oidiodendron maius* isolated from mycorrhizal roots of *Vaccinium myrtillus* growing in heavily contaminated soil was investigated in the presence of zinc ions [40] and the authors found strong specificity of HM tolerance for each fungal strain. In the presence of increasing concentrations of Zn salts (especially at higher ion concentrations), better performance of the mentioned isolate was observed in comparison with isolates from unpolluted soils. Differential resistance was also observed among populations of *H. ericae* isolated from *Calluna vulgaris* from natural heathland soils and mine-site soils contaminated with AsO_4^{3-} and Cu^{2+} [41]. *H. ericae* populations from the mine sites demonstrated resistance to AsO_4^{3-} compared with the heathland population; the mine-site populations produced significant growth at the highest AsO_4^{3-} concentrations (4.7 $mol.m^{-3}$), whereas growth of the heathland population was almost completely inhibited. All isolates produced identical responses to increasing copper concentrations with no differences observed between mine-site and heathland fungal populations. Thus, *H. ericae* on the contaminated sites has developed an adaptive resistance to arsenate whereas resistance to copper appears to be constitutive.

3.3. ECTOMYCORRHIZAL FUNGI

Negative effects of heavy metals on the development of EcM symbiosis have been repeatedly reported, both *in vitro* as well as in association with host plants. Decreasing abundance and diversity of Ectomycorrhiza morphotypes with increasing concentration of heavy metals was observed in a vicinity of the fertiliser factory [42]. Similarly, lower average number of EcM root tips in *Pinus sylvestris* was reported from site polluted with acid rain and with high aluminium availability than from a reference unpolluted site [43]. On the contrary, no evidence for a reduction of the genetic variation of subpopulations of the EcM fungus *Suillus luteus* caused by HM contamination was revealed in another field study [44]. High chromium and nickel concentrations were reported to decrease EcM colonisation by *Pisolithus tinctorius* in *Eucalyptus urophylla* [45]. Five different heavy metals were shown to negatively influence nitrogen acquisition efficacy of ectomycorrhizal birch seedlings [46]. On the contrary, no adverse effects of lead exposure on development of association of two EcM fungi with Norway spruce were found in another study [47]. Increased concentrations of cadmium and zinc reduced EcM colonisation of Scots pine with *Paxillus involutus* and EcM colonisation was found to be more sensitive to elevated HM levels than host *Pinus sylvestris* itself [48]. The authors also showed that both metals negatively influenced not only root colonisation but also dispersal of EcM fungi from colonised roots towards non-mycorrhizal roots. This was prominent mainly in the case of cadmium, which suppressed cross-colonisation of non-mycorrhizal roots with mycelium from roots already colonised by EcM fungi to a greater extent than the colonisation of already EcM roots. However, it should be noted that HM tolerance of ectomycorrhizal fungus in pure culture and symbiotic state does not always correlate. For example, strains of *Paxillus involutus* and *Laccaria bicolor* exhibited similar HM tolerance in pure culture, but in association with Norway spruce roots only the colonisation by *L. bicolor* but not by *P. involutus* was decreased by cadmium [49].

Similarly as for the other mycorrhizal types, a strong inter- and intraspecific variation in HM tolerance has been reported for the EcM fungi [46, 50, 51]. The isolates of EcM fungi originating from polluted soils with high selection pressure were repeatedly shown to perform better in the presence of increasing concentrations of heavy metals when compared to isolates from unpolluted soils [50, 52, 53, 54]. For example, experiments screening HM tolerance of ectomycorrhizal strains of *Amanita muscaria*, *Paxillus involutus*, *Pisolithus tinctorius*, *Suillus bovinus*, *S. luteus* and *Thelephora terrestris* isolated from polluted or unpolluted soils revealed that the strains originating from unpolluted soils were strongly inhibited by HM amendment into growth media, whereas most of the strains isolated from sporocarps growing in polluted soils exhibited tolerance to HMs. Some of these strains were able to grow at zinc concentrations up to 1 $mg.g^{-1}$ in medium [55]. The authors supposed EcM strains to be naturally selected for HM tolerance in polluted soil, but at the same time noted the occurrence of HM sensitive EcM strains at polluted sites, which was attributed to unequal distribution of HMs in polluted soils. However, HM tolerance of different geographical strains of the EcM fungi from polluted and unpolluted sites is not always predetermined by their origin, as it was the case of different *Paxillus involutus* strains screened for alleviation of aluminium stress [56].

4. Effect of mycorrhizal fungi on xenobiotics uptake and tolerance of the host plants

4.1. ARBUSCULAR MYCORRHIZAL PLANTS

Inoculation with AM fungi has been repeatedly reported to modify HM uptake by host plants, however, the results are contradictory [4, 11]. In general, AM plants are more efficient than non-mycorrhizal in the acquisition of micronutrients such as copper, iron, manganese and zinc when available at low concentrations [57, 58, 59, 60, 61]. However, when grown in excess of micronutrients or in soils contaminated by HMs with unknown biological function such as cadmium, chromium or lead, both AM-mediated increases [62, 63, 64, 65, 66] and decreases [57, 60, 67, 68] of HM concentrations in plant tissues were observed, depending strongly on plant-fungus combination and cultivation conditions.

Protective role of AM fungi for plants growing in HM contaminated soils together with lower HM concentrations in plant tissues were frequently reported for inoculated plants [59, 69, 70], although the opposite observations, i.e. growth inhibition connected with higher HM concentrations were also shown [62, 66, 71]. Growth stimulation induced by inoculation with AM fungi is not always connected with lower HM concentrations in plant tissues, for example alleviation of zinc phytotoxicity in mycorrhizal plants without any effect of inoculation on Zn concentrations was observed [72]. In some cases, AM can be of fundamental importance for plant survival in heavily contaminated substrate: non-inoculated plants of several species (maize, barley, alfalfa) were not able to survive long-term without inoculation with an isolate from the rhizosphere of a metallophyte species *Viola calaminaria* [21]. When inoculated with reference fungal isolate of the same species from unpolluted soil, the positive effect on plant growth and HM uptake was less pronounced [73]. Differences between AM fungal isolates of different origin were reported also in other study [74] where AM fungi from a zinc contaminated site were more effective in increasing *Adropogon gerardii* biomass at higher levels of Zn in the soil whereas plant growth at lower levels of soil Zn was better with mycorrhizal fungi from an uncontaminated site. In some cases, positive effect of AM inoculation on host plant growth is more pronounced under conditions of HM contamination than in control conditions or under slight contamination levels [75, 76]. It was observed that only an obligately mycotrophic species *Andropodon gerardii* but not facultatively mycotrophic *Festuca arundinacea* benefited from AM in uncontaminated soil whereas both species benefited from inoculation under HM contamination [77]. Protective effect of AM can vary not only among plant species but even varieties [69]. Recently, also indigenous soil bacteria were shown to significantly modify the effect of AM on HM uptake by host plants [78, 79]. Therefore, fungal isolates should be thoroughly screened prior to the final inoculation of a particular host plant species not only in greenhouse experiments, but also directly on contaminated sites where AM fungi inocula interact with the whole microbial community.

The data on the uptake of organic xenobiotics by AM plants are very scarce. Amendment of soil with up to 5% of a heavily PAH polluted soil (8.1 g PAHs kg^{-1}) reduced growth of ryegrass but AM colonisation with a non-adapted fungus (*G. mosseae* BEG69) was not affected [34]. At 5 g of PAHs kg^{-1} only mycorrhizal plants survived. The fate of PAHs in the rhizosphere and mycorrhizosphere of ryegrass inoculated with *G. mosseae* BEG69 was investigated in soil spiked with 5 $g.kg^{-1}$ of anthracene or with 1 $g.kg^{-1}$ of a mixture of eight PAHs [80]. Proportion of the total PAH amount that was taken up to plant tissues or adsorbed to roots was negligible and major part of PAH dissipation in the rhizosphere was due to biodegradation or biotransformation. The authors found no difference between inoculated and non-inoculated ryegrass in PAH dissipation, however, shoot concentrations of PAHs and their adsorption to roots were lower in mycorrhizal than in non-mycorrhizal plants.

4.2. ERICOID MYCORRHIZAL PLANTS

Results of the studies confirmed mycorrhizal colonisation with the ErM fungi as an important factor positively influencing the resistance to heavy metals. Ericaceous plant species themselves display HM resistance [81], but it is supposed that successful colonisation of certain HM polluted sites requires adaptations from both host plant and ericoid mycorrhizal fungus [12]. In general all reports on experiments with the ErM fungi showed lower content of heavy metals in shoots of mycorrhizal plants as compared to non-mycorrhizal plants but often increased HM accumulation in the roots. Seedlings of *Vaccinium macrocarpon* colonised by *Hymenoscyphus ericae* showed increased tolerance to lead expressed in reduced translocation of Pb to the shoots when compared to non-mycorrhizal seedlings [38]. In another study, two races of *Calluna*, one from a HM polluted site, the other from an unpolluted natural heathland, were compared under mycorrhizal and non-mycorrhizal conditions in sand cultures with different levels of copper and zinc [82]. The colonisation by ErM fungi lead to significant reduction of the HM content in shoots of experimental plants. The ErM fungi from the study were Cu resistant regardless of the site of their origin. Mycorrhizal endophytes isolated from *Calluna vulgaris*, *Vaccinium macrocarpon* and *Rhododendron ponticum* were reported to protect their host plants against metal toxicity when grown together in sand with addition of copper (concentrations from 0 to 75 $mg.l^{-1}$) and zinc (concentrations from 0-150 $mg.l^{-1}$) [83]. All ErM fungal endophytes were able to grow even at the highest concentrations of both elements in pure cultures. Mycorrhizal plants also showed at least some growth in all treatments, whereas non-mycorrhizal plants failed to grow at all but the lowest Cu and Zn concentrations. Mycorrhizal plants had lower concentrations of metals in their shoots than non-mycorrhizal but higher metal concentrations in roots. The role of the ErM colonisation in the regulation of iron uptake was investigated in a study where ericaceous plants were exposed to various Fe concentrations corresponding to those occurring in the extracts from heathland soil (0-144 $\mu g.ml^{-1}$) [39]. The authors observed very high affinity of mycorrhizal roots of *V. macrocarpon* and *C. vulgaris* for absorption of Fe at low concentrations, which was not observed for non-mycorrhizal roots. They suggested the involvement of a hydroxamate siderophore in the absorption of Fe by mycorrhizal plants at low external concentrations. At higher concentrations, the presence of fungal endophyte decreased Fe uptake to shoots resulting in lower concentrations of Fe in shoots comparing to non-mycorrhizal plants. A strain of *H. ericae* decreased Fe content in shoots of *Vaccinium macrocarpon* and reduced HM toxicity symptoms of the host plant [84].

4.3. ECTOMYCORRHIZAL PLANTS

Regarding heavy metal uptake by host plants, EcM fungi can decrease the uptake of range of elements, e.g. aluminium [85], cadmium [50], copper [86], lead [87] and zinc [50] and thus alleviate HM toxicity to the host plants. For example, the seedlings of birch, pine and spruce were reported to be less susceptible to toxic concentrations of zinc, copper, nickel and aluminium when formed associations with several EcM fungi [88]. In another study, *Paxillus involutus* ameliorated the toxicity of cadmium and zinc to Scots pine seedlings in terms of root length [89]. Even though cadmium inhibited ectomycorrhiza formation in seedlings, colonisation with *P. involutus* decreased cadmium and zinc transport to the plant shoots and also altered the ratio of zinc transported to the roots and shoots, with higher amount of cadmium retained in the roots of the seedlings. Investigations on the effect of *Paxillus involutus*, *Suillus luteus* and *Thelephora terrestris* on the copper resistance of *Pinus sylvestris* revealed that although *Suillus luteus* was more sensitive to increased copper concentrations than the other two fungi, it prevented HM accumulation in the needles of *Pinus sylvestris*; this ability was absent in *T. terrestris* [86]. Alleviating effect of the EcM fungi on HM toxicity was observed across wide range of fungal species, therefore, it seems to be universal feature: protective effects of *Thelephora terrestris*, *Laccaria laccata*, *Scleroderma citrinum*, *Paxillus involutus*, *Suillus luteus* and *S. bovinus* against cadmium in *Pinus sylvestris* seedlings were demonstrated [50].

On the other hand, some studies report that EcM fungi do not limit uptake of heavy metals by their host plants, moreover, they can increase such an uptake, which was the case of zinc in *Pinus sylvestris* inoculated with *Thelephora terrestris* [54]. Increased uptake of aluminium by host plants mediated by EcM fungi was show as well [90]. EcM fungi can affect HM stress in host plants differently according to the metal screened: *Laccaria proxima*, *Lactarius hibbardae*, *L. rufus* and *Scleroderma flavidum* increased tolerance of *Betula papyrifera* to different concentrations of nickel but negative growth effect of ectomycorrhizal fungi were observed at elevated levels of copper [91]. Also the concentration of the metal in the medium plays a significant role: regulation of zinc uptake by EcM fungi in *Pinus sylvestris* was dependent on the concentration of metal in the mycorrhizosphere [92]. At low Zn concentrations, the EcM fungi increased its uptake, whereas at high external Zn concentrations, the EcM fungi were able to maintain shoot Zn concentrations at relatively low levels. This effect of EcM fungi was hypothesised to be involved also in uptake of other heavy metals and it might explain also contradictory data reported on HM uptake [93].

Little information is available on EcM effects on host tolerance to organic pollutants. Tolerance of pine seedlings to m-toluate was unaltered regardless of the presence or absence of the EcM fungus *Suillus bovinus* or biodegradative bacteria in microcosms containing expanded clay and growth media [94]. On the contrary, fungus tolerance was significantly increased when grown in symbiosis with pine: it was able to withstand two-order higher concentration of m-toluate as compared to the pure culture on agar plates (0.02 vs. 2.0 $mg.kg^{-1}$). Fungal survival on agar was increased in a co-culture with the degradative bacterial strain.

5. Possible mechanisms of the interaction of mycorrhizal fungi and their host plants with xenobiotics

5.1. HEAVY METALS

Most of the references on potential mechanisms of plants-mycorrhizae-xenobiotics interaction are again related to heavy metals. It is generally acknowledged that mycorrhizal amelioration of metal toxicity is achieved via several physiological processes rather than through a single mechanism and that participation of single mechanisms in metal detoxification is metal- and species-specific [4, 95, 96].

Firstly, extraradical fungal mycelium radiating from the roots and exploiting non-rhizosphere soil enables the mycorrhizal plants to capture nutrients, mainly immobile ones such as phosphorus, zinc, copper and ammonium, also from sources non-available to non-mycorrhizal plants. Due to better nutrition, growth stimulation is often observed in mycorrhizal plants, sometimes connected with lower concentrations of HMs in plant biomass [e.g. 68, 97]. However, this "growth dilution effect" cannot explain all the experimental results and also another, direct mechanisms must be involved.

As the main mechanism, immobilisation of HMs in both intra- and extraradical fungal structures and resulting restriction of metal transfer into plant tissues has been suggested [4, 98]. In accordance with this theory, it was shown that inoculation with different EcM strains decreased zinc and cadmium uptake into pine shoots and the fungus producing more extensive extraradical mycelium showed greater effect on overcoming metal toxicity and lowering shoot metal concentrations [50, 54]. Non-mycorrhizal pine seedlings were reported to transport more copper to above-ground parts, however, roots and extraradical mycelium of inoculated plants accumulated up to two times more metal than non-mycorrhizal roots [99]. Similarly, EcM colonisation decreased Zn and Cd transport to the plant shoots and influenced the proportion of Zn transported to the roots and shoots, with a higher proportion retained in the roots of the seedlings [100]. Recently, accumulation of large amounts of aluminium at the fungal mantle and in areas with the Hartig net was described in the pine roots colonised by an EcM fungus [101]. Increased root but decreased shoot HM concentrations were also reported for three ericoid mycorrhizal plant species [83] and the authors hypothesised that fungus provided adsorptive surfaces for binding of metals and thus facilitated exclusion of metals from the shoots. Concerning AM fungi, metal immobilisation in fungal tissues and lower metal partitioning to the shoots of inoculated plants was also repeatedly demonstrated [57, 102]. For example, root-to-shoot ratio of Cd concentrations of 3.15 was reported in mycorrhizal clover plants compared to only 1.66 in non-mycorrhizal plants [67]. However, retention of the HMs in fungal structures does not seem to be a universal phenomenon as inoculation was shown to lead to the completely opposite results in some cases and significantly enhanced translocation of HMs from roots to shoots of mycorrhizal plants was observed [68, 103]. Retention of HMs in fungal structures inside colonised roots was supported also by several microanalytical studies. Accumulation of metals within intracellular AM hyphae, mainly in phosphate-rich materials in the vacuoles, was reported in the mycorrhizal roots of the fern *Pteridium aquilinum* from a contaminated soil [104]. A massive accumulation of heavy metals in inner cortical cells of AM colonised maize roots where arbuscules and intraradical hyphae were located was also confirmed using different microbeam techniques [73]. Further, a high cadmium and zinc sorption capacity of the ERM of AM fungi in comparison with other microorganisms was shown [105] and a metal tolerant *Glomus mosseae* isolate showed significantly higher binding capacity than non-tolerant isolate of the same species. Similarly, a very high cation exchange capacity for lead absorption was reported for the fungal mycelium of two EcM fungi [106]. Different HMs can be detoxified *via* different mechanisms, as shown by a study where Cd was complexed predominantly extracellularly in the Hartig net hyphae, whereas zinc was sequestered also in the cytosol of mantle hyphae [107]. Binding of cadmium onto fungal cell walls as well as accumulation of Cd in the vacuolar compartments was described for *in vitro* grown EcM fungus *Paxillus involutus* [108]. Cytoplasmatic sequestration of metals may be realised *via* different thiol compounds, including reduced glutathione, phytochelatins and metallothioneins that are essential components of HM detoxification pathways in various organisms. Using a specific histochemical staining, induction of protein-bound thiolate clusters in EcM hyphae in response to cadmium was detected [109] and the sulphur-rich clusters were hypothesised to be derived from metallothionein or metallothionein-like proteins. Later, a metallothionein-like protein was detected in two tolerant EcM fungi in response to copper exposure and this compound associated with as much as 90% of the total copper [110]. Recently, a drastically increased production of the Cd-modulated compound most probably related to a metallothionein was detected in Cd-exposed EcM mycelium of *Paxillus involutus* [111]. In contrast, neither phytochelatins nor metallothioneins were detected in EcM fungus exposed to Cd, in spite of increased glutathione levels [112] and the authors concluded that rather glutathione and glutamylcystein were involved in metal detoxification. Similarly, increased contents of glutathione and lack of phytochelatin induction in two different EcM fungi exposed to cadmium were reported recently [111, 113]. As far as AM fungi, increased synthesis of metallothioneins by mycorrhizal plants was also proposed [114, 115], however, higher synthesis of peptides containing thiol groups was not revealed in roots of AM-inoculated plants exposed to high Cd and Cu concentrations, in spite of increased concentrations of cysteine, glutathione and glutamycysteine. Recently, a novel metallothionein-like gene that was selectively up-regulated by Cu exposure was identified in the ERM of AM fungi [116].

Apart from metal binding onto cell walls, accumulation of copper in the mucilaginous outer hyphal wall zone of the ERM of AM fungi was reported [117] and the authors suggested an employment of extracellular glycoprotein glomalin into HM detoxification. Similarly in EcM fungi, zinc was shown to be bound not only to extraradical hyphal walls, but also to extrahyphal slime polymers [53]. In accordance with this finding, a frequent excretion of extracellular mucilaginous substances by EcM fungal cultures exposed to higher metal concentrations was observed [118]. The correlation between excretion of loosely adhering extrahyphal slime and zinc tolerance of fungal strains and the degree of amelioration of zinc toxicity to mycorrhizal host plants was reported also for ErM fungi [119]. A significantly higher production of oxalic acid by EcM pine seedlings than non-mycorrhizal controls was reported in response to aluminum exposure [120] and it was assumed that ectomycorrhizae produce or stimulate the roots to secrete organic acids that bind Al and prevent its absorption. Study on a Zn and Cd tolerant strain of ErM fungus *Oidiodendron maius* showed that the increase in concentration of Zn ions in cultivation medium induced a shift in the spectrum of proteins secreted by the fungus towards more basic, low molecular weight polypeptides [121]. Among other products, two superoxide dismutase isoforms were found as well. These are known to act in HM response in plants, animals and microorganisms and the authors hypothesised that they could play similar role also in ErM fungi. As another option how mycorrhizal plants may influence metal uptake, modification of metal solubility mediated by changes in pH of soil solution was suggested [122] as higher pH values and lower zinc concentrations in soil solution were observed in mycorrhizal treatments.

Interesting mechanism of eliminating arsenate toxicity was reported for the ErM fungi growing in arsenate-polluted soils. Arsenate is analogical to phosphate and can be transported across the plasma membrane *via* the phosphate co-transporter [123]. The ErM accumulating arsenate in their hyphae would thus transport phosphorus together with arsenate into plant tissues resulting in negative consequences to host plants. But *Hymenoscyphus ericae* strains from arsenic-polluted sites had the ability to reduce arsenate to arsenite and excrete the latter from their hyphae [124, 125]. By this way, *H. ericae* actively accumulated phosphate while eliminating arsenate and transported phosphate to the host plant. Ericaceous host plants appear to lack altered phosphate-arsenate uptake system and seem to be dependent on their mycobionts regarding alleviation of arsenate toxicity [125].

Also other mechanisms how mycorrhizal fungi cope with elevated concentrations heavy metals in their environment have been proposed (e.g. chelating by pigments with metal binding ability such as melanin, enzymes resistant to inactivation by metal binding, modified uptake systems at the plasmalemma, ability of membranes to withstand attack from oxygen radicals), their role, however, remains unravelled [12, 96, 126].

5.2. ORGANIC COMPOUNDS

It was shown that dissipation of condensed PAHs might be enhanced in the presence of AM fungi in the soil spiked with a mixture of different PAHs [127]. These results were consecutively verified in pot experiments conducted in two different industrially polluted soils [128]. In general, following explanations were suggested by the authors to elucidate the contribution of AM to PAH degradation: i) mycorrhiza modifies root enzyme activity, exudation and architecture in a manner that stimulates PAH degradation, either by root-derived enzymes or by rhizosphere microorganisms, ii) mycorrhizal colonisation affects root surface or rhizosphere soil properties that act on PAH availability through adsorption. Several examples of above mentioned capabilities of AM fungi can be found in the literature. Mycorrhiza was reported to enhance amount of hydrogen peroxide in the roots [129] and to stimulate oxidoreductase activities in the roots and the rhizosphere [130]. These mechanisms may contribute to increase of PAH dissipation associated with mycorrhizal roots. Mycorrhizal colonisation also modifies root exudation both quantitatively and qualitatively [131, 132], which could have further effects on the composition and activity of microbial communities in the rhizosphere [133, 134]. This mycorrhiza-associated microflora may be more effective in organics degradation in comparison with that related to non-mycorrhizal roots. Accumulation of phenolics in the roots or rhizosphere soil of mycorrhizal plants could induce degradation of more complex aromatic compounds [135].

Effects of AM on root longevity and proportion of higher order lateral roots with short life span was documented for several plant species [136]. Accumulation of root debris together with the ability of mycorrhizal plants to enrich the soil in organic matter [137] may contribute to enhanced PAH adsorption in plant rhizosphere. Finally, AM colonisation results in formation of an extensive network of extraradical mycelium, which can modify surrounding environment by extrusion of glycoproteins [138] and extracellular enzymes [139]. Hyphal biomass together with the two latter substances can serve as substrates for microbial growth. Nutrients derived from extraradical hyphae of AM fungi were hypothesised to drive co-metabolic degradation of PAHs within small soil pores where PAHs are spatially unavailable to roots [127].

Most of the ectomycorrhizal fungi have some limited ability to use polymers such as lignin and cellulose as substrates for their growth [2], however, their ability to degrade polymers of this kind was supposed to be much lower as compared to wood decomposers or even of some ericoid mycorrhizal fungi [140, 141, 142]. As it was recently reviewed, most of the EcM fungi screened for degradation of persistent organic pollutants are able to transform these compounds [143]. On the other hand, most reports concern degradation in pure cultures, rather than in symbiosis with plants. It has been demonstrated that EcM fungi can degrade e.g. trinitrotoluene [144], dichlorphenol [145], atrazine [146] and several 3-5 ring PAHs [147]. Only limited PAH degrading abilities of EcM fungi and decreased mineralisation of some PAHs in microcosms with EcM pine seedlings vs. unplanted microcosms inoculated with ectomycorrhizosphere soil were documented [148]. On the contrary, mineralisation of dichlorphenol was stimulated when EcM fungi were cultivated in symbiosis with pine than when grown in absence of the host [145].

Also degradation of lower chlorinated PCBs by eight of 13 studied species of EcM fungi was reported, however, only two species were able to degrade 4 and 5 chlorinated biphenyls [149]. While EcM fungi could sequentially hydroxylate a halogenated biphenyl ring (parent compound is then more polar and bioavailable), they were unable to cleave the ring [150]. However, hydroxylation of the biphenyl ring can be an important initial metabolic step facilitating degradation of PCBs by other rhizosphere organisms that have the capacity to degrade compounds further [143]. In this context it can be hypothesised that the presence of EcM fungi may negate the need for the presence of co-substrates. Although the tolerance of EcM fungi to different aromatic compounds is lower as compared to white rot fungi and depends on the compound type, its external concentration and the fungal species, the degradative capabilities of different fungi varied between species but not generally between the biotrophic and saprotrophic fungi [151].

6. Conclusion

Application of mycorrhizal fungi as supportive agents for phytoremediation can generally eliminate or reduce some known limitation of phytoremediation processes. Mycorrhizal fungi can help plants to acquire nutrients more effectively, increase their tolerance to edaphic stress and change accumulation of pollutants in plant tissues.

Reduced HM concentrations in plant tissues of mycorrhizal plants, together with an amplified barrier against metal translocation from plant roots to shoots, are favourable for phytostabilisation that is aimed at prevention of contamination spreading into the surroundings. The presence of AM fungi leading to decreased HM concentrations in plant shoots can be also an important factor for quality and safety of plants which enter food chains such as forage crops, medicinal herbs and vegetables. For example, tobacco was found to accumulate significantly less cadmium in the leaves when inoculated with selected AM fungi [8].

On the other hand, enhanced HM concentrations in the shoots of mycorrhizal plants induced by some mycorrhizal fungal isolates represent optimal conditions for phytoextraction technique. In some cases, elimination of present AM fungi populations (in particular if they involve strains decreasing HM translocation to the shoots) might be recommended before phytoextraction beginning. For example, application of the fungicide benomyl detrimental to mycorrhiza was shown to significantly decrease root colonisation and simultaneously to increase Pb concentrations in plant shoots [153]. Mycorrhizal fungi may be crucial also for re-vegetation efforts after heavy metal removal as the rate of site re-vegetation may be accelerated when AM fungi are present in soil. However, little is known about mycorrhiza functioning under conditions imposed by particular metal remediation protocols. First investigations have appeared showing that the quantity and species composition of glomalean propagules and the functioning of AM symbiosis could be significantly influenced by phytoextraction treatments (the choice of plant species, i.e. non-mycotrophic vs. mycotrophic, soil supplements etc.) [7]. Recently, negative effects of synthetic chelates used for chelate-induced HM phytoextraction such as EDTA on AM development were also described [102, 152].

Based on the intended phytoremediation strategy, the appropriate management of native fungi and/or application of artificial inocula should be chosen. For the introduction of artificial inocula, it seems to be essential to formulate specific products (mixtures of strains compatible with target plants and environment) rather than to use generic products all over the scale of edaphic conditions. However, it should considered that the effect of mycorrhizal inoculation may interact e.g. with fertilisation regime as demonstrated for the grasses grown in mine tailings containing high levels of zinc: plant growth was best after inoculation combined with nitrogen and phosphorus fertilisation, whereas neither mycorrhiza nor fertilisation alone had any effect on plant biomass [77]. Therefore, mycorrhizal inoculation cannot be considered as panacea and should be combined with other practices such as appropriate fertilisation or soil amendments to maximise re-vegetation success.

To conclude, an extension of knowledge on the involvement of mycorrhizal fungi in phytoremediation should still be achieved. Great attempt should be also undertaken to increase awareness of potential users of mycorrhizal inoculants regarding all possible functions and impacts of mycorrhiza applications in phytoremediation processes.

Acknowledgement

Presentation of this contribution was supported by the grant 526/04/0996 of the Grant Agency of the Czech Republic and by the Research Centre for Bioindication and Revitalisation funded by the Ministry of Education, Youth and Sports of the Czech Republic (grant 1M67985939).

References

[1] Cunningham, SD; Berti, WR and Huang, JW (1995) Phytoremediation of contaminated soils. Trends Biotechnol 12: 393-397

[2] Smith, SE and Read, DJ (1997) Mycorrhizal symbiosis. Academic Press, London.

[3] Donnelly, PK and Fletcher, JS (1994) Potential use of mycorrhizal fungi as bioremediation agents. In: Anderson T.A. (ed.) Bioremediation through Rhizosphere Technology. ACS Symposium series No. 563, American Chemical Society, pp. 93-99

[4] Leyval, C; Turnau, K and Haselwandter, K (1997) Effect of heavy metal pollution on mycorrhizal colonization and function: physiological, ecological and applied aspects. Mycorrhiza 7: 139-153

[5] Gaur, A and Adholeya, A (2004) Prospects of arbuscular mycorrhizal fungi in phytoremediation of heavy metal contaminated soils. Curr Sci 86: 528-534

[6] Schüssler, A; Schwarzott, D and Walker, C (2001) A new fungal phylum, the *Glomeromycota*: phylogeny and evolution. Mycol Res 105: 1413-1421

[7] Pawlowska, TE; Chaney, RL; Chin, M and Charvat, I (2000) Effects of metal phytoextraction practices on the indigenous community of arbuscular mycorrhizal fungi at a metal-contaminated landfill. Appl Env Microbiol 66: 2526-2530

[8] Janoušková, M; Pavlíková, D; Macek, T and Vosátka, M (2005) Arbuscular mycorrhiza decreases cadmium phytoextraction by transgenic tobacco with inserted metallothionein. Plant Soil 272: 29-40

[9] Read, DJ (1996) The structure and function of the ericoid mycorrhizal root. Ann Bot 77: 365-374

[10] Cairney, JWG (2000) Evolution of mycorrhiza systems. Naturwissenschaften 87: 467-475

[11] Meharg, AA and Cairney, JWG (2000) Co-evolution of mycorrhizal symbionts and their hosts to metal-contaminated environments. Adv Ecol Res 30: 69-112

[12] Cairney, JWG and Meharg, AA (2003) Ericoid mycorrhiza: a partnership that exploits harsh edaphic conditions. Eur J Soil Sci 54: 735-740

[13] Shaw, G and Read, DJ (1989) The biology of mycorrhiza in the Ericaceae. XIV. Effects of iron and aluminium on the activity of acid phosphatase in the ericoid endophyte *Hymenoscyphus ericae* (Read) Korf and Kernan. New Phytol 113: 529-533

[14] Wilkinson, DM and Dickinson, NM (1995) Metal resistance in trees: the role of mycorrhizae. Oikos 72: 298-300

[15] Boyle, M and Paul, EA (1988) Vesicular-arbuscular mycorrhizal associations with barley on sewage-amended plots. Soil Biol Biochem 20: 945-948

[16] Leyval, C; Singh, BR and Joner, EJ (1995) Occurrence and infectivity of arbuscular mycorrhizal fungi in some Norwegian soils influenced by heavy metals and soil properties. Water Air Soil Poll 84: 203-216

[17] del Val, C; Barea, JM and Azcón-Aguilar, C (1999) Assessing the tolerance to heavy metals of arbuscular mycorrhizal fungi isolated from sewage sludge-contaminated soils. Appl Soil Ecol 11: 261-269

[18] Ietswaart, JH; Griffioen, WAJ and Ernst, WHO (1992) Seasonality of VAM infection in three populations of *Agrostis capillaris* (Graminae) on soil with or without heavy metal enrichment. Plant Soil 139: 67-73

[19] Weissenhorn, I; Mench, M and Leyval, C (1995) Bioavailability of heavy metals and arbuscular mycorrhizas in a sewage sludge amended sandy soil. Soil Biol Biochem 27: 287-296

[20] Pawlowska, TE; Blaszkowski, J and Rühling, A (1996) The mycorrhizal status of plants colonizing a calamine soil mound in southern Poland. Mycorrhiza 6: 499-505

[21] Hildebrandt, U; Kaldorf, M and Bothe, H (1999) The zinc violet and its colonization by arbuscular mycorrhizal fungi. J Plant Physiol 154: 709-717

[22] Turnau, K; Ryszka, P; Gianinazzi-Pearson, V and van Tuinen, D (2001) Identification of arbuscular mycorrhizal fungi in soils and roots of plants colonizing zinc wastes in southern Poland. Mycorrhiza 10: 169-174

[23] Weissenhorn, I; Leyval, C and Berthelin, J (1993) Cd-tolerant arbuscular mycorrhizal (AM) fungi from heavy-metal polluted soils. Plant Soil 157: 247-256

[24] Bartolome-Esteban, H and Schenck, NC (1994) Spore germination and hyphal growth of arbuscular mycorrhizal fungi in relation to soil aluminium saturation. Mycologia 86: 217-226

[25] Kelly, CN; Morton, JB and Cumming, JR (2005) Variation in aluminium resistance among arbuscular mycorrhizal fungi. Mycorrhiza 15: 193-201

[26] Gildon, A and Tinker, PB (1983) Interactions of vesicular-arbuscular mycorrhizal infection and heavy metals in plants. I. The effects of heavy metals on the development of vesicular-arbuscular mycorrhizas. New Phytol 95: 247-261

[27] Weissenhorn, I and Leyval, C (1995) Root colonization of maize by a sensitive and a Cd-tolerant *Glomus mosseae* and cadmium uptake in sand culture. Plant Soil 175: 233-238

[28] Malcová, R; Rydlová, J and Vosátka, M (2003) Metal-free cultivation of *Glomus* sp. BEG 140 isolated from Mn-contaminated soil reduces tolerance to Mn. Mycorrhiza 13: 151-157

[29] Hepper, CM and Smith, GA (1976) Observations on the germination of *Endogone* spores. Trans Br Mycol Soc 66: 189-194

[30] Hepper, CM (1979) Germination and growth of *Glomus caledonius* spores: the effects of inhibitors and nutrients. Soil Biol Biochem 11: 269-277

[31] McIlveen, WD and Cole, H (1978) Influence of zinc on development of the ectomycorrhizal fungus *Glomus mosseae* and its mediation of phosphorus uptake by *Glycine max* "Amsoy 71". Agric Environm 4: 245-256

[32] Batkhuugyin, E; Rydlová, J and Vosátka, M (2000) Effectiveness of indigenous and non-indigenous isolates of arbuscular mycorrhizal fungi in soils from degraded ecosystems and man-made habitats. Appl Soil Ecol 14: 201-211

[33] Arines, J; Porto, ME and Vilariño, A (1992) Effect of manganese on vesicular-arbuscular mycorrhizal development in red clover plants and on soil Mn-oxidizing bacteria. Mycorrhiza 1: 127-131

[34] Leyval, C and Binet, P (1998) Effect of polyaromatic hydrocarbons in soil on arbuscular mycorrhizal plants. J Environ Qual 27: 402-407

[35] Joner, EJ and Leyval, C (2001) Influence of arbuscular mycorrhiza on clover and ryegrass grown together in a soil spiked with polycyclic aromatic hydrocarbons. Mycorrhiza 10: 155-159

[36] Liu, SL; Luo, YM; Cao, ZH; Wu, LH; Ding, KQ and Christie, P (2004) Degradation of benzo[a]pyrene in soil with arbuscular mycorrhizal alfalfa. Environ Geochem Hlth 26: 285-293

[37] Rydlová, J; Stárek, L; Macková, M; Vosátka, M and Macek, T (2005) Interactions of arbuscular mycorrhizal fungi and PCB degrading bacteria in two soils contaminated with PCB. Biodegradation (submitted)

[38] Hashem, AR (1990) *Hymenoscyphus ericae* and the resistance of *Vaccinium macrocarpon* to lead. Trans Myc Soc Jap 31: 345-353

[39] Shaw, G; Leake, JR; Baker, AJM and Read, DJ (1990) The biology of mycorrhiza in the *Ericaceae*. XVII. The role of mycorrhizal infection in the regulation of iron uptake by ericaceous plants. New Phytol 115: 251-258

[40] Martino, E; Turnau, K; Girlanda, M; Bonfante, P and Perotto, S (2000) Ericoid mycorrhizal fungi from heavy metal polluted soils: their identification and growth in the presence of zinc ions. Mycol Res 104: 338-344

[41] Sharples, JM; Meharg, AA; Chambers, SM and Cairney, WG (2001) Arsenate resistance in the ericoid mycorrhizal fungus *Hymenoscyphus ericae*. New Phytol 151: 265-270

[44] Stankenkevičiene, D and Pečiulyte, D (2004) Functioning of ectomycorrhizas and soil microfungi in deciduous forests situated along a pollution gradient next to a fertilizer factory. Pol J Environ Stud 13: 715-721

[45] Rudawska, M; Kieliszewska-Rokicka, B and Leski, T (1996) Effect of acid rain and aluminium on the mycorrhizas of *Pinus sylvestris*, in Azcón-Aguillar C and Barea JM (Eds.), Mycorrhizas in integrated systems. From genes to plant development. European Commission, Luxembourg, pp. 452-454

[46] Muller, LAH; Lambaerts, M; Vangronsveld, J and Colpaert, JV (2004) AFLP-based assessment of the effects of environmental heavy metal pollution on the genetic structure of pioneer populations of *Suillus luteus*. New Phytol 164: 297-303

[47] Aggangan, NS; Dell, B and Malajczuk, N (1989) Effects of chromium and nickel on growth of the ectomycorrhizal fungus *Pisolithus tinctorius* and formation of ectomycorrhizas on *Eucalyptus urophylla* Blake. S.T. Geoderma 84: 15-27

[48] Blaudez, D; Jacob, C; Turnau, K; Colpaert, JV; Ahonen-Jonnarth, U; Finlay, R; Botton, B and Chalot, M (2000) Differential reponses of ectomycorrhizal fungi to heavy metals *in vitro*. Mycol Res 104: 1366-1371

[49] Jentschke, G; Fritz, E and Godbold, DL (1991) Distribution of lead in mycorrhizal and non-mycorrhizal Norway spruce seedlings. Physiol Plantarum 81: 417-422

[50] Hartley, J; Cairney, JWG and Meharg, A (1999) Cross-colonization of Scots pine (*Pinus sylvestris*) seedlings by the ectomycorrhizal fungus *Paxillus involutus* in the presence of inhibitory levels of Cd and Zn. New Phytol 142: 141-149

[51] Jentschke, G; Winter, S and Godbold, DL (1999) Ectomycorrhizas and cadmium toxicity in Norway spruce seedlings. Tree Physiol 19: 23-30

[52] Colpaert, JV and van Assche, JA (1993) The effects of cadmium on ectomycorrhizal *Pinus sylvestris* L. New Phytol 123: 325-333

[53] Vodnik, D; Byrne, AR and Gogala, N (1998) The uptake and transport of lead in some ectomycorrhizal fungi in culture. Mycol Res 102: 953-958

[54] Egerton-Warburton, LM and Griffin, BJ (1995) Differential responses of *Pisolithus tinctorius* isolates to aluminium *in vitro*. Can J Bot 73: 1229-1233

[55] Denny, HJ and Wilkins, DA (1987) Zinc tolerance in *Betula* spp. 4. The mechanism of ectomycorrhizal amelioration of zinc toxicity. New Phytol 106: 545-553

[56] Colpaert, JV and van Assche, JA (1992) Zinc toxicity in ectomycorrhizal *Pinus sylvestris*. Plant Soil 143: 201-211

[57] Colpaert, JV and van Assche, JA (1987) Heavy metal resistance in some ectomycorrhizal fungi. Funct Ecol 1: 415-421

[58] Rudawska, M and Leski, T (1998) Aluminium tolerance of different *Paxillus involutus* Fr. strains originating from polluted and non-polluted sites. Acta Soc Bot Pol 67: 115-122

[59] Dehn, B and Schüepp, H (1989) Influence of VA mycorrhizae on the uptake and distribution of heavy metals in plants. Agr Ecosyst Environ 29: 79-83

[60] El-Kherbawy, M; Angle, JS; Heggo, A and Chaney, RL (1989) Soil pH, rhizobia, and vesicular-arbuscular mycorrhizae inoculation effects on growth and heavy metal uptake of alfalfa (*Medicago sativa* L.). Biol Fertil Soils 8: 61-65

[61] Heggo, A; Angle, JS and Chaney, RL (1990) Effects of vesicular-arbuscular mycorrhizal fungi on heavy metal uptake by soybeans. Soil Biol Biochem 22: 865-869

[62] Loth, FG and Höfner, W (1995) Einfluss der VA-Mykorrhiza auf die Schwermetallaufnahme von Hafer (*Avena sativa* L.) in Abhängigkeit vom Kontaminationsgrad der Böden. Z Pflanzenernaehr Bodenk 158: 339-345

[63] Liu, A; Hamel, C; Hamilton, RI; Ma, BL and Smith, DL (2000) Acquisition of Cu, Zn, Mn and Fe by mycorrhizal maize (*Zea mays* L.) grown in soil at different P and micronutrient levels. Mycorrhiza 9: 331-336

[64] Killham, K and Firestone, MK (1983) Vesicular arbuscular mycorrhizal mediation of grass response to acidic and heavy metal depositions. Plant Soil 72: 39-48

[65] Carneiro, MAC; Siqueira, JO and Moreira, FMD (2001) Establishment of herbaceous plants in heavy metal contaminated soils inoculated with arbuscular mycorrhizal fungi. Pesqui Agropecu Bras 36: 1443-1452

[66] Jamal, A; Ayub, N; Usman, M and Khan, AG (2002) Arbuscular mycorrhizal fungi enhace zinc and nickel uptake from contaminated soil by soybean and lentil. Int J Phytorem 4: 205-221

[67] Liao, JP; Lin, XG; Cao, ZH; Shi, YQ and Wong, MH (2003) Interactions between arbuscular mycorrhizae and heavy metals under sand culture experiment. Chemosphere 50: 847-853
[68] Yu, X; Cheng, J and Wong, MH (2005) Earthworm-mycorrhiza interaction on Cd uptake and growth of ryegrass. Soil Biol Biochem 37: 195-201
[69] Joner, EJ and Leyval, C (1997) Uptake of ^{109}Cd by roots and hyphae of a *Glomus mosseae/Trifolium subterraneum* mycorrhiza from soil amended with high and low concentrations of cadmium. New Phytol 135: 353-360
[70] Weissenhorn, I; Leyval, C; Belgy, G and Berthelin, J (1995) Arbuscular mycorrhizal contribution to heavy metal uptake by maize (*Zea mays* L.) in pot culture with contaminated soil. Mycorrhiza 5: 245-251
[71] Rivera-Becerril, F; Calantzis, C; Turnau, K; Caussanel, JP; Belimov, AA; Gianinazzi, S; Strasser, RJ and Gianinazzi-Pearson, V (2002) Cadmium accumulation and buffering of cadmium-induced stress by arbuscular mycorrhiza in three *Pisum sativum* L. genotypes. J Exp Bot 53: 1177-1185
[72] Chen, BD; Li, XL; Tao, HQ; Christie, P and Wong, MH (2003) The role of arbuscular mycorrhiza in zinc uptake by red clover growing in a calcareous soil spiked with various quantities of zinc. Chemosphere 50: 839-846
[73] Dosskey, MG and Adriano, DC (1993) Trace element toxicity in VA mycorrhizal cucumber grown on weathered coal fly ash. Soil Biol Biochem 25: 1547-1552
[74] Dueck, TA; Visser, P; Ernst, WHO and Schat, H (1986) Vesicular-arbuscular mycorrhizae decrease zinc-toxicity to grasses growing in zinc-polluted soil. Soil Biol Biochem 18: 331-333
[75] Kaldorf, M; Kuhn, AJ; Schröder, WH; Hildebrandt, U and Bothe, H (1999) Selective element deposits in maize colonized by a heavy metal tolerance conferring arbuscular mycorrhizal fungus. J Plant Physiol 154: 718-728
[76] Shetty, KG; Hetrick, BAD and Schwab, AP (1995) Effects of mycorrhizae and fertilizer amendments on zinc tolerance of plants. Environ Pollut 88: 307-314
[77] Davies, FT; Puryear, JD; Newton, RJ; Egilla, JN and Grossi, JAS (2001) Mycorrhizal fungi enhance accumulation and tolerance of chromium in sunflower (*Helianthus annuus*). J Plant Physiol 158: 777-786
[78] Chen, BD; Liu, Y; Shen, H; Li, XL and Christie, P (2004a) Uptake of cadmium from an experimentally contaminated calcareous soil by arbuscular mycorrhizal maize (*Zea mays* L.). Mycorrhiza 14: 347-354
[79] Hetrick, BAD; Wilson, GWT and Figge, DAH (1994) The influence of mycorrhizal symbiosis and fertilizer amendments on establishment of vegetation in heavy metal mine soil. Environ Pollut 86: 171-179
[80] Vivas, A; Barea, JM and Azcon, R (2005) Interactive effect of *Brevibacillus brevis* and *Glomus mosseae*, both isolated from Cd contaminated soil, on plant growth, physiological mycorrhizal fungal characteristics and soil enzymatic activities in Cd polluted soil. Environ Pollut 134: 257-266
[81] Vivas, A; Vörös, A; Biró, B; Barea, JM; Ruiz-Lozano, JM and Azcón, R (2003) Beneficial effects of indigenous Cd-tolerant and Cd-sensitive *Glomus mosseae* associated with a Cd-adapted strain of *Brevibacillus* sp. in improving plant tolerance to Cd contamination. Appl Soil Ecol 24: 177-186
[82] Binet, P; Portal, JM and Leyval, C (2000) Fate of polycyclic aromatic hydrocarbons (PAH) in rhizosphere and mycorrhizosphere of ryegrass. Plant Soil 227: 207-213
[83] Marrs, RH and Bannister, P (1978) The adaptation of *Calluna vulgaris* (L.) Hull to contrasting soil types. New Phytol 81: 753-761
[84] Bradley, R; Burt, AJ and Read, DJ (1981) Mycorrhizal infection and resistance to heavy metal toxicity in *Calluna vulgaris*. Nature 292: 335-337
[85] Bradley, R; Burt, AJ and Read, DJ (1982) The biology of mycorrhiza in the *Ericaceae*. VIII. The role of mycorrhizal infection in heavy metal resistance. New Phytol 91: 197-209
[86] Hashem, AR (1995) The role of mycorrhizal infection in the tolerance of *Vaccinium macrocarpon* to iron. Mycorrhiza 5: 451-454
[87] Hentschel, E; Godbold, DL; Marschner, P; Schlegel, H and Jentschke, G (1993) The effect of *Paxillus involutus* Fr. on aluminium sensitivity of Norway spruce seedlings. Tree Physiol 12: 379-390
[88] van Tichelen, KK; van Straelen, T and Colpaert, JV (1999) Nutrient uptake by intact mycorrhizal *Pinus sylvestris* seedlings: a diagnostic tool to detect copper toxicity. Tree Physiol 19: 189-196
[89] Dixon, RK (1988) Response of ectomycorrhizal *Quercus rubra* to soil cadmium, nickel and lead. Soil Biol Biochem 20: 555-559

[90] Cumming, JR and Weinstein, LH (1990) Aluminum-mycorrhizal interactions in the physiology of pitch pine seedlings. Plant Soil 125: 7-18

[91] Hartley-Whitaker, J; Cairney, JWG and Meharg, AA (2000) Sensitivity to Cd or Zn of host and symbiont of ectomycorrhizal *Pinus sylvestris* L. (Scots pine) seedlings. Plant Soil 218: 31-42

[92] Schier, G and McQuattie, C (1995) Effect of aluminium on the growth, anatomy, and nutrient content of ectomycorrhizal and nonmycorrhizal eastern white pine seedlings. Can J For Res 25: 1252-1262

[93] Jones, MD and Hutchinson, TC (1986) The effect of mycorrhizal infection on the response of *Betula papyrifera* to nickel and copper. New Phytol 102: 429-442

[94] Bucking, H and Heyser, W (1994) The effect of ectomycorrhizal fungi on Zn uptake and distribution in seedlings of *Pinus sylvestris* L. Plant Soil 167: 203-212

[95] Khan, AG; Kuek, C; Chaudhry, TM; Khoo, CS and Hayes, WJ (2000) Role of plants, mycorrhizae and phytochelators in heavy metal contaminated land reclamation. Chemosphere 41: 197-207

[96] Sarand, I; Timonen, S; Koivula, T; Peltola, R; Haahtela, K; Sen, R and Romantchuk, M (1999) Tolerance and biodegradation of m-toluate by Scots pine, a mycorrhizal fungus and fluorescent pseudomonads individually and under associative conditions. J Appl Microbiol 86: 817-826

[97] Hartley, J; Cairney, JWG and Meharg, AA (1997) Do ectomycorrhizal fungi exhibit adaptive tolerance to potentially toxic metals in the environment? Plant Soil 189: 303-319

[98] Jentschke, G and Godbold, DL (2000) Metal toxicity and ectomycorrhizas. Physiol Plantarum 109: 107-116

[99] Kucey, RMN and Janzen, HH (1987) Effects of VAM and reduced nutrient availability on growth and phosphorus and micronutrient uptake of wheat and field beans under greenhouse conditions. Plant Soil 104: 71-78

[100] Godbold, DL; Jentschke, G; Winter, S and Marschner, P (1998) Ectomycorrhizas and amelioration of metals stress in forest trees. Chemosphere 36: 757-762

[101] van Tichelen, KK; Colpaert, JV and Vangronsveld J (2001) Ectomycorrhizal protection of *Pinus sylvestris* against copper toxicity. New Phytol 150: 203-213

[102] Hartley-Whitaker, J; Cairney, JWG and Meharg, AA (2000) Toxic effects of cadmium and zinc on ectomycorrhizal colonization of Scots pine (*Pinus sylvestris* L.) from soil inoculum. Environ Toxicol Chem 19: 694-699

[103] Moyer-Henry, K; Silva, I; Macfall, J; Johannes, E; Allen, N; Goldfarb, B and Rufty, T (2005) Accumulation and localization of aluminium in root tips of loblolly pine seedlings and the associated ectomycorrhiza *Pisolithus tinctorius*. Plant Cell Environ 28: 111-120

[104] Chen, B; Shen, H; Li, X; Feng, G and Christie, P (2004) Effects of EDTA application and arbuscular mycorrhizal colonization on growth and zinc uptake by maize (*Zea mays* L.) in soil experimentally contaminated with zinc. Plant Soil 261: 219-229

[105] Citterio, S; Prato, N; Fumagalli, P; Aina, R; Massa, N; Santagostino, A; Sgorbati, S and Berta, G (2005) The arbuscular mycorrhizal fungus *Glomus mosseae* induces growth and metal accumulation changes in *Cannabis sativa* L. Chemosphere 59: 21-29

[106] Turnau, K; Kottke, I and Oberwinkler, F (1993) Element localization in mycorrhizal roots of *Pteridium aquilinum* (L.) Kuhn collected from experimental plots treated with cadmium dust. New Phytol 123: 313-324

[107] Joner, EJ; Briones, R and Leyval, C (2000) Metal-binding capacity of arbuscular mycorrhizal mycelium. Plant Soil 226: 227-234

[108] Marschner, P; Jentschke, G and Godbold, DL (1998) Cation exchange capacity and lead sorption in ectomycorrhizal fungi. Plant Soil 205: 93-98

[109] Frey, B; Zierold, K and Brunner, I (2000) Extracellular complexation of Cd in the Hartig net and cytosolic Zn sequestration in the fungal mantle of *Picea abies – Hebeloma crustaliniforme* ectomycorrhizas. Plant Cell Environ 23: 1257-1265

[110] Blaudez, D; Botton, B and Chalot, M (2000) Cadmium uptake and subcellular compartmentation in the ectomycorrhizal fungus *Paxillus involutus*. Microbiology 146: 1109-1117

[111] Morselt, AFW; Smits, WTM and Limonard, T (1986) Histochemical demonstration of heavy metal tolerance in ectomycorrhizal fungi. Plant Soil 96: 417-420

[112] Howe, R; Evans, RL and Ketteridge, SW (1997) Copper-binding proteins in ectomycorrhizal fungi. New Phytol 135: 123-131

[113] Courbot, M; Diez, L; Ruotolo, R; Chalot, M and Leroy, P (2004) Cadmium-responsive thiols in the ectomycorrhizal fungus *Paxillus involutus*. Appl Environ Microb 70: 7413-7417

[114] Galli, U; Meier, M and Brunold, C (1993) Effects of cadmium on non-mycorrhizal and mycorrhizal Norway spruce seedlings [*Picea abies* (L.) Karst.] and its ectomycorrhizal fungus *Laccaria laccata* (Scop. Ex Fr.) Bk. & Br.: Sulphate reduction, thiols and distribution of the heavy metal. New Phytol 125: 837-843

[115] Ott, T; Fritz, E; Polle, A and Schützendübel, A (2002) Characterisation of antioxidative systems in the ectomycorrhiza-building basidiomycete *Paxillus involutus* (Bartsch) Fr. and its reaction to cadmium. FEMS Microbiol Ecol 42: 359-366

[116] Galli, U; Schüepp, H and Brunold, C (1994) Heavy metal binding by mycorrhizal fungi. Physiol Plantarum 92: 364-368

[117] Galli, U; Schüepp, H and Brunold, C (1995) Thiols of Cu-treated maize plants inoculated with the arbuscular-mycorrhizal fungus *Glomus intraradices*. Physiol Plantarum 94: 247-253

[118] Lafranco, L; Bolchi, A; Ros, EC; Ottonello, S and Bonfante, P (2002) Differential expression of a metallothionein gene during the presymbiotic versus the symbiotic phase of an arbuscular mycorrhizal fungus. Plant Physiol 130: 58-67

[119] Gonzalez-Chavez, C; D'Haen, J; Vangronsveld, J and Dodd, JC (2002) Copper sorption and accumulation by the external mycelium of different *Glomus* ssp. (arbuscular mycorrhizal fungi) isolated from the same polluted soil. Plant Soil 240: 287-297

[120] Tam, PCF (1995) Heavy metal tolerance by ectomycorrhizal fungi and metal amelioration by *Pisolithus tinctorius*. Mycorrhiza 5: 181-187

[121] Denny, HJ and Ridge, I (1995) Fungal slime and its role in the mycorrhizal amelioration of zinc toxicity to higher plants. New Phytol 130: 251-257

[122] Ahonen-Jonnarth, U; van Hess, PAW; Lundström, US and Finlay, RD (2000) Organic acids produce by mycorrhizal *Pinus sylvestris* exposed to elevated aluminium and heavy metal concentrations. New Phytol 146: 557-567

[123] Martino, E; Franco, B; Piccoli, G; Stocchi, V and Perotto, S (2002) Influence of zinc ions on protein secretion in a heavy metal tolerant strain of the ericoid mycorrhizal fungus *Oidiodendron maius*. Mol Cell Biochem 231: 179-185

[124] Christie, P; Li, X and Chen, B (2004) Arbuscular mycorrhiza can depress translocation of zinc to shoots of host plants in soils moderately polluted with zinc. Plant Soil 261: 209-217

[125] Meharg, AA and Macnair, MR (1992) Suppression of the high affinity phosphate uptake system: a mechanism of arsenate tolerance in *Holcus lanatus* L. J Exp Bot 43: 519-524

[126] Sharples, JM; Meharg, AA; Chambers, SM and Cairney, JWG (2000) Evolution: Symbiotic solution to arsenic contamination. Nature 404: 951-952

[127] Sharples, JM; Meharg, AA; Chambers, SM and Cairney, JWG (2000) Mechanism of arsenate resistance in the ericoid mycorrhizal fungus *Hymenoscyphus ericae*. Plant Physiol 124: 1327-1334

[128] Perotto, S and Martino, E (2001) Molecular and cellular mechanisms of heavy metal tolerance in mycorrhizal fungi: what perspectives for bioremediation? Minerva Biotecnol 13: 55-63

[129] Joner, EJ; Johansen, A; Loibner, AP; de la Cruz, MA; Szolar, OHJ; Portal, JM and Leyval, C (2001) Rhizosphere effects on microbial community structure and dissipation and toxicity of polycyclic aromatic hydrocarbons (PAHs) in spiked soil. Environ Sci Technol 35: 2773-2777

[130] Joner, EJ and Leyval, C (2003) Rhizosphere gradients of polycyclic aromatic hydrocarbon dissipation in two industrial soils and the impact of arbuscular mycorrhiza. Environ Sci Technol 37: 2371-2375

[131] Salzer, P; Corbiere, H and Boller, T (1999) Hydrogene peroxide accumulation in *Medicago truncatula* roots colonized by the arbuscular mycorrhiza-forming fungus *Glomus intraradices*. Planta 208: 319-325

[132] Criquet, S; Joner, EJ; Léglize, P and Leyval, C (2000) Anthracene and mycorrhiza affect the activity of oxidoreductases in the roots and the rhizosphere of lucerne (*Medicago sativa* L.). Biotechnol Lett 22: 1733-1737

[133] Mada, RJ and Bagyaraj, DJ (1993) Root exudation from *Leucaena leucocephala* in relation to mycorrhizal colonization. World J Microb Biot 9: 342-344

[134] Jones, DL; Hodge, A and Kuzyakov, Y (2004) Plant and mycorrhizal regulation of rhizodeposition. New Phytol 163: 459-480

[135] Secilia, J and Bagyaraj, DJ (1987) Bacteria and actinomycetes associated with pot cultures of vesicular-arbuscular mycorrhizas. Can J Microbiol 33: 1069-1073

[136] Kandeler, E; Marschner, P; Tscherko, D; Gahoonia, TS and Nieksen, NE (2002) Microbial community composition and functional diversity in the rhizosphere of maize. Plant Soil 238: 301-312

[137] Vierheilig, H; Gagnon, H; Strack, D and Maier, W (2000) Accumulation of cyclohexanone derivates in barley, wheat and maize roots in response to inoculation with different arbuscular mycorrhizal fungi. Mycorrhiza 9: 291-293

[138] Hooker, JE and Atkinson, D (1996) Arbuscular mycorrhizal fungi-induced alteration to tree-root architecture and longevity. Z Pflanzenernaehr Bodenk 159: 229-234

[139] Quintero-Ramos, M; Espinoza-Victoria, D; Ferrera-Cerrato, R and Bethlenfalvay, GJ (1993) Fitting plants to soil through mycorrhizal fungi – mycorrhiza effects on plant growth and soil organic matter. Biol Fert Soils 15: 103-106

[140] Wright, SF and Upadhyaya, A (1998) A survey of soils for aggregate stability and glomalin, a glycoprotein produced by hyphae of arbuscular mycorrhizal fungi. Plant Soil 198: 97-107

[141] Joner, EJ; van Aarle, IM and Vosátka, M (2000) Phosphatase activity of extraradical arbuscular mycorrhizal hyphae: A review. Plant Soil 226: 199-210

[142] Trojanowski, J; Haider, K and Hüttermann, A (1984) Decomposition of 14C-labelled lignin, holocellulose amd lignocellulose by mycorrhizal fungi. Arch Microbiol 139: 202-206

[143] Haselwandter, K; Bobleter, O and Read, DJ (1990) Degradation of 14C-labelled lignin and dehydroplymer of coniferyl alcohol by ericoid and ectomycorrhizal fungi. Arch Microbiol 153: 352-354

[144] Cairney, JWG and Burke, RM (1994) Fungal enzymes degrading plant cell walls: their possible significance in ectomycorrhizal symbiosis. Mycol Res 98: 1345-1356

[145] Meharg, AA and Cairney, JWG (2000) Ectomycorrhizas – extending the capabilities of rhizosphere remediation? Soil Biol Biochem 32: 1475-1484

[146] Meharg, AA; Cairney, JWG and Maguire, N (1997) Mineralization of 2,4-dichlorphenol by ectomycorrhizal fungi in axenic culture and in symbiosis with pine. Chemosphere 34: 2495-2504

[147] Meharg, AA; Dennis, GR and Cairney, JWG (1997) Biotransformation of 2,4,6-trinitrotoluen (TNT) by ectomycorrhizal basidiomycetes. Chemosphere 35: 513-521

[148] Donnelly, PK; Entry, JA and Craftword, DL (1993) Degradation of atrazine and 2,4-dichlorphenoxyacetic acid by mycorrhiozal fungi at three nitrogen concentrations *in vitro*. Appl Environ Microbiol 59: 2642-2647

[149] Braun-Lüllemann, A; Hüttermann, A and Majcherczyk, A (1999) Screening of ectomycorrhizal fungi for degradation of polycyclic aromatic hydrocarbons. Appl Microbiol Biot 53: 127-132

[150] Genney, DR; Alexander, IJ; Killham, K and Meharg, AA (2004) Degradation of the polycyclic aromatic hydrocarbon (PAH) fluorene is retarded in a Scots pine ectomycorrhizosphere. New Phytol 163: 641-649

[151] Donnelly, PK and Fletcher, JS (1995) PCB metabolism by ectomycorrhizal fungi. B Environ Contam Tox 54: 507-513

[152] Green, NA; Meharg; AA; Till, C; Troke, J and Nicholson, JK (1999) Degradation of 4-fluorbiphenyl by mycorrhizal fungi as determined by F-19 nuclear magnetic resonance spectroscopy and C-14 radiolabelling analysis. Appl Environ Microbiol 65: 4021-4027

[153] Dittmann, J; Heyser, W and Bucking, H (2002) Biodegradation of aromatic compounds by white rot and ectomycorrhizal fungal species and the accumulation of chlorinated benzoic acid in ectomycorrhizal pine seedlings. Chemosphere 49: 297-306

[154] Jurkiewicz, A; Orłowska, E; Anielska, T; Godzik, B and Turnau, K (2004) The influence of mycorrhiza and EDTA application on heavy metal uptake by different maize varieties. Acta Biol Cracov Bot 46: 7-18

[155] Hovsepyan, A and Greipsson, S (2004) Effect of arbuscular mycorrhizal fungi on phytoextraction by corn (*Zea mays*) of lead-contaminated soil. Int J Phytorem 6: 305-321

ASSESSING RISKS AND CONTAINING OR MITIGATING GENE FLOW OF TRANSGENIC AND NON-TRANSGENIC PHYTOREMEDIATING PLANTS

TON ROTTEVEEL[1], HANI AL-AHMAD[2] AND JONATHAN GRESSEL[2]
[1]Plant Protection Service, NL-6700 HC Wageningen, The Netherlands and [2]Plant Sciences, Weizmann Institute of Science, Rehovot, Israel
Fax +972-8-934-4181, E-mail: Jonathan.Gressel@weizmann.ac.il

1. Introduction - needs for preventing gene flow in phytoremediating species

Plants have been used to correct human error over the ages. The few species capable of revegetating Roman lead and zinc mine tailings in Wales [2] taught us that there are a limited number of species that can withstand toxicants: some by exclusion, and others that can withstand toxic wastes after they have been taken up. Plants with the latter type mechanism are of interest for phytoremediation. Ideally, one might consider that it is best to use the species that naturally take up particular toxic wastes, but these are often slow growing (e.g. mosses, lichens, or the *Thlaspi* species that take up heavy metals) [3] or may have a potential to be weedy. If the desired wild species do not exist locally, there may be a reticence or legal issues about introducing them into the ecosystem, toxic as it may be, due to fear that the plants or their genes may spread to other areas.

Two types of multi-cut species are used, with the cut material burnt to extract the heavy metals or to oxidize the organic wastes: herbaceous species such as *Brassica juncea* and *Spartina* spp. (cord grasses), which are most efficient at dealing with surface wastes, and trees such as *Populus* spp., for dealing with deeper wastes [4]. *Brassica juncea* (Indian mustard) wild type had been used commercially, because it grows rapidly, and is easy to cultivate as a crop, but especially because of its inherent ability to take up some heavy metals. This ability has been enhanced by mutant selection (in tissue culture) for heavy metal resistance [5], from *Thlaspi* by protoplast fusion (along with many other genes) [6], but it was better yet to transgenically transfer genes leading to enhanced glutathione content [7, 8] to make the necessary phytochelatins.

A single cropping of *B. juncea* does not clean up a toxic site; many growth cycles are required, with multiple harvests and natural reseeding. *B. juncea,* even more than its close relative *B. napus (*oilseed rape) is not fully domesticated, and the multiple cycles of cropping would allow the possibility of selecting for feral forms that may persist or crossing the genes into related *Brassica* species, or cultivated varieties of Indian mustard. Thus, gene containment and/or mitigation seem necessary to prevent volunteers

M. Mackova et al. (eds.), Phytoremediation Rhizoremediation, 259–284.

from becoming feral and to prevent crossing into related species. Similarly, many oppose introducing transgenic or non-transgenic phytoremediating tree species such as poplars unless they can be prevented from establishing outside of the contaminated area or from hybridising with related native or introduced species.

The herbaceous plants, shrubs, and trees used for phytoremediation pose certain biological risks, whether transgenic or not. Many of the species are semi-domesticated and introduced from habitats far removed from the site requiring phytoremediation. Such species pose a risk of becoming established in the contaminated site after the contaminant is remediated, and also pose a risk of spread to adjacent areas, displacing native or other desirable species, or hybridising with other varieties of the same species or even other varieties of the same species. *Spartina* [9, 10] and *Populus* [11, 12] are often proposed for phytoremediation, yet they commonly form hybrids with other species in their genera. In the case of *Spartina,* the results were devastating when the new world *Spartina alterniflora* crossed with the European *S. maritime* around 1870, the hybrids massively displaced all other native species from the ecosystem [13]. Populus species easily form hybrids [14], and native species could easily be displaced by hybrids. An added concern is that transgenes in the phytoremediation species may introgress into related species. If a non-transgenic species poses a risk, the addition of specific transgenes can actually reduce the risk.

We describe below the molecular tools that can be used to contain gene flow within the bioremediation site, and separately, molecular mitigation tools that can prevent establishment of such transgenes should they leak out of the phytoremediation site, which are appropriate for non-transgenic and transgenic bioremediating species alike. Molecular solutions to gene flow problems for non-transgenic phytoremediation species may sound oxymoronic in the present climate surrounding transgenics. Still, if the scientifically determined risk of spread of a phytoremediating species outweighs the utility of the species for phytoremediation, such molecular solutions should be sought to allow effective phytoremediation while preventing gene flow.

Genes can flow from bioremediation sites in three forms – seeds carried by various vectors, vegetative propagules, and pollen. Typically pollen is thought of as the source of gene movement, but even without human intervention seeds carrying an undesirable trait can move large distances; e.g. maternally inherited triazine resistance in *Solanum nigrum* has moved 20 km per year from a single site – the distance a bird flies from eating berries to defecating [15]. Some species can move long distances as vegetative propagules, e.g. feral forms of asexually propagated Jerusalem artichoke (*Helianthus tuberosus*) have become widely spread in Europe along riverbanks [16]. The number 1 and 16th Worst Weeds of the World, *Cyperus* spp., are primarily spread asexually [17].

This review will not cover the toxicological risks of the pollutants sequestered in or vaporized from plants used for phytoremediation, or the toxicological risks of not phytoremediating a contaminated site.

2. Assessing the likelihood of risks

Species used for phytoremediation are often no ordinary agricultural species, nor are they used in agricultural contexts. In their natural surroundings any species occupies a specific niche that determines its occurrence in space, time and function of the ecosystem concerned. Genetic modification might alter a niche, and does certainly so when phytoremediation traits are attached to species formerly sensitive to the chemicals involved, and equally certain if genetic modification changes fitness. However, the basic biological traits will be the same as for any plant species.

In this section we examine the interaction between different factors that might produce an ecological risk situation. First the factors are indicated and briefly described as we did for the biotechnology derived herbicide resistant plants [18]. However, the situation is clearly more complicated since phytoremediation traits always operate under natural conditions compared to herbicide resistance, which requires that the herbicide be applied to be operational.

For contaminated sites it might be thought that there is no risk as the purpose of the use of phytoremediation plants is cleaning of the site by being grown there. However, there are many rare, sometimes endemic species that are specialised on growing on (heavy metal) containing soils. An example is *Viola calaminaria,* which is endemic in the Netherlands, Belgium and Western Germany, and completely restricted to zinc, lead, and/or cadmium containing soils [19]. Such a rare and endangered species may easily be out competed by an engineered if phytoremediating species are used on, or invade the *Viola* habitats, and in this way a loss of biodiversity may occur. Conversely, this species may be used to mine the genes for zinc phytoremediation, to be transformed into faster growing species, for use elsewhere.

The theoretical case of a remediation species becoming invasive on uncontaminated sites is currently a remote possibility but cannot completely ruled out, especially as they can flow via pollen to related species, forming hybrid swarms.

2.1. DECISION TREES FOR ASSESSING THE LEVEL OF RISK

Here we combine the different factors through the use of decision trees. We want to state clearly that a decision tree is not a quantitative tool producing a quantified risk. It is an aid in risk evaluation providing unbiased guidance in indicating hazards attached to a certain species. When a hazard is indicated, more detailed and quantified data acquisition will often be necessary.

Different levels of certainty apply to the various factors used. For instance: invasiveness is highly unpredictable whereas the presence or absence of vegetative propagation is not.

The keys are layered; first assessing the "biological hazard" (Key 1), which is equal for engineered and non-engineered species, and then we examine the extent to which measures aimed at containment and mitigation affect the total risk (Key 2). Finally, the keys are designed to assist in determining whether the growing conditions might trigger identified hazards into real risks, both on contaminated sites (Key 3) and in an uncontaminated environment (Key 4).

2.1.1 Key 1 Assessing basic hazards imposed by biology.

1\. Invasiveness
1a. The plant has no known invasive characters:

Basic Biological Hazard low, go to Key 2A
1b. The plant is a known invasive, go to 2
2. The plant is a known invasive
2a. The plant has no sexual propagation (Most species have a sexual propagation pathway. However, some species may almost never propagate sexually, or when cultivated, propagation might be entirely vegetative), go to 3
2b. The plant has sexual propagation, go to 4
3. Known invasive, without sexual propagation
3a. The plant has a proven capacity for efficient natural long range dissemination:
Basic Biological Hazard high: go to Key 2C
3b. The plant has little capacity for efficient natural long range dissemination; dissemination takes place with the aid of people:
Basic Biological Hazard medium: go to Key 2B
4. Known invasive with sexual propagation
4a. The species is cross-pollinating - go to 5
4b. The species is self- pollinating - go to 6
5. Cross pollinating invasive species
5a. The plant has a proven capacity for efficient natural long range dissemination:
Basic Biological Hazard very high: go to Key 2D
5b. The plant has little capacity for efficient natural long range dissemination; dissemination takes place with the aid of people:
Basic Biological Hazard high: go to Key 2C
6. Self pollinating invasive species
6a. The plant has a proven capacity for efficient natural long range dissemination
Basic Biological Hazard high: go to Key 2C
6b. The plant has little capacity for efficient natural long range dissemination; dissemination takes place with the aid of people:
Basic Biological Hazard medium: go to Key 2B

2.1.2 Key 2A Assessing hazards imposed by containment and mitigation measures - Basic Biological hazard low.

1. Presence of added genetic containment and mitigation measures
1a. No measures new to the species added:
Biological hazard low, go to Key 3B and Key 4
1b. New measures have been added aimed at containment and/or mitigation:
Biological Hazard very low; go to Key 3A and Key 4

2.1.3 Key 2B Assessing hazards imposed by containment and mitigation measures - Basic Biological hazard medium.

1. Presence of added genetic containment and mitigation measures
1a. No measures new to the species added. (If containment or mitigation genes present in the species gene pool are used in cultivar breeding, no new possibilities are introduced in the species, and hence the hazard to the environment is estimated as equal to the basic biological hazard): Biological Hazard medium, go to Key 3C and Key 4

1b. New measures have been added aimed at containment and/or mitigation, go to 2
2. Decreasing gene flow
2a. The plant has added genes, not in the parent, enhancing containment, go to 3
2b. The plant has added genes, not present in the parent, enhancing mitigation, go to 5
3. New containment genes present
3a. Containment genes added at random, go to 4
3b. Containment genes present as tandem constructs:
Hazard very low, go to Key 3A and Key 4
4. Random containment genes present, presence of mitigation genes
4a. The plant has added genes engineered in tandem, enhancing mitigation:
Hazard very low, go to Key 3A and Key 4
4b. The plant has added genes incorporated at random, enhancing mitigation:
Hazard very low, go to Key 3A and Key 4
4c. The plant has no mitigation genes added: Hazard low, go to Key 3b and Key 4
5. No containment genes present, presence of mitigation genes
5a. The plant has added genes engineered in tandem, enhancing mitigation:
Hazard very low: go to Key 3A and Key 4
5b. The plant has added genes incorporated at random, enhancing mitigation:
Hazard low, go to Key 3B and Key 4

2.1.4 Key 2C Assessing hazards imposed by containment and mitigation measures - Basic Biological hazard high.

1. Presence of added genetic containment and mitigation measures
1a. No measures new to the species added:
Biological hazard high: go to Key 3D and Key 4
1b. New measures have been added aimed at containment and/or mitigation, go to 2
2. Decreasing gene flow
2a. The plant has added genes, not in the parent, enhancing containment, go to 3
2b. The plant has added genes, not in the parent species, enhancing mitigation, go to 6
3. New containment genes present
3a. Containment genes added at random, go to 4
3b. Containment genes present as tandem constructs, go to 5
4. Random containment genes present, presence of mitigation genes
4a. The plant has added genes engineered in tandem, enhancing mitigation:
Hazard very low, go to Key 3A and Key 4
4b. The plant has added genes incorporated at random, enhancing mitigation:
Hazard very low, go to Key 3A and Key 4
4c. The plant has no mitigation genes added. Hazard medium, go to Key 3C and Key 4
5. Tandem containment genes and mitigation genes present
5a. The plant has added genes engineered in tandem, enhancing mitigation:
Hazard very low, go to Key 3A and Key 4
5b. The plant has added genes incorporated at random, enhancing mitigation:
Hazard very low, go to Key 3A and Key 4
5c. No mitigation genes present: Hazard low, go to Key 3B and Key 4
6. No containment genes present, presence of mitigation genes

6a. The plant has added genes engineered in tandem, enhancing mitigation:
Hazard very low: go to Key 3A and Key 4
6b. The plant has added genes incorporated at random, enhancing mitigation:
Hazard medium, go to Key 3C and Key 4

2.1.5 Key 2D Assessing hazards imposed by containment and mitigation measures - Basic Biological hazard very high.

1. Presence of added genetic containment and mitigation measures
1a. No novel measures species added:
Biological hazard very high: go to Key 3E and Key 4
1b. New measures have been added aimed at containment and/or mitigation, go to 2
2. Decreasing gene flow
2a. The plant has added genes, not in the parent, enhancing containment, go to 3
2b. The plant has added genes, not in the parent species, enhancing mitigation, go to 6
3. New containment genes present
3a. Containment genes added at random, go to 4
3b. Containment genes present as tandem constructs, go to 5
4. Random containment genes present, presence of mitigation genes
4a. The plant has added genes engineered in tandem, enhancing mitigation:
Hazard very low, go to Key 3a and 4
4b. The plant has added genes incorporated at random, enhancing mitigation:
Hazard low, go to Key 3B and 4
4c. The plant has no mitigation genes added: Hazard high, go to Key 3D and Key 4
5. Tandem containment genes present, presence of mitigation genes
5a. The plant has added genes engineered in tandem, enhancing mitigation:
Hazard very low, go to Key 3A and Key 4
5b. The plant has added genes incorporated at random, enhancing mitigation:
Hazard very low, go to Key 3A and Key 4
6. No containment genes present, presence of mitigation genes
6a. The plant has added genes engineered in tandem, enhancing mitigation:
Hazard low: go to Key 3A and Key 4
6b. The plant has added genes incorporated at random, enhancing mitigation:
Hazard high, go to Key 3D and Key 4

2.1.6 Key 3A Assessing risks to contaminated sites - Biological hazard very low.

Contaminated site refers to all sites contaminated with the compound(s) for which the plant may be used in cleaning up. There still may be unwanted side-effects to some naturally occurring contaminated sites because they are inhabited by rare and protected wild species that evolved to withstand the contamination. These species may be out-competed by the "bioremediating" species.

1a. The species (transgenic or not) needs management to survive at the site, go to 2
1b. The species (transgenic or not) survives at a site without management, go to 4
2. Species only surviving under management
2a. The species poses a risk to higher trophic levels on the site, go to 3
2b. The species poses no risk to higher trophic levels on the site:

Risk very low for contaminated sites, go to Key 4

3. Species survives under management, posing risk to higher trophic levels on the site

3a. To species confined to this specific contaminated environment:
Risk medium for contaminated sites, go to Key 4

3b. To species not confined to this specific contaminated environment:
Risk low for contaminated sites, go to Key 4

4. Species survives without management

4a. Species unable to spontaneously invade a contaminated site, go to 5

4b. Species able to invade and dominate the contaminated site, go to 7

5. Species survives but unable to invade contaminated sites

5a. The species poses a risk to higher trophic levels on the site, go to 6

5b. The species poses no risk to higher trophic levels on the site:
Risk very low for contaminated sites, go to Key 4

6. Species survives, but unable to spontaneously invade, yet poses a risk to higher trophic levels on the site

6a. To species confined to this specific contaminated environment:
Risk medium for contaminated sites, go to Key 4

6b. To species not confined to this specific contaminated environment.
Risk low for contaminated sites, go to Key 4

7. Species survives and able to spontaneously invade contaminated sites

7a. The species poses a risk to higher trophic levels on the site, go to 8

7b. The species poses no risk to higher trophic levels on the site:
Risk very low for contaminated sites, go to Key 4

8. Species survives, can spontaneously invade, and poses a risk to higher trophic levels on the site

8a. To species confined to this specific contaminated environment:
Risk medium for contaminated sites, go to Key 4

8b. To species not confined to this specific contaminated environment:
Risk low for contaminated sites, go to Key 4

2.1.7 Key 3B Assessing risks to contaminated sites - Biological hazard low.

1a. The species (transgenic or not) needs management to survive at a site, go to 2

1b. The species (transgenic or not) survives at a site without management, go to 4

2. Species only survives under management

2a. The species poses a risk to higher trophic levels on the site, go to 3

2b. The species poses no risk to higher trophic levels on the site:
Risk very low for contaminated sites, go to Key 4

3. Species survives under management posing risk to higher trophic levels on the site

3a. To species confined to this specific contaminated environment:
Risk medium for contaminated sites, go to Key 4

3b. To species not confined to this specific contaminated environment:
Risk low for contaminated sites, go to Key 4

4. Species survives without management

4a. Species unable to spontaneously invade a contaminated site, go to 5

4b. Species can invade and dominate a contaminated site, go to 7

5. Species survives but unable to spontaneously invade contaminated sites
5a. The species poses a risk to higher trophic levels on the site, go to 6
5b. The species poses no risk to higher trophic levels on the site:
Risk very low for contaminated sites, go to Key 4
6. Species survives, but unable to spontaneously invade and poses risk to higher trophic levels on the site
6a. To species confined to this specific contaminated environment:
Risk medium for contaminated sites, go to Key 4
6b. To species not confined to this specific contaminated environment:
Risk low for contaminated sites, go to Key 4
7. Species survives and able to spontaneously invade contaminated sites
7a. The species poses a risk to higher trophic levels on the site, go to 8
7b. The species poses no risk to higher trophic levels on the site:
Risk very low for contaminated sites, go to Key 4
8. Species survives and can spontaneously invade, and poses a risk to higher trophic levels on the site
8a. To species confined to this specific contaminated environment:
Risk medium for contaminated sites, go to Key 4
8b. To species not confined to this specific contaminated environment.
Risk low for contaminated sites, go to Key 4

2.1.8 Key 3C Assessing risks to contaminated sites - Biological hazard medium.

1a. The species (transgenic or not) needs management to survive at a site, go to 2
1b. The species (transgenic or not) survives at a site without management, go to 4
2. Species only survives under management
2a. The species poses a risk to higher trophic levels on the site, go to 3
2b. The species poses no risk to higher trophic levels on the site:
Risk very low for contaminated sites, go to Key 4
3. Species survives only under management, but poses a risk to higher trophic levels on the site
3a. To species confined to this specific contaminated environment:
Risk medium for contaminated sites, go to Key 4
3b. To species not confined to this specific contaminated environment.
Risk low for contaminated sites, go to Key 4
4. Species survives without management
4a. Species unable to spontaneously invade contaminated site, go to 5
4b. Species can spontaneously invade and dominate a contaminated site, go to 7
5. Species survives but unable to spontaneously invade contaminated sites
5a. The species poses a risk to higher trophic levels on the site, go to 6
5b. The species poses no risk to higher trophic levels on the site:
Risk very low for contaminated sites, go to Key 4
6. Species survives, yet unable to spontaneously invade, but poses risk to higher trophic levels on the site
6a. To species confined to this specific contaminated environment:
Risk medium for contaminated sites, go to Key 4

6b. To species not confined to this specific contaminated environment:
Risk low for contaminated sites, go to Key 4
7. Species survives and is able to spontaneously invade contaminated sites
7a. The species poses a risk to higher trophic levels on the site, go to 8
7b. The species poses no risk to higher trophic levels on the site:
Risk very low for contaminated sites, go to Key 4
8. Species survives, can spontaneously invade and poses risk to higher trophic levels on the site
8a. To species confined to this specific contaminated environment:
Risk medium for contaminated sites, go to Key 4
8b. To species not confined to this specific contaminated environment:
Risk low for contaminated sites, go to Key 4

2.1.9 Key 3D Assessing risks to contaminated sites - Biological hazard high.

1a. The species (transgenic or not) needs management to survive at a site, go to 2
1b. The species (transgenic or not) survives at a site without management, go to 4
2. Species only survives under management
2a. The species poses a risk to higher trophic levels on the site, go to 3
2b. The species poses no risk to higher trophic levels on the site:
Risk very low for contaminated sites, go to Key 4
3. Species survives only under management yet poses a risk to higher trophic levels on the site
3a. To species confined to this specific contaminated environment:
Risk medium for contaminated sites, go to Key 4
3b. To species not confined to this specific contaminated environment:
Risk low for contaminated sites, go to Key 4
4. Species survives without management
4a. Species unable to spontaneously invade a contaminated site, go to 5
4b. Species spontaneously invades and dominates a contaminated site, go to 7
5. Species survives but unable to spontaneously invade contaminated sites
5a. The species poses a risk to higher trophic levels on the site, go to 6
5b. The species poses no risk to higher trophic levels on the site:
Risk very low for contaminated sites, go to Key 4
6. Species survives, is unable to spontaneously invade, yet poses risk to higher trophic levels on the site
6a. To species confined to this specific contaminated environment:
Risk medium for contaminated sites, go to Key 4
6b. To species not confined to this specific contaminated environment:
Risk low for contaminated sites, go to Key 4
7. Species survives and can spontaneously invade contaminated sites
7a. The species poses a risk to higher trophic levels on the site, go to 8
7b. The species poses no risk to higher trophic levels on the site:
Risk very low for contaminated sites, go to Key 4
8. Species survives, can spontaneously invade and poses a risk to higher trophic levels on the site

8a. To species confined to this specific contaminated environment:
    Risk medium for contaminated sites, go to Key 4
8b. To species not confined to this specific contaminated environment:
    Risk low for contaminated sites, go to Key 4

2.1.10 Key 3E Assessing risks to contaminated sites - Biological hazard very high.

1a. The species (transgenic or not) needs management to survive at a site, go to 2
1b. The species (engineered or not) survives at a site without management, go to 4
2. Species only survives under management
2a. The species poses a risk to higher trophic levels on the site, go to 3
2b. The species poses no risk to higher trophic levels on the site:
    Risk very low for contaminated sites, go to Key 4
3. Species survives under management posing risk to higher trophic levels on the site
3a. To species confined to this specific contaminated environment:
    Risk medium for contaminated sites, go to Key 4
3b. To species not confined to this specific contaminated environment:
    Risk low for contaminated sites, go to Key 4
4. Species survives without management
4a. Species unable to spontaneously invade contaminated site, go to 5
4b. Species can spontaneously invade and dominate a contaminated site, go to 7
5. Species survives but unable to spontaneously invade contaminated sites
5a. The species poses a risk to higher trophic levels on the site, go to 6
5b. The species poses no risk to higher trophic levels on the site:
    Risk very low for contaminated sites, go to Key 4
6. Species survives, is unable to spontaneously invade, yet poses a risk to higher trophic levels on the site
6a. To species confined to this specific contaminated environment:
    Risk medium for contaminated sites, go to Key 4
6b. To species not confined to this specific contaminated environment:
    Risk low for contaminated sites, go to Key 4
7. Species survives and can spontaneously invade contaminated sites
7a. The species poses a risk to higher trophic levels on the site, go to 8
7b. The species poses no risk to higher trophic levels on the site:
    Risk very low for contaminated sites, go to Key 4
8. Species survives, can spontaneously invade, and poses a risk to higher trophic levels on the site
8a. To species confined to this specific contaminated environment:
    Risk medium for contaminated sites, go to Key 4
8b. To species not confined to this specific contaminated environment:
    Risk low for contaminated sites, go to Key 4

2.1.11 Key 4 Assessing risks to the natural, uncontaminated environment.

1. The basic biological hazard has been estimated in Key 2 as:
1a. biological hazard very low: non-invasive species, new containment and/or mitigation genes added: Risk very low

1b. Biological hazard low, go to 2
1c. Biological hazard medium, go to 3
1d. Biological hazard high, go to 5
1e. Biological hazard very high, go to 7
2. Biological hazard low – species can invade uncontaminated environment
2a. Species unable to spontaneously invade and dominate an uncontaminated site: Risk to uncontaminated environment very low
2b. Species able to spontaneously invade and dominate an uncontaminated site: Risk to uncontaminated environment low
3. Biological hazard medium - species can invade uncontaminated environment
3a. Species unable to spontaneously invade and dominate an uncontaminated site: Risk to uncontaminated environment very low
3b. Species able to invade and dominate an uncontaminated site, go to 4
4. Invasive on uncontaminated sites: risk to trophic levels
4a. The species poses a risk to higher trophic levels Risk to uncontaminated environment high
4b. The species poses no risk to higher trophic levels: Risk to uncontaminated environment medium
5. Biological hazard high - ability to invade uncontaminated environment
5a. Species unable to spontaneously invade and dominate an uncontaminated site: Risk to uncontaminated environment very low
5b. Species able to invade and dominate an uncontaminated site, go to 6
6. Invasive on uncontaminated sites: risk to trophic levels
6a. The species poses a risk to higher trophic levels: Risk to uncontaminated environment very high
6b. The species poses no risk to higher trophic levels: Risk to uncontaminated environment high
7. Biological hazard very high - ability to invade uncontaminated environment
7a. Species unable to spontaneously invade and dominate a uncontaminated site: Risk to uncontaminated environment very low
7b. Species able to invade and dominate a uncontaminated site: Risk to uncontaminated environment very high

3. Dealing with the risks

3.1. GENE FLOW

Genes do flow in nature, not only within species, but also among related species that do not readily cross, in a process coined "diagonal" gene transfer [20] to readily distinguish between vertical gene transfer in readily crossing species and horizontal gene transfer between totally unrelated species. For example, a DNA sequence typical of hexaploid wheat, found in modified form in some progenitors of wheat, was not found in >90 accessions of *Aegilops peregrina* (syn. *Ae. variabilis*) but was found in two geographically distinct populations of that species with >99% sequence identity to wheat [21]. In agroecosystems, such inadvertent gene flow may be undesirable.

There are two general approaches to dealing with gene flow: (1) "contain" the transgenes in the novel variety so that gene inflow, gene outflow or both are precluded depending on the mechanism; (2) "mitigate" gene flow effects if there are inevitable "leaks" in the containment system, which should also prevent volunteer populations of the phytoremediation species from establishing and/or reaching maturity so that they cannot evolve into problems. Most discussions so far have dealt with "containing" gene flow from managed ecosystems to "natural" ecosystems with less on "mitigation" of the effects of gene flow after it has occurred [22-27]. Only recently has discussion begun dealing with gene flow within the agroecosystems, both on preventing and mitigating endo-feral (evolution within the biotype) and exo-feral (evolution of less domesticated forms by crossing with wild or weedy forms) dedomestication of species as volunteer weeds [28]. Containment and mitigation are discussed below in the general context of bi-directional containment as well as mitigation.

3.2. CONTAINING GENE FLOW

Several molecular mechanisms have been suggested for containing gene flow (i.e., to prevent gene flow between the phytoremediating species and relatives), especially by pollen, ignoring the other routes of sporophyte propagule (seeds and asexual parts) movement, especially transgenes within the phytoremediating species (i.e., to prevent outflow to related species), or to mitigate the effects of transgene flow once it has occurred [20, 22, 26, 29]. It is more important to prevent gene flow from the phytoremediating species to outside the contaminated site than to prevent influx into the phytoremediation site, as the phytoremediating species should be most fit to live the contaminated site. Even though the hybrids may be the same in either direction, the likelihood of such a hybrid establishing on a phytoremediation site is minimal.

3.2.1 Containment by targeting genes to a cytoplasmic genome

The most widely discussed containment possibility is to integrate the transgene of choice in the plastid or mitochondrial genomes [30-32]. There are good reasons to engineer phytoremediating genes into chloroplasts besides the presumed biosafety. The chloroplasts are often the targets of environmental contaminants and need protection. Additionally, many genes of value come from bacteria with similar codon usage as chloroplasts. Such genes often need to be re-engineered to plant codon usage before inserting into the nuclear genome [33]. Indeed the bacterial genes *mer*A/*mer*B that convert organomercurials into elemental mercury (which is later volatilised) were successfully introduced into chloroplasts of tobacco [34]. Still, the same genes were active in *Arabidopsis* when the *merB* was augmented with a peptide that targeted the gene product into the endoplasmic reticulum, despite the bacterial codon usage differences [35].

The opportunity of gene outflow is limited due to the predominantly maternal inheritance of these genomes in many, but far from all species. This is presently an arduous technology, which so far is limited to a few species. It does not preclude the outside species from pollinating the bioremediating species, and then acting as the recurrent pollen parent, but this is less of a problem on a bioremediation site than off site.

The claim of strict maternal inheritance of plastome-encoded traits [32, 36, 37] was not substantiated. Tobacco [38] and other species [39] often have between a 10^{-3}–10^{-4} frequency of pollen transfer of plastid inherited traits. Pollen transmission of plastome traits can only be easily detected using both large samples and selectable genetic markers. A large-scale field experiment utilized a *Setaria italica* (foxtail or birdseed millet) with chloroplast-inherited atrazine resistance (bearing a nuclear dominant leaf marker) crossed with five different male sterile herbicide susceptible lines. Chloroplast-inherited resistance was pollen transmitted at a frequency of 3×10^{-4} in >780,000 hybrid offspring [40]. At this transmission frequency, the probability of transgene movement via plastomic gene flow is orders of magnitude greater than by spontaneous nuclear genome mutations. Thus, chloroplast transformation is probably unacceptable for preventing transgene outflow, unless stacked with additional mechanisms, and as noted above, will not at all impede gene inflow. Maliga [32] discounts the relevance of the findings with tobacco and *Setaria* as being due to an origin of the plastids from interspecific (closely related) cytoplasmic substitution, where pollen transmission barriers can break down [41]. *Setaria viridis,* the wild progenitor of *Setaria italica* is biologically con-specific with it [42]. There are two problems with this denigration of the relevance of pollen movement of plastome encoded genes: 1) it is just such interspecific movement that could be a problem between phytoremediating species and related species; 2) he [32] ignores the discussion in Darmency *et al.* [39] of cases of intraspecific transmission of plastomic traits by pollen at about the same frequency, within the same species, as reported above between species.

3.2.2 Male sterility coupled with transplastomic traits

A novel additional combination that considerably lowers the risk of plastome gene outflow within a field (but not gene influx from related strains or species) can come from utilizing male sterility with transplastomic traits [40]. Introducing plastome-inherited traits into varieties with complete male sterility would vastly reduce the risk of transgene flow, except in the small isolated areas required for line maintenance. Such a double failsafe containment method might be considered sufficient where there are highly stringent requirements for preventing gene outflow to interbreeding species adjacent to the phytoremediation sites. Plastome-encoded transgenes for non-selectable traits (e.g. for phytoremediation) could be transformed into the chloroplasts together with a trait such as tentoxin or atrazine resistance as a selectable plastome marker. With such mechanisms to further reduce out-crossing risk, plastome transformation can possibly meet the initial expectations.

3.2.3 Genetic use restriction technologies and recoverable block of function

Other molecular approaches suggested for transgene containment include: seed sterility, utilizing the genetic use restriction technologies (GURT) ('terminator gene') [43, 44], and recoverable block of function (RBF) [45] to prevent transgene flow. Such proposed technologies control both the gene influx of exo-ferality and endo-feral volunteer seed dispersal, but theoretically if the controlling element of the transgene is silenced, expression would occur, rendering a critical defect in principle and practice. The frequency of loss of such controlling elements is yet unclear, as there have been no large-scale field trials to test this.

3.2.4 Repressible seed lethal technologies

An impractical technology has been proposed to use a "repressible seed-lethal system" [46]. The seed-lethal trait and its repressor must be simultaneously inserted at the same locus on homologous chromosomes in the hybrid used for phytoremediation (in our specific case), to prevent recombination (crossing over), a technology that is not yet workable in plants. The hemizygote transgenic seed lethal parent of the hybrid cannot reproduce by itself, as its seeds are not viable. If the hybrid could be made, half the progeny would not carry the seed lethal trait (or the trait of interest linked to it) and they would have to be culled, which would not be easy without a marker gene. A containment technology should leave no viable volunteers with the transgene, but this complex technology would kill only 25% of the progeny and 50% would be like the hybrid parents and 25% would contain just the repressor. Thus, the repressor can cross from the volunteers to related weeds, and so can the trait of choice linked with the lethal, and viable hybrid plants could form. The death of a quarter of the seeds in all future generations is inconsequential to plants that copiously produce seed, as long as the transgenic trait provides some selective advantage.

In summary, none of the above containment mechanisms is absolute, but the risk could be reduced by stacking a combination of containment mechanisms, compounding the infrequency of gene introgression. Still, even at very low frequencies of gene transfer, once gene transfer occurs, the new bearer of the transgene could disperse throughout the population if it has just a small fitness advantage.

3.2.5 Transient transgenics

It is possible to insert certain phytoremediation traits encoding transgenes on RNA viruses or in endomycorrhizae that are expressed in the plant, but are not carried through meiosis into reproductive cells, and thus there will be no gene flow via seeds or pollen. Attempts had been made to use endophytes to carry useful genes into plants by pressure-infiltrating the endophytes into seeds [47, 48]. The advantage of the technology was that it was not variety specific, such that indigenous species or varieties can be used. There are endophytes that naturally participate in phytoremediation processes, e.g. the *Methylobacterium* sp. that inhabits poplars and degrades explosives [49]. Genes from this or similar species can and have been engineered into other endophytic bacteria, with quite promising results [50, 51].

The same or other infection procedures could be used to introduce phytoremediation traits by disarmed plant disease viruses as the vector. The possibility that such a procedure might work was borne out in many cases with dicots showing that they express virus-encoded genes, e.g. [52]. It was possible to infect *Arabidopsis* with tobacco etch virus carrying the *bar* gene; the gene was fully expressed in the plants [53].

Cucurbits artificially infected with an attenuated zucchini yellow mosaic potyvirus containing the same transgene s were resistant in the field [54]. An NTPII carrying wheat streak mosaic virus was used to infect various grains, and the gene was expressed (immunologically) [55]. The virus carrying the genes was expressed in the roots following leaf infection, though not in all tissues.

Considerable technological obstacles of infection of the phytoremediating species will have to be worked out. While no gene flow from the plants is expected, endophytic bacteria are prone to horizontal gene transfer among themselves, an issue which bacterial biosafety experts will have to consider. There are biosafety issues relating to the mode of disarming to be considered, and it must be demonstrated that there is no gene introgression from the virus to the plant chromosomes, as well as no-extra-nuclear transmission of the virus through ovules or pollen in very large numbers of individuals. It is necessary to transfect the phytoremediating species every generation, which may be easier with perennials, such as poplar, and it may be more cumbersome with annual species.

3.3. PREVENTING ESTABLISHMENT BY TRANSGENIC MITIGATION

If a transgene confers even a small fitness disadvantage, the less fit transgenic volunteers and their own or hybrid progeny should only be able to exist as a very small proportion of the population. Therefore, it should be possible to mitigate volunteer establishment and gene flow by lowering the fitness of transgene recipients below the fitness of competitors, so that the volunteer or hybrid offspring will reproduce with considerably less success than its non-transgenic competitors. A concept of "transgenic mitigation" (TM) was proposed [22], in which mitigator genes are linked or fused to the desired primary transgene. Thus, a transgene with a desired trait is directly linked to a transgene that decreases fitness in volunteers (Fig. 1). TM could also be used as a stand-alone procedure with non-transgenic phytoremediating species to reduce the fitness advantage of hybrids and their rare progeny, and thus substantially reduce the risk of exo-feral hybrid volunteer persistence.

This TM approach is based on the premises that: 1) tandem constructs act as tightly linked genes, and their segregation from each other is exceedingly rare; 2) the gain of function dominant or semi-dominant TM traits chosen are neutral or favourable to phytoremediating species, but deleterious to volunteer progeny and their hybrids due to a negative selection pressure; and 3) individuals bearing even mildly harmful TM traits will be kept at very low frequencies in volunteer/hybrid populations because strong competition with their own wild type or with other species should eliminate even marginally unfit individuals, and prevent them from persisting in the field population [22].

Thus, it was predicted that if the primary gene(s) for phytoremediation advantage being engineered into a phytoremediating species or a crop will not persist in future generations if it is flanked by TM gene(s), such as genes (for crops) encoding dwarfing, strong apical dominance to prevent tillering (in grains) or multi-heading (in crops like sunflowers), determinate growth, non-bolting genes, uniform seed ripening, non-shattering, anti-secondary dormancy. When they are in such a tandem construct, the overall effect would be deleterious to the volunteer progeny and to hybrids. Indeed a

TM gene such as anti-shattering should decrease re-seeding, and thus the number of initial volunteers. With crops or phytoremediating species there is typically a small amount of shattering due to imperfect harvesting equipment, which may leave a few seeds behind. Because the TM genes will reduce the competitive ability of the rare hybrids, they should not be able to compete and persist in easily measurable or biologically significant frequencies in agroecosystems [20, 22].

Once TM genes are isolated, the actual cost of cloning them into TM constructs is minimal, compared to the total time and effort in producing a transgenic phyto-remediating species. The cost is even inconsequential in systems where biolistic co-transformation allows introducing genes into the same site such that the tandem construct is made by the plant.

3.3.1 Demonstration of Transgenic Mitigation in tobacco and oilseed rape

We used tobacco (*Nicotiana tabacum*) as a model plant to test the TM concept: a tandem construct was made containing an *ahas*^{R} (acetohydroxy acid synthase) gene for herbicide resistance as the primary desirable gene of choice, and the dwarfing Δ*gai* (gibberellic acid-insensitive) truncated gene as a mitigator [23].

Dwarfing would be disadvantageous to the rare weeds introgressing the TM construct, as they could no longer compete, but is desirable in many crops, preventing lodging and producing less stem with more leaves. The dwarf and herbicide resistant TM transgenic hybrid tobacco plants (simulating a TM introgressed hybrid) were more reproductive than the wild type when cultivated alone (without herbicide). They formed many more flowers than the wild type when cultivated by themselves, which is indicative of a higher harvest index. Conversely, the TM transgenics were weak competitors and highly unfit when co-cultivated with the wild type in ecological simulation of competition. The inability to achieve flowering on the TM plants in the competitive situation resulted in zero reproductive fitness of the TM plants grown in an equal mixture with the wild type at typical field spacing of plants resulting from seed rain of volunteer weeds [23].

From the data above it is clear that transgenic mitigation should be advantageous to a phytoremediation species growing alone, while disadvantageous to a hybrid with it living in the competitive environment of the phytoremediation site, or off site. If a rare pollen grain bearing tandem transgenic traits bypasses containment, it must compete with multitudes of wild type pollen to produce a hybrid. Its rare progeny must then compete with more fit wild type cohorts during self-thinning and establishment. Even a small degree of unfitness encoded in the TM construct would bring about the elimination of the vast majority of progeny in all future generations, as long as the primary gene provides no selective advantage that counterbalances the unfitness of the linked TM gene. Most phytoremediating genes have a drag, not an increased fitness off the phytoremediation site. We have inserted the same construct into oilseed rape and have tested the selfed progeny, as well as hybrids with the weed *Brassica campestris*

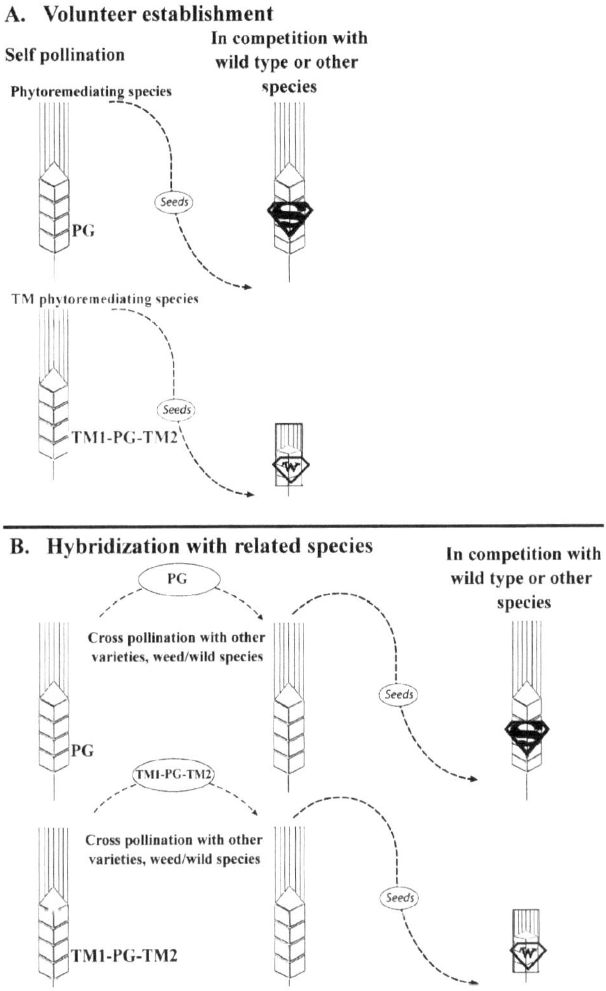

Fig. 1. Transgenic Mitigation to prevent establishment of (A) volunteers and (B) hybrids between phytoremediation species and relatives. The phytoremediation species bears desirable transgenes coupled in tandem with transgenes encoding traits that are neutral or positive for the phytoremediating species, but render volunteers or hybrids unfit to compete outside of cultivation. Source: From ref. [1], with permission of Springer Verlag.

x *B. rapa*. When cultivated alone, the dwarf transgenic oilseed rape grew at almost the same rate as the transgenic (Fig. 2A), but produced twice as much seed as the non-transgenic isoline (Fig. 2C). When the TM transgenic oilseed rape plants were co-cultivated in competition with the wild type, they were unable to grow normally (Fig. 2B), and hardly set seed (Fig. 2C) because they were so unfit to reproduce.

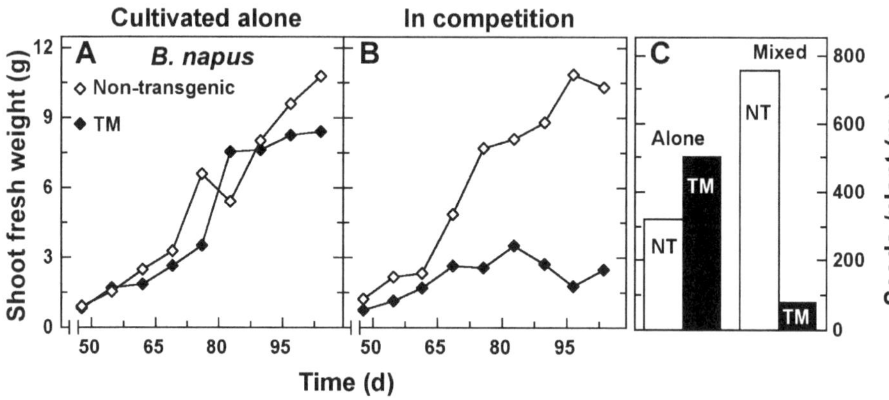

Fig. 2. Suppression of B. growth and C. seed yield of TM (transgenic mitigator) bearing oilseed rape plants carrying a dwarfing gene in tandem with a herbicide resistance gene (closed symbols and bars) when in competition with non-transgenic plants (open symbols and bars)), and A. near-normal growth of the transgenics and C. much higher seed yield of the transgenics when cultivated separately without herbicide at 3 cm spacing in a biocontainment screenhouse. (Unpublished data: Al-Ahmad and Gressel, 2005).

The rare hybrid offspring from escaped pollen bearing transgenic mitigator genes would not pose a dire threat, especially to wild species outside fields, as the amount of pollen reaching the pristine wild environment would only be at a minuscule fraction of the pollen from the wild type. This is dependent on the distance, source size, and on fertility barriers. Large-scale cultivation creates large pollen sources, and in theory a wild population having its niche on "the edge of agriculture" with coincident pollen shed could be swamped. There has been pollen flow, but no swamping with native DNA of wheat sporadically appearing in a ruderal *Aegilops* sp. [21]. Presently, there are no well documented cases where fertility barriers do not prevent more than the formation of a few infertile hybrids near the borders, as well as the rare introgressions, as have been happening for time immemorial. Any unfit hybrids and their rare backcross offspring containing transgenes linked to TM genes should still be eliminated. Further large-scale field studies will be needed with crop/weed pairs to continue to evaluate the positive implications of risk mitigation.

3.3.2 Risk that introgression of TM traits will affect relatives of the phytoremediating species

A model by Haygood *et al.* [56] claims to "prove" the premise that "demographic swamping" by transgenes would cause "migrational meltdown" of wild species related to the crop or phytoremediating species, especially if the introgressed genes confer unfitness. This proposition that recurrent gene flow from crops or phytoremediating species, even TM gene flow, could affect wild relatives deserves some discussion, as it negates the concept of transgenic mitigation.

They claim that their model demonstrates that recurrent gene flow from transgenic crops or phytoremediating species with less fit genes will cause wild populations to shrink. Firstly, conventional crops already belie this possibility. There are few if any major domesticated crops that are fit to live in a wild ecosystem, so their normal genes should confer a modicum of unfitness. Such crop x wild hybrids continually form, yet no evidence is presented that demographic swamping has occurred due to recurrent gene flow from the crops or phytoremediating species, nor could we locate any published data to that effect. Indeed, considerable evidence has been presented that many crops exist near their wild or weedy progenitors, without causing the extinction of the progenitors, despite gene flow.

There are other mundane yet fatal flaws in their model based on shaky premises and assumptions not borne out by plant biology. Three problematic issues that seem to invalidate the relevance of their model for the vast majority of conceivable crop or phytoremediating species/wild species systems, are discussed below:

- to get the level of swamping that they [56] discuss, the wild relative and the phytoremediating species would have to live in the same ecosystem. There are typically geographic separations between phytoremediation ecosystems and wild ecosystems, with the extent of pollen flow decreasing exponentially with distance between them – usually to a low asymptote due to wind currents or insects not fully following simple physics. There should always be far more wild pollen in the wild ecosystems, so hybridisation events in the wild from crop pollen will be rare, even with masses of pollen occurring within the agroecosystem. Thus their basic assumption of transgenic pollen swamping wild type pollen in the wild is invalid. Indeed, even when they assume an enormous 10% of hybridisations in the wild each generation coming from transgenic pollen, according to their model it will take about 20 generations of recurrent pollination for the unfit allele to become fixed in half the population, and 50 generations for an unfit gene to asymptotically reach 80% of the population. As discussed below, their other assumptions leading to these numbers are also off target, so it should actually take much longer;
- they assume synchronous flowering, no self-fertilization, and no genetic or other barriers to cross-fertilization; indeed, this negates the definition of speciation. It is exceedingly rare for pollen from one species to fertilize another species without any genetic barrier in the wild relative. Of the species mentioned in [57], this might only occur with con-specific wild sunflowers, which might fit this criterion, but even in this case there are genomic deterrents to introgression (as reviewed in [26]. The flow of genes between con-specific rice and red (weedy)-rice does not fit their assumptions because they are cleistogamous, predominantly self-fertilizing before the flowers open, and the amount of outcrossing possible is very low. Of course weedy rice is not a wild species (by definition), so it too is not really relevant to their case. There are fertilization barriers of different chromosome numbers, non-homology etc, which limit fertilization of wild relatives of oilseed rape and wheat, so they are outside the models;
- their models assume animal-type replacement rates – a few progeny per mating, where lower fitness can indeed become fixed. Most wild relatives

> of phytoremediating herbaceous or tree species produce copious amounts of seed to replace parents. Hundreds to thousands typically germinate in the area occupied by a parent and the process of self-thinning is ferociously competitive, eliminating less fit individuals. Our experimental data show that at realistic seed output and seeding rates, unfit individuals are eliminated or remain at a low frequency, just as unfit mutations are maintained in populations at some low frequency (the relative fitness multiplied by the mutation frequency).

Their conclusion that "the most striking implication of this model is the possibility of thresholds and hysteresis, such that a small increase in (unfit gene) immigration can lead to fixation of a disfavoured crop allele….." [56] flies in the face of evolutionary evidence, and decades of classic and contemporary field data showing that only near-neutral genes exist in pockets of the evolutionary landscape of plants, and blatantly unfit plant genes are not known to exist in such pockets unless all the fit genes are somehow removed. Just as endogenous unfavored gene mutations exist in the wild at a frequency lower than the mutation rate, transgenes from phytoremediating species that have a fitness penalty will exist in the wild at a rate lower than the immigration rate. As discussed above, the immigration rate to the wild is perforce very low. Unfit genes are eliminated from populations of plants that produce large numbers of seeds, whereas the genes could be fixed in populations of animals with few progeny. When a model contradicts reams of data, it is more likely than not that the model is invalid.

Haygood *et al.* [56] further contend that their model would work if the phytoremediating species were heterozygous for the unfit gene (and many transgenic hybrids have the transgene in a single parent and are thus hemizygous). The data in Fig. 2 clearly demonstrate that when even half of the backcross progeny contain a TM construct, they cannot compete with their non-transgenic sibs, let alone the wild type. Part of the problem may be that Haygood *et al.* [56] (p. 1880 column 2) "assume (that) the number of plants surviving to maturity does not vary from one generation to the next", a questionable assumption for unfit phenotypes when they must compete with fit cohorts and other species.

In summary, where might their model have some validity? Even though, despite their claims, the model has limited validity for the "wild" ecosystems, the model might be valid for a few weeds (not wild species) related to phytoremediating species. Weeds are man-made domesticated species (of a sort), and they are dependent on human controlled agro-ecosystems. These systems change continuously, which leads to continual shifting weed populations with an ever-changing composition. Over time new species invade, and old species go extinct, adapt, or are once more confined to their original natural environments. This is the nature of agriculture itself. It is likely that weeds that are evolutionarily threatened by the flow of unfit genes would evolve exclusionary mechanisms that block extinction;, e.g., they could evolve a shift to predominant self-fertilization that would protect them from transgenic pollen bearing unfit genes. The model of Haygood *et al.* [56] may be right for certain animal systems but irrelevant for the vast majority of plant systems. They fail to mention specific plant systems where their model might be valid. Indeed, the species that naturally phytoremediate mine sites (for the last 2000 years in the case of Roman sites) are so unfit to compete off of mine sites that the heavy metal resistant genes are not found in the same species of wind-pollinated grasses a few cm from the edge of mine tailings [58]. Some pollen flowed, but the hybrid offspring cannot compete with wild-type offspring.

3.3.3 Following transgene flow to volunteers and feral forms

Using the various containment and mitigation strategies it should be possible to keep transgene "leaks" below risk thresholds, which have to be specified by science-based regulators on a case-to-case basis. As the numbers of transgenic species being released is increasing, and the problems of monitoring for such genes increases geometrically, we suggested that a uniform biobarcode™ system be used, where a small piece of non-coding DNA having uniform recognition sites are at the ends (for single PCR primer pair amplification) with an assigned variable region in between. Thus, PCR-automated sequencing could be used to determine the origin of "leaks", contamination, liability, as well as intellectual property violations [59].

4. Special transgenic mitigation genes for phytoremediation

As more genes become isolated and their properties elucidated, it appears that many might be specifically utilizable to contain and mitigate gene flow in plants used for phytoremediation. Some genes that can be used for containment might be better used for mitigation. For example, various *Populus* species have been genetically engineered and field-tested out of doors for heavy metal tolerance or for metabolising halogenated hydrocarbons, as well as male sterility, and lack of fertility [60], but necessarily linked in tandem, so the traits can segregate. Male sterility and lack of fertility can prevent gene outflow, albeit typically leaky. Thus, some pollen bearing the phytoremediation traits can escape to the wild, and some pollen from the wild can fertilize the few flowers appearing on a tree. In the case of vegetatively propagated species such as poplars, male sterility can be coupled with female sterility, which will prevent pollen from nearby related species from effectively pollinating the phytoremediating poplar. Additionally, floral ablation can be used (no pollination in either direction) can be used, as described in a review of the earlier literature [61]. A presently used cytotoxin gene under the control, of a PTD flower promoter imparts "high levels" of floral ablation in poplar, a species commonly used for phytoremediation [62], with complete loss of flower buds in some lines tested in the greenhouse, in plants also engineered for early flowering. Whether they are leaky and allow some flowering as plants mature is being tested in field trials now in progress (S.H. Strauss, Oregon State Univ., pers. comm. 2004). If the infertility is not 100% and the genes are just used for containment, i.e., not engineered in a tandem construct with the phytoremediation genes, the infertility genes can segregate from the phytoremediation genes in further generations, giving fertile plants with the phytoremediation traits. If the same infertility genes are engineered in a tandem construct or in such a way that they will be linked *in planta* (as happens with most biolistic co-transformants), the two sets of traits will remain linked, and the rare escapee bearing infertility and phytoremediation will remain “mitigated”, i.e., in a perennially low proportion of the population.

Some traits are appropriate containing/mitigating both tree, shrub, and herbaceous phytoremediating plants, for example: the overexpression of a cytokinin oxidase [63], which reduces the levels of isopentenyl and zeatin type cytokinins. This in turn leads to phenotypes with far reduced shoot systems (unfitness to compete) but with faster growing more extensive root systems [64], all the better for extracting toxic wastes.

Irreversible sterility is best for trees and shrubs that can be vegetatively propagated, reversible male sterility is better for herbaceous species, as it allows seed production, as described below.

4.1. CONTAINMENT/MITIGATION FOR HERBACEOUS PHYTOREMEDIATION AGENTS

Mitigating genes should easily prevent or delay flowering in rosette type herbaceous species such as the *Brassica* spp. that are two phase species, where the vegetative material is harvested, and flowering (bolting) is detrimental. This could easily be effected by preventing gibberellic acid biosynthesis [65], either in a TM construct and/or by permanent mutation of the kaurene oxidase gene using a chimeraplastic gene conversion system [66], a system that as yet is hard to use in plants. Kaurene oxidase suppression would require the use of gibberellic acid to ‘force’ flowering for seed production. There should be a concomitant biosafety requirement that seed production areas be far removed from areas where weedy or other feral or wild relatives grow to prevent pollen transfer.

Delaying of bolting and flowering by using a different transgene has recently been demonstrated. Curtis *et al.* [67] engineered a fragment of the *GIGANTEA* gene, the gene encoding a protein that is part of the photoperiod recognition system, into radish using an antisense approach. Bolting was considerably delayed, and thus seed production could come about without reversal mechanisms if seed producers waited long enough. If despite all isolation distances, a TM construct or a mutant in a seed production area introgresses with a wild species, the progeny will also be delayed, i.e., the transgenic hybrid would be non-competitive with cohorts.

4.2. SPECIAL CONTAINMENT/MITIGATION GENES FOR PHYTOREMEDIATING TREES

In forestry, the possibility of gene flow is especially problematic as the duration until long-term implications of gene movement become apparent can be longer than human lifetimes. The introgression of traits from these species to wild populations has been extensively discussed by [20, 68] and thus containment/mitigation requirements should be stringent. Some phytoremediating species such as the poplars are vegetatively propagated and thus flowers and seeds are not important – indeed may provide a metabolic/genetic drag. Such phytoremediating trees can be vegetatively propagated, and if sterile, besides possibly higher yield and biosafety, allergy-causing pollen clouds and messy fruits would be prevented. An ideal gene for doing this is barnase under the

T29 tapetum-specific promoter [69]. The ribonuclease is only produced in the tapetum and prevents pollen formation with no other ill effects.

If one has an important phytoremediating species in which transgenics are exceedingly worthwhile, yet the risks of cultivation too great, one could envisage using a pollen sterility system coupled with flower drop, as described above and the crop could be propagated by artificial seed, e.g., artificially encased somatic embryos produced in mechanized tissue culture systems. As noted above, such genes are being tested [60], but whether in tandem with phytoremediation traits, or separate is not clear.

Poplar height is under control of gibberellic acid, just as it is with herbaceous species [70]. The GAI and related dwarfism genes are thus being tested in poplar to ascertain whether the shorter, fatter trees concept cited will grow any faster and be less competitive under competition. So far a field trial has been growing for one year and the researches at Oregon State University have many short, fattish trees (size varies from 1/3 to 2m)...but it will take several more years to ascertain the capacity to mitigate (Steven Strauss, personal communication, 2004). They believe that better genes or more specific promoters may be needed to really make the concept work. The professional foresters are quite sceptical, given that tall and straight trees is what they have been taught to seek all their careers (Steven Strauss, personal communication, 2004).

Another approach by scientists at Oji Paper Company in Japan for an analogous situation has been announced (in a news release) [71]. They engineered *Eucalyptus* to withstand very acid soils, and graft non-transgenic rapidly growing *Eucalyptus* on the transgenic acid-tolerant rootstock. There can be no transgene flow from these plants, unless suckers or shoots form on the rootstocks. Similar grafting approaches could be used with many bioremediating tree species.

5. Concluding remarks

Systems exist that can theoretically preclude a phytoremediating species from becoming established outside the contaminated area being treated, whether by containing gene flow or by preventing the establishment of hybrids by mitigation. There is evidence that some of these systems are efficient in crops, and there is no reason they could not be used in phytoremediating species, where a risk of transgene flow is perceived. Thus, if a risk of establishment is discerned using the enabling decision tree proved above, such a risk should not preclude developing transgenic phytoremediation species – it should stimulate the imagination to devise and test systems to deal with the potential problems.

Acknowledgements

The second and last authors' research on transgenic mitigation was supported by the Levin Foundation, by INCO–DC contract no. ERB IC18 CT 98 0391, and by a bequest from Israel and Diana Safer. This chapter is a heavily augmented and updated version of Gressel and Al-Ahmad [1], and updated portions from that review are included with the permission of the Springer Verlag.

References

[1] Gressel, J and Al-Ahmad, H (2005) Assessing and managing biological risks of plants used for bioremediation, including risks of transgene flow. Z Naturforsch C, 60: 154-165
[2] Smith, RAH and Bradshaw, AD (1979) Use of metal tolerant plant populations for the reclamation of metalliferous wastes. J Appl Ecol 16: 595-603
[3] Kramer, U; Smith, RD; Wenzel, WW; Raskin, I and De, S (1997) The role of metal transport and tolerance in nickel hyperaccumulation by *Thlaspi goesingense* Halacsy. Plant Physiol 115: 1641-1650
[4] Pilon-Smits, E and Pilon, M (2002) Phytoremediation of metals using transgenic plants. Crit Rev Plant Sci 21: 439-456
[5] Schulman, RN; Salt, DE and Raskin, I (1999) Isolation and partial characterization of a lead-accumulating *Brassica juncea* mutant. Theor Appl Genet 99: 398-404
[6] Dushenkov, S; Skarzhinskaya, M; Glimelius, K; Gleba, D and Raskin, I (2002) Bioengineering of a phytoremediation plant by means of somatic hybridization. Intl J Phytoremed 4: 117-126
[7] Zhu, YL; Pilon-Smits, EAH; Tarun, AS; Weber, SU; Jouanin, L and Terry, N (1999) Cadmium tolerance and accumulation in Indian mustard is enhanced by overexpressing gamma-glutamylcysteine synthetase. Plant Physiol 121: 1169-1177
[8] Bennett, LE; Burkhead, JL; Hale, KL; Terry, N; Pilon, M and Pilon-Smits, EAH (2003) Analysis of transgenic Indian mustard plants for phytoremediation of metal-contaminated mine tailings. J Envir Qual 32: 432-440
[9] Weis, JS and Weis, P (2004) Metal uptake, transport and release by wetland plants: implications for phytoremediation and restoration. Envir Intl 30: 685-700
[10] Rugh, CL; Gragson, GM; Meagher, RB and Merkle, SA (1998) Toxic mercury reduction and remediation using transgenic plants with a modified bacterial gene. Hortscience 33: 618-621
[11] Bittsanszky, A; Komives, T; Gullner, G; Gyulai, G; Kiss, J; Heszky, L; Radimszky, L and Rennenberg, H (2005) Ability of transgenic poplars with elevated glutathione content to tolerate zinc(2+) stress. Envir Intl 31: 251-254
[12] Schoenmuth, BW and Pestemer, W (2004) Dendroremediation of trinitrotoluene (TNT) - Part 2: Fate of radio-labelled TNT in trees. Envir Sci Pollution Res 11: 331-339
[13] Gray, AJ; Marshall, DF and Raybould, AF (1991) A century of evolution in *Spartina angelica*. Adv Ecol Res 21: 1-62
[14] Kowarik, I (2005) Urban Ornamentals Escaped from Cultivation, in J Gressel, Ed., Crop Ferality and Volunteerism, pp. 97-121, CRC Press, Boca-Raton
[15] Stankiewicz, M; Gadamski, G and Gawronski, SW (2001) Genetic variation and phylogenetic relationships of triazine-resistant and triazine-susceptible biotypes of *Solanum nigrum* - analysis using RAPD markers. Weed Res 41: 287-300
[16] Bervillé, A; Muller, M-H; Poinso, B and Serieys, H (2005) Ferality : risks of gene flow between sunflower and other *Helianthus* species, in J Gressel, Ed., Ferality and volunteerism in plants, pp. 209-230, CRC Press, Boca-Raton
[17] Holm, LG; Plucknett, JD; Pancho, LV and Herberger, JP (1977) The World's Worst Weeds, Distribution and Biology. University Press of Hawaii (Vol. pp. 609) Honolulu
[18] Gressel, J and Rotteveel, T (2000) Genetic and ecological risks from biotechnologically-derived herbicide resistant crops: decision trees for risk assessment. Plant Breeding Rev 18: 251-303
[19] Bizoux, JR; Brevers, F; Meerts, P; Graitson, E and Mahy, G (2004) Ecology and conservation of Belgian populations of *Viola calaminaria*, a metallophyte with a restricted geographic distribution. Belgian J Bot 137: 91-104
[20] Gressel, J (2002) Molecular biology of weed control, Taylor and Francis, London
[21] Weissmann, S; Feldman, M and Gressel, J (2003) Evidence for sporadic introgression of a DNA sequence from polyploid wheat into *Aegilops peregrina* (*Ae. variabilis*), in Proceedings - 10th International Wheat Genetics Symposium, Paestum (Italy), 539-542
[22] Gressel, J (1999) Tandem constructs: preventing the rise of superweeds. Trends Biotech 17: 361-366
[23] Al-Ahmad, H; Galili, S and Gressel, J (2004) Tandem constructs to mitigate transgene persistence: tobacco as a model. Mol Ecol 13: 697-710
[24] Al-Ahmad, H; Galili, S and Gressel, J (2005) Poor competitive fitness of transgenically mitigated tobacco in competition with the wild type in a replacement series. Planta (in press)

[25] Ellstrand, NC (2003) Dangerous liaisons – when cultivated plants mate with their wild relatives, Johns Hopkins University Press, Baltimore MD

[26] Stewart, CN; Halfhill, MD and Warwick, SI (2003) Transgene introgression from genetically modified crops to their wild relatives. Nature Rev Genet 4: 806-817

[27] Jenszewski, E; Ronfort, J and Chevre, AM (2003) Crop-to-wild gene flow, introgression, and possible fitness effects of transgenes. Envir Biosafety Res 2: 9-24

[28] Gressel, J and Al-Ahmad, H (2005) Molecular mitigation of ferality, in J Gressel, Ed., Crop Ferality and Volunteerism, pp. 371-388, CRC Press, Boca-Raton

[29] Daniell, H (2002) Molecular strategies for gene containment in transgenic crops. Nat Biotechnol 20: 581-586

[30] Khan, MS and Maliga, P (1999) Fluorescent antibiotic resistance marker for tracking plastid transformation in higher plants. Nat Biotechnol 17: 910-915

[31] Maliga, P (2002) Engineering the plastid genome of higher plants. Curr Opin Plant Biol 5: 164-172

[32] Maliga, P (2004) Plastid transformation in higher plants. Annu Rev Plant Biol 55: 289-313

[33] Tian, JL; Shen, RJ and He, YK (2002) Sequence modification of merB gene and high organomercurial resistance of transgenic tobacco plants. Chinese Sci Bull 47: 2084-2088

[34] Ruiz, ON; Hussein, HS; Terry, N and Daniell, H (2003) Phytoremediation of organomercurial compounds via chloroplast genetic engineering. Plant Physiol 132: 1344-1352

[35] Bizily, SP; Kim, T; Kandasamy, MK and Meagher, RB (2003) Subcellular targeting of methylmercury lyase enhances its specific activity for organic mercury detoxification in plants. Plant Physiol 131: 463-471

[36] Daniell, H; Datta, R; Varma, S; Gray, S and Lee, SB (1998) Containment of herbicide resistance through genetic engineering of the chloroplast genome. Nat Biotechnol 16: 345-348

[37] Bock, R (2001) Transgenic plastids in basic research and plant biotechnology. J Mol Biol 312: 425-438

[38] Avni, A and Edelman, M (1991) Direct selection for paternal inheritance of chloroplasts in sexual progeny of *Nicotiana*. Mol Gen Genet 225: 273-277

[39] Darmency, H (1994) Genetics of herbicide resistance in weeds and crops, in SB Powles and JAM Holtum, Eds., Herbicide Resistance in Plants: Biology and Biochemistry, pp. 263-298, Lewis, Boca-Raton

[40] Wang, T; Li, Y; Shi, Y; Reboud, X; Darmency, H and Gressel, J (2004) Low frequency transmission of a plastid encoded trait in *Setaria italica*. Theor Appl Genet 108: 315-320

[41] Kiang, A-S; Connolly, V; McConnell, DJ and Kavanagh, TA (1994) Paternal inheritance of mitochondria and chloroplasts in *Festuca pratensis-Lolium perenne* intergeneric hybrids. Theor Appl Genet 87: 681-688

[42] Darmency, H (2005) Incestuous relations of foxtail millet (*Setaria italica*) with its parents and cousins, in J Gressel, Ed., Crop Ferality and Volunteerism, pp. 81-96, CRC Press, Boca-Raton

[43] Oliver, MJ; Quisenberry, JE; Trolinder, NLG and Keim, DL (1998) Control of plant gene expression. US Patent 5,723,765

[44] Crouch, ML (1998) How the terminator terminates: An explanation for the non-scientist of a remarkable patent for killing second generation seeds of crop plants http://www.bio.indiana.edu/people/terminator/html, The Edmonds Institute, Edmond WA, USA

[45] Kuvshinov, V; Koivu, K; Kanerva, A and Pehu, E (2001) Molecular control of transgene escape from genetically modified plants. Plant Sci 160: 517-522

[46] Schernthaner, JP; Fabijanski, SF; Arnison, PG; Racicot, M and Robert, LS (2003) Control of seed germination in transgenic plants based on the segregation of a two-component genetic system. Proc Natl Acad Sci USA 100: 6855-6859

[47] Fahey, JW and Anders, J (1995) Delivery of beneficial clavibacter microorganisms to seeds and plants. US Patent 5,415,672

[48] Tomasino, SF; Leister, RT; Dimock, MB; Beach, RM and Kelly, JL (1995) Field performance of *Clavibacter xyli* subsp. *cynodontis* expressing the insecticidal protein gene *cryIA(c)* of *Bacillus thuringiensis* against european corn borer in field corn. Biol Control 5: 442-448

[49] Van Aken, B; Yoon, JM and Schnoor, JL (2004) Biodegradation of nitro-substituted explosives 2,4, 6-trinitrotoluene, hexahydro-1,3,5-trinitro-1,3,5-triazine, an octahydro-1,3,5,7-tetranitro-1,3, 5-tetrazocine by a phytosymbiotic *Methylobacterium* sp. associated with poplar tissues (*Populus deltoides* x *nigra* DN34). Appl Environ Microbiol 70: 508-517

[50] Barac, T; Taghavi, S; Borremans, B; Provoost, A; Oeyen, L; Colpaert, JV; Vangronsveld, J and van der Lelie, D (2004) Engineered endophytic bacteria improve phytoremediation of water-soluble, volatile, organic pollutants. Nat Biotechnol 22: 583-588
[51] Newman, LA and Reynolds, CM (2005) Bacteria and phytoremediation: new uses for endophytic bacteria in plants. Trends Biotechnol 23: 6-8
[52] Koo, M; Bendahmane, M; Lettieri, GA; Paoletti, AD; Lane, TE; Fitchen, JH; Buchmeier, MJ and Beachy, RH (1999) Protective immunity against murine hepatitis virus (MHV) induced by intranasal or subcutaneous administration of hybrids of tobacco mosaic virus that carries an MHV epitope. Proc Natl Acad Sci USA 96: 7774-7779
[53] Whitham, SA; Yamamoto, ML and Carrington, JC (1999) Selectable viruses and altered susceptibility mutants in *Arabidopsis thaliana*. Proc Natl Acad Sci USA 96: 772-777
[54] Shiboleth, YM; Arazi, T; Wang, Y and Gal-On, A (2001) A new approach for weed control in a cucurbit field employing an attenuated potyvirus-vector for herbicide resistance. J Biotech 92: 37-46
[55] Choi, I-R; Stenger, DC; Morris, TJ and French, R (2000) A plant virus vector for systemic expression of foreign genes in cereals. Plant J 23: 547-555
[56] Haygood, R; Ives, AR and Andow, DA (2003) Consequences of recurrent gene flow from crops to wild relatives. Proc Roy Soc London B 270: 1879-1886
[57] Gressel, J, Ed. (2005) Crop Ferality and Volunteerism, CRC Press
[58] Bradshaw, AD (1982) Evolution of heavy metal resistance - an analogy for herbicide resistance? In HM LeBaron and J Gressel, Eds., Herbicide Resistance in Plants, pp. 293-307, Wiley, New York
[59] Gressel, J and Ehrlich, G (2002) Universal inheritable barcodes for identifying organisms. Trends Plant Sci 7: 542-544
[60] USDA-APHIS (2004) Search Results of the Field Test Release Permits Database for the U.S., http://www.nbiap.vt.edu/cfdocs/fieldtests3.cfm
[61] Meilan, R; Brunner, AM; Skinner, JS and Strauss, SH (2001) Modification of flowering in transgenic trees, in N Morohoshi and A Komamine, Eds., Molecular breeding of woody plants, pp. 247-256, Elsevier, Amsterdam
[62] Skinner, JS; Meilan, R; Ma, CP and Strauss, SH (2003) The *Populus PTD* promoter imparts floral-predominant expression and enables high levels of floral-organ ablation in *Populus, Nicotiana* and *Arabidopsis*. Molec Breeding 12: 119-132
[63] Bilyeu, KD; Cole, JL; Laskey, JG; Riekhof, WR; Esparza, TJ; Kramer, MD and Morris, RO (2001) Molecular and biochemical characterization of a cytokinin oxidase from maize. Plant Physiol 125: 378-386
[64] Werner, T; Motyka, V; Laucou, V; Smets, R; Van Onckelen, H and Schmulling, T (2003) Cytokinin-deficient transgenic *Arabidopsis* plants show multiple developmental alterations indicating opposite functions of cytokinins in the regulation of shoot and root meristem activity. Plant Cell 15: 2532-2550
[65] Hedden, P and Kamiya, Y (1997) Gibberellin biosynthesis: Enzymes, genes and their regulation. Annu Rev Plant Physiol Plant Molec Biol 48: 431-460
[66] Zhu, T; Mettenburg, K; Peterson, DJ; Tagliani, L and Baszczynski, CL (2000) Engineering herbicide-resistant maize using chimeric RNA/DNA oligonucleotides. Nat Biotechnol 18: 555-558
[67] Curtis, IS; Nam, HG; Yun, JY and Seo, K-H (2002) Expression of an antisense GIGANTEA (GI) gene fragment in transgenic radish causes delayed bolting and flowering. Transgenic Res 11: 249-256
[68] Llewellyn, DJ (2000) Herbicide tolerant forest trees, in SM Jain and SC Minocha, Eds., Molecular Biology of Woody Plants (Vol. 2), pp. 439-466, Kluwer, Dordrecht
[69] Mariani, C; Debeuckeleer, M; Truettner, J; Leemans, J and Goldberg, RB (1990) Induction of male sterility in plants by a chimeric ribonuclease gene Nature 347: 737-741
[70] Busov, VB; Meilan, R; Pearce, DW; Ma, C; Rood, SB and Strauss, SH (2003) Activation tagging of a dominant gibberellin catabolism gene (GA 2-oxidase) from poplar that regulates tree stature. Plant Physiol 132: 1283-1291
[71] Anonymous (2004) Oji grafts natural eucalyptus onto gene-altered eucalyptus. Nikkei Report, Aug 6, see: http://www.agbioworld.org/newsletter_wm/index.php?caseid=archive&newsid=2209

HUMAN EXPOSURE ASSESSMENT FOR FOOD – ONE EQUATION FOR ALL CROPS IS NOT ENOUGH

STEFAN TRAPP[1] AND ALES KULHANEK[2]
[1]*Environment & Resources DTU, Technical University of Denmark, DK-2800 Kongens Lyngby, E-mail: stt@er.dtu.dk*
[2]*Department of Environmental Chemistry, Institute of Chemical Technology, Technická 5, Prague 160 00, Czech Republic*

Keywords: Benzo(a)pyrene, bioconcentration, health risk, models, plant uptake, soil

1. Introduction

Several risk assessment tools for contaminated soils have been developed, i.e. the Contaminated Land Exposure Assessment Model – CLEA (UK), CETOX (Denmark), CSOIL (The Netherlands), the Soil Screening Guidance – SSG (USA) or the (more general) European Union System for the Evaluation of Substances – EUSES (EU). Each of these tools uses a different approach for the calculation of the transfer into food [1]. The methods differ, and the estimation of transfer into food has a very high uncertainty when calculating human exposure. Another shortcoming is that only one type of crop, usually green vegetables, is considered. However, green vegetables represent only a small fraction of vegetable food, compared to bread, potatoes, juice and other beverages. Besides this, the uptake into leaves does not necessarily correlate with uptake into other plant parts. The uptake and transport behavior of neutral organic compounds can be adequately described with the available theory [2]. Plant specific models (leafy vegetables, root vegetables, fruits from trees) have been developed [3, 4, 5]. The results from crop-specific exposure assessment can not only provide more detailed information for risk management, they may also lead to different conclusions [6].

The objective of this chapter is to gain an insight into the underlyimg equations of the models, to show the similarities, but also the differences, and to compare the outcome to the empirical equation of Travis & Arms [7]. Crop-specific models were developed by describing the basic processes of convective or diffusive uptake, chemical equilibrium between plant tissue (roots, wood, leaves, tubers) and surrounding soil or air, as well as fluxes inside the plants. This mechanistic principle of building is similar for all crop-specific models. However, processes and the parameterization depend on the type of crop (Figure 1). The models are briefly described below but for more detailed descriptions and the limits of applicability we refer the reader to the recent report of Samsøe-Petersen et al. [8] or published original work [3, 4, 5].

M. Mackova et al. (eds.), Phytoremediation Rhizoremediation, 285–300.

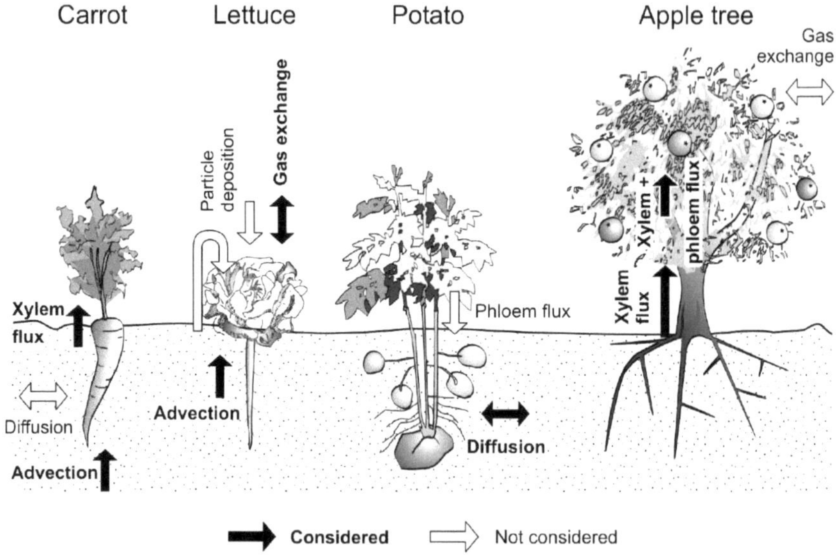

Figure 1. Principles of crop-specific plant uptake models.

Example compound: Benzo(a)pyrene (BaP) is used as example for the calculations. BaP is a polycyclic aromatic hydrocarbon (PAH). It is one of the strongest carcinogens registered. Its sources are incomplete burning processes from heating, traffic, industry, and smoking. It occurs practically everywhere in the environment. BaP is very lipophilic (characterized by the partition coefficient between n-octanol and water K_{OW}, log K_{OW} = 6.13) and semivolatile (characterized by the partition coefficient between air and water, K_{AW}, also known as dimensionless Henry's Law Constant, $K_{AW} = 1.39 \times 10^{-5}$) [9].

1. Chemical equilibrium

1.1. SORPTION TO SOIL MATRIX

The natural bulk soil consists of soil matrix, soil solution and soil gas. The distribution coefficient between soil matrix and water K_d (L water/kg soil) of organic chemicals is related to the fraction of organic carbon in soil OC (kg/kg):

$$C_M / C_W = K_d = OC \times K_{OC}$$

C_W is the equilibrium concentration in the aqueous solution (mg/L), C_M is the concentration sorbed to the soil matrix (mg/kg). It follows the expression to calculate the dissolved concentration of a chemical in soil (mg/L) from the total concentration in soil C_{Soil} (mg/kg, wet weight):

$$\frac{C_W}{C_{Soil}} = \frac{\rho_{wet}}{OC \times K_{OC} \times \rho_{dry} + P_W + K_{AW} \times P_A} = K_{WS} = \frac{1}{K_{SW}} \quad (1)$$

ρ_{wet} is the density of the wet soil (kg/L), ρ_{dry} is the density of the dry soil, and P_W and P_A are the volume fractions of water and air in the soil (L/L) (data see Table 1). K_{AW} is the partition coefficient between air and water (also named dimensionless Henry's Law constant) and K_{OC} is the partition coefficient between organic carbon and water and can be estimated from [10]:

$$\log K_{OC} = 0.81 \log K_{OW} + 0.1$$

Example calculation. The concentration of BaP in wet soil is 1 mg/kg. What is the concentration in soil solution?

$$K_{OC} = 10^{0.81 \log KOW + 0.1} = 10^{0.81 \times 6.13 + 0.1} = 116\,225$$

$$C_W = \frac{\rho_{wet}}{OC \times K_{OC} \times \rho_{dry} + P_W + K_{AW} \times P_A} \times C_{soil}$$

$$= \frac{1.95\text{kg/L}}{0.02\text{g/g} \times 116225 L/kg \times 1.6 kg/L + 0.35 L/L + 0.1 L/L \times 1.39 \times 10^{-5}} \times 1 mg/kg$$

$$= 0.0005 \text{ mg/L}$$

Table 1. Typical soil data.

Parameter	Symbol	Value	Unit
Soil wet density	ρ_{wet}	1.95	kg/L
Organic carbon content	OC	0.02	kg/kg
Soil pore water	P_W	0.35	L/L
Soil gas pores	P_A	0.1	L/L
Soil dry density	ρ_{dry}	$= \rho_{wet} - P_W$	kg/L

1.2. ROOT CONCENTRATION FACTOR RCF

Briggs *et al.* [11] macerated barley roots and made shaking experiments in water with chemicals of different K_{OW}. They expressed the result as "root concentration factor" RCF (L/kg):

$$RCF = \frac{\text{concentration in roots (mg/kg)}}{\text{concentration in water (mg/L)}}$$

The RCF increased with K_{OW}. The fit curve between RCF and K_{OW} was

$$\log (RCF - 0.82) = 0.77 \log K_{OW} - 1.52 \quad \text{or} \quad RCF = 0.82 + 0.03\, K_{OW}^{0.77}$$

The RCF can be rewritten as K_{RW} (L/kg), which describes the equilibrium partitioning between root concentration C_R (mg/kg fresh weight) and water C_W (mg/L). The partitioning occurs into the water, the lipid and the gas phase of the root:

$$K_{RW} = W_R + L_R\, a\, K_{OW}^{\ b} + P_A(\text{root})\, K_{AW} \qquad (2)$$

K_{OW} is the equilibrium partition coefficient between n-octanol and water, W and L are water and lipid content of the plant root, 'b' for roots is 0.77, 'a' = $1/\rho_{octanol}$ = 1.22. Partitioning into the gas phase of the root, P_A(root), is usually negligible.

Table 2. Water and lipid content of carrots.

Parameter	**Symbol**	**Value**	**Unit**
Carrot water content	W_R	0.89	L/kg
Carrot lipid content (*)	L_R	0.025	kg/kg
Carrot air pores	P_A(carrot)	0.05	L/kg

(*) includes all lipid-like compounds, also waxes like suberin and cutin.

Example Root concentration factor of BaP:

BaP by Briggs' equation

$$\log (RCF - 0.82) = 0.77 \log K_{OW} - 1.52 = 0.77 \times 6.13 - 1.52 = 3.20$$

$$RCF - 0.82 = 10^{3.20} = 1584.8$$

$$RCF = 0.82 + 1584.8 = 1586 \text{ L kg}^{-1}$$

or

$$RCF = 0.82 + 0.03\, K_{OW}^{0.77} = 0.82 + 0.03 \times 1\,348\,963^{0.77} = 1576 \text{ L kg}^{-1}$$

By equation 2

K_{RW} = 0.89 L kg^{-1} + 0.025 kg kg^{-1} x 1.22 L kg^{-1} x 1 348 9630.77+ 0.05 L kg^{-1} x 1.39 x 10^{-5} = 1602 L kg^{-1}

2. Dynamic root uptake model for neutral lipophilic organics

The "carrot model" calculates uptake into root with the transpiration water [4]. The change of chemical mass in roots = + flux in with water – flux out with water

$$\frac{dm_R}{dt} = C_W \times Q - C_{Xy} \times Q$$

m_R is the mass of chemical in roots, Q is the transpiration stream (L/d), C_{XY} is the concentration in the xylem (mg/L) at the outflow of the root. Diffusive uptake is not considered (the carrot is peeled!). From mass, we get the concentration by dividing through the mass of the root M:

$$\frac{d(C_R \times M)}{dt} = \frac{dm_R}{dt} = C_W \times Q - C_{Xy} \times Q$$

If growth is exponential, and the ratio Q/M (transpiration to plant mass) is constant, the growth by exponential dilution can be considered by a first-order growth rate k (d^{-1}):

$$\frac{dC_R}{dt} = C_W \times Q / M - C_{Xy} \times Q / M - k \times C_R$$

If the xylem sap is in equilibrium with the root, the concentration $C_{Xy} = C_R/K_{RW}$. Then,

$$\frac{dC_R}{dt} = C_W \times Q / M - C_R / K_{RW} \times Q / M - k \times C_R$$

Setting this to steady-state ($dC_R/dt = 0$) gives us

$$C_W \times Q / M = C_R \times (\frac{Q}{K_{RW} \times M} - k)$$

And we solve for the concentration in the root C_R:

$$C_R = \frac{Q}{\frac{Q}{K_{RW}} + kM} C_W$$

The ratio of the concentration in soil water C_W to that in bulk soil C_{Soil} is K_{WS}, and for the bioconcentration factor BCF (ratio of concentrations in plant and soil) between carrot and bulk soil follows:

$$BCF = \frac{C_R}{C_{Soil}} = \frac{C_R}{C_W} \times K_{WS} = \frac{Q}{\frac{Q}{K_{RW}} + kM} \times K_{WS} \qquad (3)$$

The model approach is strictly limited to non-ionising compounds, log $K_{OW} > 1$. The parameterization of the model is for 1 m^2 soil, with 1 kg roots, a transpiration of 1 L d^{-1} and a root growth rate of 0.1 d^{-1}. This parameterization is not idealized, the values are realistic (Table 3).

Table 3. More data for carrots.

Parameter	Symbol	Value	Unit
Transpiration stream	Q	1	L/d
Root mass	M	1	Kg
1st order growth rate	k	0.1	d^{-1}

Example calculation for BaP.

From before: $K_{WS} = 0.0005$ kg/L, $K_{RW} = 1602$

$$BCF = \frac{Q}{\frac{Q}{K_{RW}} + kM} \times K_{WS} = \frac{1\,L/d}{1\,L/d \big/ 1602\,L/kg + 0.1\,d^{-1} \times 1\,kg} \times 0.0005\,kg/L = 0.005\,kg/kg$$

2.1. TRANSLOCATION UPWARDS

The water, which is taken up by the roots, does not stay there but is translocated in the xylem to the leaves and evaporates from the stomata. Only 1-2% is taken up into the plant cells. Chemicals, which are dissolved in the "transpiration stream" (= the xylem sap), can be moved upwards, too. The 'Transpiration Stream Concentration Factor' TSCF is defined as the concentration ratio between xylem sap and external solution (water).

$$TSCF = C_{Xy}/C_W$$

The TSCF can be calculated from the root model. The xylem sap consists of water, and with $C_{Xy} = C_R / K_{RW}$ follows

$$\frac{C_{Xy}}{C_W} = \frac{Q/K_{RW}}{\frac{Q}{K_{RW}} + kM} \qquad (4a)$$

Alternatively, empirical relations exist. The TSCF is related to the K_{OW} [11] by a bell-shaped (Gaussian) curve:

$$TSCF = 0.784 \times \exp\left\{\frac{-(\log K_{OW} - 1.78)^2}{2.44}\right\} \tag{4b}$$

For poplar trees, a similar relation was found by [12] (corrected by authors):

$$TSCF = 0.756 \times \exp\left\{\frac{-(\log K_{OW} - 2.50)^2}{2.58}\right\} \tag{4c}$$

It is recommended to use Briggs' equation (4b) [11] for herbal plants and the green vegetables model (experiments were done with the grass barley), and Burken & Schnoors equation (4c) [12] for woody plants and the fruit tree model (experiments were done on poplars).

The calculated TSCF (4a) is equivalent to the empirically derived fit curves of Briggs *et al.* (4b) and Burken and Schnoor (4c), but only as long as log K_{OW} is not <1.5 (Figure 2). At lower values, reduced uptake was found in the experimentally determined TSCF-curves, possibly due to decreasing membrane permeability for polar compounds. These experiments were done in hydroponic solution. The TSCF depends on diffusion processes, and the root surface area is a limiting factor for exchange [13]. In soil, roots will - differently from hydroponic solution - form root hairs, which increase largely the surface area. Therefore, the uptake limitation for polar compounds does probably not exist. Experiments to clarify this are lacking so far.

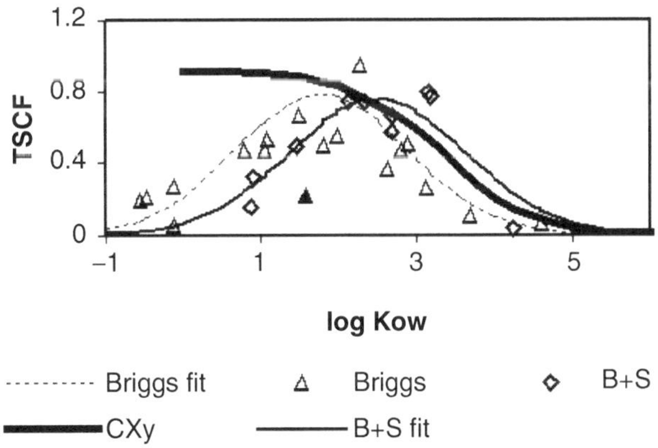

Figure 2. Comparison of empirical and calculated TSCF; 'Briggs fit' is eq. 4b; 'Δ Briggs' is original data from Briggs et al. [11]. 'B+S' is original data from Burken and Schnoor [12]; 'CXY' is calculated with eq. 4a; 'B+S fit' is eq. 4c.

Example calculation TSCF.

$$\frac{C_{Xy}}{C_W} = \frac{Q / K_{RW}}{\frac{Q}{K_{RW}} + kM} = \frac{1L/d \big/ 1602L/kg}{1L/d \big/ 1602L/kg + 0.1d^{-1} \times 1kg} = 0.0062L/L$$

$$\text{TSCF} = 0.784 \times \exp\left\{\frac{-(6.13\text{-}1.78)^2}{2.44}\right\} = 0.00034\ L/L$$

$$\text{TSCF} = 0.756 \times \exp\left\{\frac{-(6.13\text{–}2.50)^2}{2.58}\right\} = 0.0046\ L/L$$

All three models predict low translocation of BaP, the root model's theoretical curve is in between the empirical predictions.

3. Fruit tree model

Based on Trapp *et al.* [14]. Trees differ from herbs by their woody stem. Wood is composed of cellulose, hemicellulose and lignin. Lignin is a giant macromolecule with lipophilic sorption properties. The ratio between a chemical's concentration in wood and its concentration in the water is named "K_{Wood}" (mg chemical per g dry wood to mg chemical per mL water). The log K_{Wood} is significantly correlated to the log K_{OW} of the chemical and the following regression was established [15]:

Oak: $\log K_{Wood} = -0.27 + 0.632 \log K_{OW}$

The tree model considers chemical flux into stem via xylem, flux out of the stem after equilibration, dilution by growth and first order metabolism. Exchange with air and bark and phloem flow are (in this version) neglected. The transport of the compound is considered as a passive transport, taking place with the transpired water from the xylem of the roots:

$$\text{influx per time} = Q \times C_{Xy} = Q \text{ x } C_W \text{ x TSCF}$$

where Q is the transpiration stream (m^3/year) and C_{Xy} is the concentration of the compound in the xylem sap, which is TSCF x C_W. The chemical is transported out of the stem with the water transpired from the xylem of the stem into leaves and fruits:

$$\text{loss per time} = Q \times C_{Stem} / K_{Wood}$$

where C_{Stem} is the concentration of the chemical in the stem (mg/kg). Then, the mass balance of the stem is

$$\frac{dm_{Stem}}{dt} = + Q \times C_W \times TSCF - Q \times C_{Stem} / K_{Wood}$$

and by replacing mass with concentration, and with exponential growth rate k_G:

$$\frac{dC_{Stem}}{dt} = \frac{Q \times C_W \times TSCF}{M} - \frac{Q \times C_{Stem}}{K_{Wood} \times M} - k_G \times C_{Stem}$$

The steady-state solution (dC/dt = 0) for the concentration in the stem is

$$C_{Stem}(t = \infty) = \frac{C_{Xy} \times \frac{Q}{M}}{\frac{Q}{M \times K_{Wood}} + k_G} \qquad (5)$$

Uptake into fruits. Transport from stem into fruits can occur both with the phloem and xylem sap. It may be assumed that the water flow into fruits is about 20 times the dry matter content. The calculated water flow per kg fruit (fresh weight) is:

$$Q_F = dw \times 20$$

where Q_F is the water flow (L) into 1 kg of fruit, and dw is the dry weight fraction of the fruit. The amount of chemical m (mg) transported into the fruit can then be calculated from the concentration in the stem by assuming chemical equilibrium between wood, phloem and xylem:

$$m = Q_F \times C_{Stem} / K_{Wood}$$

The chemical concentration in the fruit C_F (mg/kg fresh weight) is obtained by dividing the chemical mass m by the weight of the fruit M_F (1 kg):

$C_F = m / M_F$ or in one step:

$$\frac{C_F}{C_{Soil}} = \frac{Q_F}{K_{Wood} \times M_F} \times \frac{C_{Stem}}{C_{Soil}} \qquad (6)$$

Example calculation for an apple tree stand with 100 tons/ha dry wood and 300 mm/year ($3000\ m^3\ ha^{-1}\ year^{-1}$) transpiration, a growth rate of 0.01 $year^{-1}$, and the water flux for apples (3.12 L water per kg fruit). Results we have from before: Dissolved concentration in soil $C_W = 0.0005\ mg\ L^{-1}$; TSCF [12] = $0.0046\ L\ L^{-1}$.

$$\log K_{Wood} = -0.266 + 0.632 \log K_{OW} = 3.61;\ K_{Wood} = 4057.$$

Concentration in stem (Equation 5)

$$\frac{C_{Stem}}{C_{Soil}} = \frac{0.0005 \times 4.6 \times 10^{-3} \times \dfrac{3000}{100}}{\dfrac{3000}{100 \times 4057} + 0.01} = 0.004\ \frac{\text{mg/kg fresh plant}}{\text{mg/kg wet soil}}$$

Concentration in apples (Equation 6); Water flux into apples:

$$Q_F = dw \times 20 = 0.156 \times 20 = 3.12\ L/kg$$

$$\frac{C_F}{C_{Soil}} = \frac{Q_F}{K_{Wood} \times M_F} \times \frac{C_{Stem}}{C_{Soil}} = \frac{3.12L}{4057 L/kg \times 1kg} \times 0.004\ kg/kg = 3.1 \times 10^{-6} kg/kg$$

4. TGD-model for leafy vegetables

Based on Trapp and Matthies [3]. This model was developed for the uptake of neutral organic substances into leafy vegetable or green fodder. The main objective of the model was the integration into multi-media fate models, which are used in risk assessment of new and existing chemicals. The model was adopted in the Technical Guidance Documents (TGD) of the European Union System for the Evaluation of Substances (EUSES) [10]. For leafy vegetables, there is uptake from soil via the xylem and exchange with air. The change of mass in leaves = + translocation from stem + uptake from air - loss to air

$$\frac{dm_L}{dt} = +Q \times TSCF \times C_W + C_{Air} \times g \times A - \frac{C_L \times g \times A \times \rho}{K_{LA}}$$

The differential equation for the concentration in leaves considers again additionally the dilution by exponential growth:

$$\frac{dC_L}{dt} = +\frac{Q}{M_L} \times TSCF \times C_W + \frac{C_{Air} \times g \times A}{M_L} - k_{Growth} \times C_L - \frac{C_L \times g \times A \times \rho}{K_{LA} \times M_L}$$

where L is the index for leaves. The equation can be rewritten and gives the standard linear differential equation

$$\frac{dC_L}{dt} = b - aC_L \quad \text{with the solution} \quad C_L(t) = C_L(0) \times e^{-at} + \frac{b}{a}(1 - e^{-at}) \qquad (7)$$

where

$$a = \frac{A \times g \times \rho}{K_{LA} \times M_L} + k_{Growth} \quad \text{and} \quad b = C_W \times TSCF \times Q / M_L + C_{Air} \times g \times \frac{A}{M_L}$$

The steady-state solution is

$$C_L(t = \infty) = \frac{b}{a}$$

Table 4. Parameterization of the leafy vegetables model, normalized to 1 m^2 (data taken from the original publication [3]).

Parameter	**Symbol**	**Value**	**Unit**
Shoot mass	M_L	1	kg
Leaf area	A	5	m^2
Shoot density	P	500	kg/m^3
Transpiration	Q	1	L/d
Lipid content	L	0.02	kg/kg
Water content	W	0.8	L/kg
Conductance	G	10^{-3}	$m\ s^{-1}$
Growth rate	k_L	0.035	d^{-1}
Time to harvest	T	60	d

Example calculation for benzo(a)pyrene. C_{Soil} is 1 mg/kg, C_{Air} is 1 ng m^{-3}. What is the concentration in leaves? Is the uptake from soil or from air? Data we know: C_W is 0.5 x 10^{-3} mg/L; TSCF (Briggs) = 0.00036

Loss term: $$a = \frac{A \times g \times \rho}{K_{LA} \times M_L} + k_{Growth}$$

$= 5\ m^2 \times 10^{-3} \times 86400\ m\ d^{-1} \times 500\ kg\ m^{-3} / (1.2 \times 10^9\ m^3 : m^3 \times 1\ kg) + 0.035\ d^{-1}$

$= 0.035\ d^{-1}$

Input term: $b = C_W \times TSCF \times Q / M_L + C_A \times g \times \frac{A}{M_L}$

$= 0.5 \text{ x } 10^{-3}$ mg/L x 0.00036 x 1 L/d / 1kg + 10^{-6} mg m^{-3} x 10^{-3} x 86400 m d^{-1} x 5 m^2 / 1kg

$= 0.18 \text{ x } 10^{-6}$ mg kg^{-1} d^{-1} + 0.00043 mg kg^{-1} d^{-1} = 0.00043 mg kg^{-1} d^{-1}

From an inspection of this last line it can be seen that the uptake into leaves from air is 240 times faster than from soil. The concentration in leaves at t = 60 days is:

$$C_L(t) = C_L(0) \times e^{-at} + \frac{b}{a}(1 - e^{-at})$$

$$= 0 + \frac{0.00043 \quad mg \quad kg^{-1} \quad d^{-1}}{0.0035 \quad d^{-1}} \times (1 - e^{-0.0035 \times 60}) \quad = 0.011 \quad mgkg^{-1}$$

And steady-state:

b/a = 0.00043 mg kg^{-1} d^{-1} / 0.035 d^{-1} = 0.012 mg/kg.

If the uptake is only from soil:

b/a = 0.18 x 10^{-6} mg kg^{-1} d^{-1}/0.035 d^{-1} = 5.1 x 10^{-6} mg/kg.

Note: This model is used in the EU risk assessment for new and existing chemicals (EUSES), but it lacks one important pathway: deposition of soil on leaves. Many green vegetables are contaminated by attached soil (e.g., lettuce). To consider this, there is always a default soil-to-plant transfer with particles of 1% attached soil assumed, which means a minimum BCF plant/soil of 0.01 (wet weight based):

BCF with soil = BCF model + 0.01 = 5.1 x 10^{-6} + 0.01 $\approx$ 0.01mg/kg

5. The regression of Travis and Arms (T&A)

Travis and Arms [7] established a regression that estimates uptake of neutral organic chemicals into above-ground plants. The regression has been obtained from a range of empirical data acquired by the authors from literature. Mainly data from uptake of pesticides with a log K_{OW} ranging from 1.15 to 9.35 were used but no particular type of vegetation was targeted in the original study. The form of the regression is:

$$\log BV\ (dry) = 1.588 - 0.578 \log K_{OW} \qquad (8)$$

where BV is the bioconcentration factor vegetation and is the concentration ratio between plants (dry weight) and soil (dry weight). The BV was converted into wet weight bioconcentration factors (BCF) using

$$BCF\ (wet) = BV\ (dry) \times (1 - W) \times \rho_{wet} / \rho_{dry}$$

where W is the water content of the plants, and ρ is the density of the soil.

In practice, the regression of Travis and Arms needs lesser efforts to calculate BCFs. Therefore, it might serve as an "early warning". The advantage of the crop-specific models is the possibility to adapt them to specific situations. So they might be used when the regression signals a critical situation.

BCF of BaP in carrot with the Travis and Arms regression:

$$\log BV\ (dry) = 1.588 - 0.578 \times 6.13 = -1.955 \quad \rightarrow BV\ (dry) = 0.01$$

Carrot water content; W = 0.89

$$BCF\ (wet) = BV\ (dry) \times (1 - 0.89) \times 1.95 / 1.6 = 0.134 \times BV(dry) = 0.0013$$

6. Exposure and risk assessment

For the calculation of exposure, the concentrations in the different food crops need to be multiplied with the consumption of these foodstuffs. The consumption, expressed as daily dietary intake, may vary for different groups of society (e.g., children, vegetarians) and different regions. To derive the risk associated with food consumption, toxicological values, such as acceptable daily intake, have to be compared with the actual uptake.

6.1. DAILY DIETARY INTAKE (DDI)

For the calculation of the daily dietary intake (DDI), the consumption of vegetables is multiplied with the calculated concentration in the crop types, i.e. root vegetable, fruits, potatoes and leafy vegetable. The daily dietary intake DDI is then

$$\text{DDI [mg/d]} = \sum_i \text{concentration in crop(i) [mg kg}^{-1}\text{] x mass of crop(i) consumed [kg d}^{-1}\text{]}$$

Table 5. Average consumption [g person^{-1} d^{-1}] in the Czech Republic [16].

Crop type	Example	Consumption
Leafy vegetable	Spinach, lettuce	2.4
Tree fruit	Apples, pears	62.7
Root vegetable	Carrots	13.5
Sum		78.6

The DDI with the Travis & Arms regression (equation 8), neglecting the differences in water content, is:

$$\text{DDI} = 0.0013 \text{ mg/kg x } 0.079 \text{ kg/d} = 102 \text{ ng/d}$$

This is a quite similar result to the crop-specific calculation (see example below), giving confidence in both methods. However, the crop-specific approach allows a much more situation-specific estimation. E.g., the use of a BaP-contaminated garden for growing vegetables might yield a quite considerable additional exposure to BaP, whereas the use for a grass lawn plus some apple trees would lead to a very small additional exposure of humans to BaP. "Crop-taylored management" of polluted soils is a chance to reduce the risk associated with soil pollution with very low costs.

Example calculation: Daily Dietary Intake DDI of BaP in the Czech Republic by consumption of various crops

Crop	Concentration (mg/kg)	Consumption (kg/d)	DDI (mg/d)	DDI (ng/d)
Root vegetable	0.005	13.5 x 10^{-3}	6.75 x 10^{-5}	67.5
Tree fruit	3.1 x 10^{-6}	62.7 x 10^{-3}	1.94 x 10^{-7}	0.19
Leafy vegetable	5.1 x 10^{-6}	2.4 x 10^{-3}	1.22x 10^{-8}	0.01
Leafy veg. + attached soil	0.01	2.4 x 10^{-3}	2.4 x 10^{-5}	24
Sum		0.079	9.7 x10^{-6}	91.7

6.2. ACCEPTABLE DAILY INTAKE (ADI)

"Sola dosis facit venenum" (Paracelsus, 16th century) – everything is toxic, it is only the dosis that makes a thing a poison or a remedy. An acceptable daily intake of the chemical under investigation has to be defined by human toxicologists.

A "virtually safe dose of BaP as a marker of the mixture of carcinogenic PAH in food would be in the range 0.06 to 0.5 ng BaP kg^{-1} bw d^{-1}" [17]. If we chose the lowest value of this virtually safe dose as acceptable, the acceptable daily intake (ADI) of BaP via food for an adult weighing 70 kg would be 4.2 ng d^{-1}. If the upper limit of the virtually safe dose is used, the ADI of BaP would be 35 ng d^{-1} (the unit "ng BaP kg^{-1} bw d^{-1}" means nanogram benzo(a)pyrene per kilogram bodyweight and day).

6.3. ACCEPTABLE SOIL CONCENTRATION (ASC)

To derive an acceptable soil concentration, we expect that the life-long consumption of food produced from this soil does not lead to oral intake above the virtually safe dose. Using the linearity of BCF to concentration in soil, the acceptable soil concentration (ASC) can be estimated from DDI (calculated for a concentration in soil of 1 mg/kg):

$$\text{ASC [mg kg}^{-1}\text{ wet wt]} = \{\text{ADI [mg d}^{-1}\text{]/DDI [mg d}^{-1}\text{]}\} \times 1 \text{ mg kg}^{-1} \text{ wet wt}$$

What would be a "virtually safe" or acceptable concentration of BaP in soil?

with ADI = 4.2 ng/d   ASC = 4.2 / 91.7 = 0.046 mg BaP/kg soil

with ADI = 35 ng/d   ASC = 35 / 91.7 = 0.38 mg BaP/kg soil

Note: This is only for uptake with root vegetables, tree fruits and leafy vegetables. The legal standards for BaP in soil are: Czech Republic: 0.1 mg/kg (dry weight), US-EPA 0.087 mg/kg (dry weight); Denmark: 3 mg/kg (since 2005).

7. Conclusions

The concept of crop-specific human exposure assessment for soil pollutants allows adapting to different food habits, different life-styles, different regions and different accepted risks. Models for transfer into major food crops are combined with regional food baskets to give the exposure of the population, which can be compared to tolerable risks based on health considerations. Backwards, tolerable concentrations of pollutants in soil for the agricultural production and gardening can be derived. The approach might therefore be used to derive rational soil quality standards for a large variety of chemicals with reasonable effort. It could also be the basis for an internationally harmonized procedure in soil risk assessment [18].

Acknowledgements

We are grateful for the international students funding Erasmus/Socrates awarded to Ales Kulhanek and for the founding MSM 223200003 from the government of the Czech Republic. We furthermore thank J.-P. Schwitzguébel for organising the COST 859 (and other actions). In particular, we thank Lise Samsøe-Petersen and Dorte Rasmussen, DHI Hørsholm, and the Danish EPA for support.

References

[1] Versluijs, CW; Koops, R; Kreule, P and Waitz, MFW (1998) The accumulation of soil contaminants in crop, location-specific calculation based on the CSOIL module. Part 1: Evaluation and suggestion for model development. RIVM Report No 711 701 008, Bilthoven, The Netherlands

[2] Trapp, S and Mc Farlane, JC (1995) Plant Contamination. Modeling and Simulation of Organic Chemical Processes. Boca Raton, Florida: Lewis Pub

[3] Trapp, S and Matthies, M (1995) Generic One-Compartment Model for Uptake of Organic Chemicals by Foliar Vegetation. Environ Sci Technol 29, 2333-2338; Erratum 30: 360

[4] Trapp, S (2002) Dynamic root uptake model for neutral lipophilic organics. Environ Toxicol Chem 21: 203-206

[5] Trapp, S; Rasmussen, D and Samsøe-Petersen, L (2003) Fruit Tree Model for Uptake of Organic Compounds from Soil. SAR - QSAR Environ Res 14: 17-26

[6] Kulhanek, A; Trapp, S; Sismilich, M; Janku, J and Zimova, M (2005) Cropp-Specific Human Exposure Assessment for Polycyclic Aromatic Hydrocarbons in Czech Soils. Sci Tot Environ (in press)

[7] Travis, C and Arms, A (1988) Bioconcentration in beef, milk and vegetation. Environ Sci Technol 22: 271-274

[8] Samsøe-Petersen, L; Rasmussen, D and Trapp, S (2003) Modellering af optagelse af organiske stoffer i grøntsager og frugt. Rapport til Miljøstyrelsen, Miljøprojekt Nr 765 2003

[9] Rippen, G (2002) Handbuch Umweltchemikalien. ecomed, Landsberg a.L, D

[10] EC European Commission (1996) Technical Guidance Document in Support of Commission Directive 93/67/EEC on Risk Assessment for New Notified Substances and Commission Regulation (EC) No 1488/94 on Risk Assessment for existing substances. European Commission, Office for Official Publications of the European Communities, Luxemburg, Luxemburg

[11] Briggs, GG; Bromilow, RH and Evans, AA (1982) Relationships between lipophilicity and root uptake and translocation of non-ionised chemicals by barley. Pestic Sci 13: 495-504

[12] Burken, JG and Schnoor, JL (1998) Predictive relationships for uptake of organic contaminants by hybrid poplar trees. Environ Sci Technol 32: 3379-3385

[13] Trapp, S (2000) Modeling uptake into roots and subsequent translocation of neutral and ionisable organic compounds. Pest Manag Sci 56: 767-778

[14] Trapp, S; Rasmussen, D and Samsøe-Petersen, L (2003) Fruit Tree Model for Uptake of Organic Compounds from Soil. SAR - QSAR Environ Res 14: 17-26

[15] Trapp, S; Miglioranza, KSB and Mosbæk, H (2001) Sorption of Lipophilic Organic Compounds to Wood and Implications for their Environmental Fate. Environ Sci Technol 35: 1561-1566

[16] Ruprich, J (2000) Spotrebni kos potravin pro Ceskou republiku - 2000, Expozicni faktory CR 1997. National Institute of Public Health, Prague, Czech Republic

[17] EC European Commission (2002) Health and Consumer Protection Directorate (ed.): Opinion of the scientific committee on food on the risks to human health of polycyclic aromatic hydrocarbons in food. Brussels, Belgium. http://europa.eu.int/comm/food/fs/sc/scf/index_en.html

[18] Trapp, S (2003) Harmonized European Assessment for the Transfer of Chemicals from Soil into Food. Journal of Soils and Sediments 3: 6

MIX
Papier aus verantwortungsvollen Quellen
Paper from responsible sources
FSC® C105338

If you have any concerns about our products,
you can contact us on
ProductSafety@springernature.com

In case Publisher is established outside the EU,
the EU authorized representative is:
Springer Nature Customer Service Center GmbH
Europaplatz 3, 69115 Heidelberg, Germany

Printed by Libri Plureos GmbH
in Hamburg, Germany